T0261606

Ecology of Climate Change

MONOGRAPHS IN POPULATION BIOLOGY
EDITED BY SIMON A. LEVIN AND HENRY S. HORN

A complete series list follows the index.

Ecology of Climate Change

The Importance of Biotic Interactions

ERIC POST

PRINCETON UNIVERSITY PRESS
Princeton and Oxford

Published by Princeton University Press, 41 William Street,
Princeton, New Jersey 08540
In the United Kingdom: Princeton University Press, 6 Oxford Street,
Woodstock, Oxfordshire OX20 1TW

press.princeton.edu

ISBN 978-0-691-14847-2

British Library Cataloging-in-Publication Data is available

This book has been composed in Times LT Std

Printed on acid-free paper. ∞

Printed in the United States of America

1 3 5 7 9 10 8 6 4 2

For Pernille

Tempus fugit, amor manet

Contents

Preface

Purpose, Perspective, and Scope

Reports of above-average annual temperatures and record-warm years are by now well familiar. The U.S. National Oceanic and Atmospheric Administration (NOAA) announced on January 12, 2011, that 2010 had equaled 2005 as the year of highest global mean temperatures on record. In the same report, NOAA listed a series of global climate highlights, including the fact that 2010 had experienced the greatest average global precipitation on record, but with substantial regional variation, and that it had also been the fourteenth year in a row with above-average temperatures in the United States. As this book moves into production in the summer of 2012, the report titled "State of the Climate National Overview—May 2012," released by NOAA's National Climatic Data Center, declared, "The average U.S. temperature during spring was . . . 5.2°F [2.9°C] above average, the warmest spring on record." This book is aimed at clarifying how and why climate change matters from an ecological perspective.

Despite the obvious focus this objective places on the importance of abiotic influences in ecological dynamics, the emphasis throughout this book is, perhaps somewhat counterintuitively, on the primacy of biotic interactions in shaping responses to climate change. Although the earliest work addressing specifically the ecological consequences of contemporary climate change was concerned with quantifying biological responses to changes in abiotic factors such as temperature, precipitation, and nutrient availability, more recent research has made increasingly clear the importance of interactions among organisms in determining responses to climate change across levels of organization (Gilman et al. 2010; Urban et al. 2012). Within a framework emphasizing the classic topics of enduring relevance in ecology, this book argues that interactions among organisms are not merely background noise that must be accounted for statistically or experimentally in the study of climate change but rather the determining factor in the responses of individuals, populations, communities, and ecosystems to climate change. Hence, the perspective advanced in this book holds that an appreciation of the importance of interactions among organisms will improve our understanding and study of individual-level life history or phenological responses to climate change, which most commonly have been studied purely as organism-environment interactions. We will also see that both competition among conspecifics, or density dependence, and interactions with aspecific competitors, resources, and predators determine the dynamical responses of populations to climate change. At the community level, the strength of exploitation and interference interactions exerts tremendous influence over the stability response of species assemblages to climate change. Finally, interactions at and among all these levels of organization come to bear on ecosystem responses to climate change, which traditionally have been studied within the conceptual framework of abiotic influences on variation in the availability and rates of turnover of nutrients.

THE TENSION AND FACILITATION HYPOTHESES OF BIOTIC RESPONSE TO CLIMATE CHANGE

As a central theme that illustrates the importance of organism-organism interactions in responses to climate change across levels of biological organization, we may consider a generalized tension or, alternatively, facilitation between the strength of climatic versus biotic influences on dynamics at the level of interest and the implications these influences pose for the stability properties of such dynamics. This tension (or facilitation) is formalized mathematically in individual chapters on phenological, population, and community dynamics; a heuristic overview is presented here as a means of establishing a viewpoint from which the rest of this book advances.

We may consider first the simple case in which the strength of interactions among organisms ranges along a spectrum from weak to strong (figure 0.1). In the perspective developed throughout this book, the strength of the climatic influence on interactions among organisms, which interactions also range along a spectrum from weak to strong, is expected to be related to the strength of the biotic interactions occurring among organisms, as depicted in figure 0.1. In general, we may think of the interaction between climatic and biotic influences in ecological dynamics as assuming one of two forms, either *tension* (the dashed line in figure 0.1) or *facilitation/promotion* (the solid line in figure 0.1). In this context, use of the term *tension* implies a trade-off between the strengths of climatic and biotic influences on ecological dynamics; in other words, these forces oppose each other. By contrast, use of the term *facilitation* or *promotion* implies an enhancement of the strength of climatic or biotic influences on ecological dynamics by an increase in the strength of the other; in other words, these forces enhance each other.

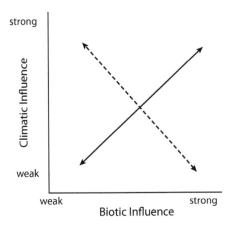

strong

Climatic Influence

weak

weak   strong
Biotic Influence

Figure 0.1. The tension (dashed line) and facilitation/ promotion (solid line) hypotheses of the relationship between the strength of biotic interactions and climatic influence in ecological dynamics. According to the tension hypothesis, in general, weak biotic interactions will be associated with strong climatic influences on ecological dynamics, while strong biotic interactions will be associated with weak climatic influences on dynamics. In contrast, according to the facilitation/ promotion hypothesis, as biotic interactions strengthen, so do climatic influences on ecological dynamics, while weak biotic interactions are associated with weak climatic influences on dynamics.

In the simple case in which biotic interactions take the form of density dependence, or intraspecific competition, weak interactions are expected to be destabilizing for dynamics, while strong interactions should be stabilizing, as will be exposed quantitatively in chapter 4. According to the tension hypothesis, when biotic interactions are weak, the role of climate in dynamics may assume greater importance, moving such dynamics toward instability (figure 0.2a). Conversely, with strong, stabilizing biotic interactions, the contribution of climate to ecological dynamics would be expected to be minimized (figure 0.2a). The tension hypothesis also recognizes, however, that the direction of causality in the tension between climatic and biotic factors in determining ecological dynamics may be reversed: strong climatic effects may weaken or override biotic influences, potentially leading to destabilization of ecological dynamics. Conversely, a weakening of any climatic influence would be expected to promote a strengthening of stabilizing biotic influences.

The use of arrows in the example illustrated in figure 0.2 and the following illustrations to depict the net effect on ecological dynamics of the interaction between the strengths of climatic and biotic influences on dynamics is intended to represent the variable nature of the expected outcomes of this interaction. Hence, when I refer to a strengthening or a weakening of climatic or biotic influences on dynamics and a tendency toward stabilization or destabilization of ecological dynamics, I am describing a variability that might be observed through time in a single population, species, or system; or across space among populations, species, or systems; or across levels of biological organization, from individuals to populations to communities, in a given location.

The nature of the interaction between the strengths of abiotic and biotic influences on ecological dynamics is, of course, reversed in the facilitation/

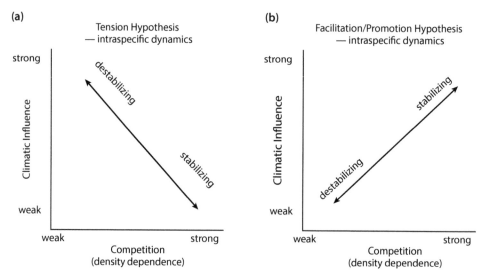

Figure 0.2. Contrasting predictions of the tension hypothesis (a) and facilitation/promotion hypothesis (b) for the stability of population dynamics. According to the tension hypothesis, strong density dependence is associated with a weak climatic influence on population dynamics, producing stable dynamics, while weak density dependence is associated with a strong climatic influence on dynamics, which leads to a destabilization of dynamics. According to the facilitation/promotion hypothesis, stable population dynamics result from the association of an increasingly strong influence of climate on dynamics with a strengthening of density dependence, while destabilization of population dynamics results from a weakening of density dependence with a weakening of the influence of climate on dynamics.

promotion hypothesis (figure 0.2b). In this case, the strengths of the biotic and abiotic influences on dynamics increase together or weaken together (figure 0.2b). Hence, in the simple example of intraspecific competition, as the strength of density dependence increases, the contribution of climatic conditions to the dynamics of that population may increase. This appears to have been the case with the population of reindeer introduced to St. Matthew Island in the U.S. state of Alaska, which, after increasing rapidly for several years, began to exhibit indications of density-dependent limitation through reduced offspring production and recruitment just before a severe winter precipitated a population crash (Klein 1968).

Turning to a consideration of interspecific interactions, we may, for instance, examine the consequences of interspecific competition, or interference interactions, for the coexistence of two species in a laterally structured community (figure 0.3). The distinctions between interference and exploitation and between laterally and vertically structured communities are discussed in greater detail in chapter 6. Here, though, according to the tension hypothesis, strong interspecific competition would be accompanied by a weak climatic influence on interference interactions between members of the competing species, in which case competitive exclusion would be expected (figure 0.3a). Conversely, weak interspecific competition may result from—or result in, depending on the direction of causality—a strong climatic influence on the interacting species, in which case stable coexistence would be expected. By contrast, the facilitation/promotion hypothesis predicts that competitive exclusion results from a simultaneous increase in the strengths of the influences of climate and competition (figure 0.3b). In this case, an increasingly strong climatic effect may have tipped the balance of competition in favor of one of the interacting species, or an adverse effect of climate on the inferior of two competitors may have been promoted by intense competition between the two interacting species.

The consequences of tension or facilitation between climatic and biotic influences on interactions among species also extend to exploitative interactions

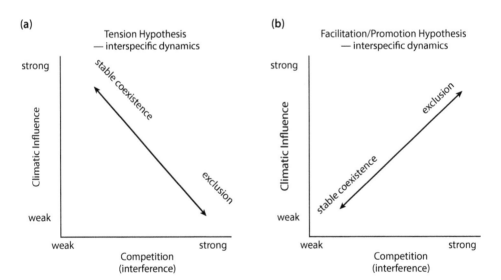

Figure 0.3. Implications of the tension hypothesis (a) and facilitation/promotion hypothesis (b) for the outcome of interspecific competition in laterally structured communities. According to the tension hypothesis, stable coexistence results from a weakening of interference in association with a strengthening of the climatic influence on one or both competitors, while competitive exclusion results from a strengthening of interference in association with a weakening of this climatic influence. In contrast, according to the facilitation/promotion hypothesis, stable coexistence should result when interference weakens in association with a weakening of the climatic influence on one or both competing species, while competitive exclusion is expected to result from a strengthening of interference with a strengthening of the climatic influence on, for example, an inferior competitor.

occurring across trophic levels in vertically structured communities (figure 0.4). In this case, however, the consequences relate to the balance between bottom-up and top-down drivers in regulating such interactions. When a tension exists between climatic and biotic influences in exploitation interactions, the dynamics across trophic levels should be regulated primarily by bottom-up interactions when the climatic influence is strong and the influence of exploitation is weak, but it should be regulated primarily by top-down interactions when climatic influences are weak and exploitation is strong (figure 0.4a). By contrast, when there is facilitation or promotion between climatic and biotic influences in exploitation interactions, bottom-up regulation should result when both the climatic influence and exploitation are weak, while top-down regulation should result when both the climatic influence and exploitation are strong (figure 0.4b). In Isle Royale National Park, for instance, wolf pack size increases with winter snowfall, leading to higher rates of moose kills, declining moose population size, and enhanced growth of balsam fir (Post, Peterson, et al. 1999).

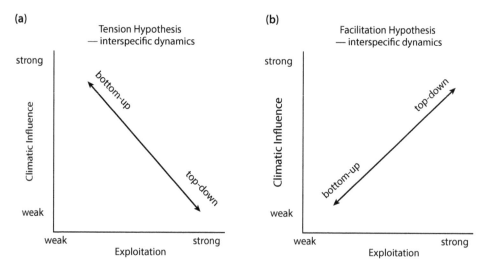

Figure 0.4. Predictions of the tension hypothesis (a) and facilitation/promotion hypothesis (b) for the directionality of regulation of trophic interactions in vertically structured communities. According to the tension hypothesis, top-down regulation is expected to occur as a result of a strengthening of exploitation interactions in association with a weakening of climatic influences on exploitation, while bottom-up regulation is expected to occur when exploitation interactions weaken with a strengthening of climatic influences. In contrast, according to the facilitation/promotion hypothesis, top-down regulation is expected to occur when a strengthening of climatic influences is accompanied by a strengthening of exploitation interactions, while bottom-up regulation should result from a weakening of exploitation interactions with a weakening of the influence of climate on exploitation.

Conceptualizing the relationships between biotic interactions and climatic influences on ecological dynamics according to the tension and facilitation/ promotion hypotheses should improve our thinking about the ecological consequences of climate change and our understanding of how and why individuals, populations, species assemblages, and ecosystems respond to climate change in variously disparate or similar ways. I will note consistencies with these two hypotheses at various points throughout this book to make their distinction clearer but will not do so in every single case, to avoid tedium.

Many important studies did not make it into this book, and some readers will no doubt take exception to the absence, or in some cases the presence, of certain case studies. At the outset, I will admit that I have not taken extensive measures to present a taxonomically, geographically, or disciplinarily balanced view of the study of ecological responses to climate change. Nor have I attempted to assemble a complete review of such responses. Omissions may motivate some authors to reemphasize and draw attention to their own work in the literature, which in the end serves the objective of this book just as well because such action focuses attention on a subject that as a whole exceeds in importance the work of any individual. At the same time, some of the ideas presented in this book will undoubtedly stir disagreement and foster debate within the scientific community, but this too should be viewed positively because it forces us as scientists to scrutinize our assumptions and explore more thoroughly the foundational principles that underlie our understanding of how nature works.

I have tried to include in this book what I believe are the most salient aspects of ecological responses to climate change and, more important, to place them in the context of the foundations of ecology as a discipline and of ecological theory as a body of knowledge. In some cases, relevance to the classic studies on which contemporary ecology has been built will be more obvious than in others, but in the latter instance the broader context of ecology as a discipline should be apparent. I have also tried to incorporate, more through structure than through outright proselytizing, some sense of the utility and importance of studying the ecology of climate change across levels of organization, from individuals to populations, to species, to species assemblages, and finally to ecosystems. I have done this out of a personal conviction that understanding what one observes at the level of one's own investigation is enhanced by understanding how that level relates to all other levels of organization. Such a perspective also serves to emphasize, and through this emphasis to help develop an understanding of, the importance of biotic interactions in ecological responses to climate change. I hope readers will see the connections I have tried to draw among the related topics herein, and perhaps realize new ones that will fuel stimulating and important research on the ecology of climate change.

Acknowledgments

The acknowledgments section of any book is invariably its shortest, yet it is the most important. Many people contributed to the successful completion of this project, but I must first thank my editor at Princeton University Press, Alison Kalett, for encouraging me to write this book. From the day she paid me a personal visit to pitch the idea for a book on the ecological consequences of climate change to the day the final manuscript was delivered, Alison has been an unwavering source of encouragement, support, and enthusiasm. I was initially reluctant to engage in this undertaking, but thanks to Alison's vision, foresight, and guidance, the process has turned out to be one of great personal reward.

Numerous colleagues provided constructive feedback on individual, and in some cases multiple, chapters, or through correspondence improved my understanding of and thinking on certain topics, and I thank them for their selflessness and generosity of time. These include Steve Beissinger, Julio Betancourt, Jessica Blois, Jedediah Brodie, John Bryant, Terry Chapin, Mads Forchhammer, Duane Froese, Tara Fulton, Russ Graham, Mark Hebblewhite, Blair Hedges, Toke Høye, David Inouye, Tony Ives, Steve Jackson, Bill Laurance, Mark McPeek, Eric Nord, Town Peterson, Bob Ricklefs, Daniel Schindler, Os Schmitz, Paddy Sullivan, Nick Tyler, Jack Williams, Chris Wilmers, and Joe Wright. I am especially grateful to Steve Beissinger, Mark McPeek, and Bob Ricklefs for entertaining my numerous questions and conjectures about the nature of biological communities and for engaging in an extensive exchange on this subject that enriched my knowledge and understanding of it. Personally, I believe the resulting chapter on community dynamics and stability under climate change developed into the most interesting and stimulating in this book, and if the reader agrees, then these three colleagues must share in its appeal. I am also grateful to Os Schmitz for guidance and detailed advice on how to proceed with this project, as well as for his personal and professional support throughout the development of my career. Two anonymous referees read the entire manuscript and offered numerous suggestions and points of critique that greatly improved its quality, scope, and scholarship. The talented pool of graduate students currently working in my lab tolerated my numerous physical and mental absences during the writing of this book and on many occasions provided substantive feedback on the ideas presented here. They include

Mike Avery, Sean Cahoon, Didem Ikis, Jeff Kerby, and David Watts. Some of the data from my own work that are represented in a few of the chapters in this book were collected with the help of these students, and with the help of previous students, assistants, friends, and family members who have worked with me in Greenland since 1993, including Tom Adams, Jesper Bahrenscheer, Pernille Sporon Bøving, Todd Costello, Nell Herrmann, Toke Høye, Syrena Johnson, Megan MacArthur, Christian Pedersen, Ieva Perkons, Mason Post, Taylor Rees, Henning Thing, and Tyler Yenter.

Parts of my research reported on in this book were supported by the Department of Biology at Penn State University, the U.S. National Science Foundation, and the National Geographic Society's Committee for Research and Exploration. The field components of my research would not have been possible without logistical support from the personnel of, over the years, the 109th Airlift Wing, New York Air National Guard; PICO; VECO Polar Resources; and CH2MHILL Polar Services. I am especially indebted to the manager of Kangerlussuaq International Science Support (KISS), Basse Vaengtoft, for his many years of hospitality and assistance. I am particularly grateful to Basse for facilitating an impromptu introduction in Kangerlussuaq, during the International Polar Year, to Crown Prince Frederik of Denmark, Crown Princess Victoria of Sweden, and Crown Prince Haakon of Norway, which provided a spur-of-the-moment opportunity to informally describe to them aspects of the research reported on in this book. I also wish to extend very special and heartfelt thanks to my dear friend and close colleague, Mads Forchhammer, for a stimulating and long-lived collaboration that has provided the foundation for many of the ideas that made their way into this book.

There is one person who does not fall exclusively into any one of the foregoing categories of acknowledgment because she has had a pervasive role in my research over the past twenty-one years and in all stages of the book project. Far more important, she has provided me with the emotional support and encouragement necessary not only for the completion of this project but also for a sustained devotion to long-term research in remote settings away from family. This person is my wife, Pernille, who not only has been my longest-running collaborator but also introduced me to fieldwork in Greenland, where I got my feet wet working as her field assistant. Pernille has always believed in me when I most needed a boost, and provided a critical perspective when my ideas needed grounding. There is no greater gift than to love and be loved, and it is this foundation that underlies and sustains all aspects of my life.

I've already thanked my son, Mason, for his help and cheerful companionship during fieldwork over four seasons in Greenland. I would also like to thank my two daughters, Phoebe and Boochie, for their abundant good cheer, hugs, and expressions of love and humor, which not only kept me going during the writing and editing of this manuscript but also gave me good reason to look forward to finishing a day's work.

Finally, I wish to thank my mother, Janice Ronchetto, and my father, Douglas Post, for their love and support, and for extending to me the childhood freedom to explore the woods, ponds, and creeks of central Wisconsin, where the seeds of my interest in ecology were sown. Unfortunately, my dad did not live to see the publication of this book. However, I am happy we were able to share many conversations about it during the final year of his life. As my first biology teacher, he set me on the trajectory that ultimately led to its writing.

Eric Post

Ecology of Climate Change

CHAPTER 1

A Brief Overview of Recent Climate Change and Its Ecological Context

This chapter summarizes the most prominent abiotic components of recent climate change to establish the environmental context from which the discussion in the rest of the book proceeds. As will become clear in subsequent chapters, the rapid pace and broad geographic extent of abiotic changes reviewed here cannot be viewed in isolation for, as important as they are as drivers of ecological dynamics, in many if not most cases the ecological responses resulting from them owe to an alteration of biotic interactions. As an example we may consider the recent mass abandonment of retreating ice floes by Pacific walrus along the coast of Alaska in 2007 and 2010, which coincided with the lowest and third-lowest arctic sea ice extent recorded to date (Post and Brodie 2012). Walrus are benthic feeders: they consume mussels on the seafloor in shallow water during dives from the ice edge, which they also rest on between foraging bouts (Oliver et al. 1985; Ray et al. 2006). As the ice melts and the edge moves farther away from shore and the shallow-water environments that harbor the mussels walrus eat, it becomes energetically unprofitable for these large marine mammals to try to remain in proximity to the ice edge, for they would have to travel progressively longer distances to reach the shallow-water seabeds (Ray et al. 2010). The rate of annual sea ice loss over the Arctic has been on the order of 56,000 km^2 each year since 1979 (figure 1.1). As suggested later in this chapter, loss of sea ice does not represent simply an abiotic response to recent warming; it also represents the loss of critical habitat necessary for the survival and reproduction of many species, and it should probably be considered the high-latitude equivalent, in terms of negative impacts on faunal species, of deforestation in the Tropics (Post and Brodie 2012).

Before I summarize the most prominent abiotic features of recent climate change, two points of clarification are necessary. First, what is meant by *recent*? In the context of this book, recent refers to climatic changes that were set in motion by the onset of the Industrial Revolution but have become most clearly manifest, in terms of global abiotic changes, since approximately the middle

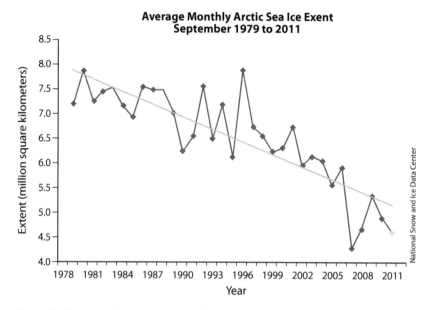

Figure 1.1. Estimates of annual minimum sea ice extent over the Arctic between 1979 and 2011. Minimum sea ice extent is estimated annually in September, after melting has peaked. The annual values are expressed as percentages of the mean for the period 1979–2000. *Image from the U.S. National Snow and Ice Data Center, annual sea ice index (http://nsidc.org/data/seaice_index/).*

of the twentieth century (Mann et al. 1998, 1999). Second, what is meant by *rapid?* It is perhaps more difficult to identify the parameters of this characteristic of recent climate change, but at the same time the notion of rapid change is probably more intuitive to ecologists than is the notion of recent change. For the purposes of this book, and in keeping with most of the examples adduced herein, *rapid* refers to changes occurring on annual to decadal timescales. It will no doubt be tempting to argue that decadal-scale climatic variation is most certainly not rapid in the ecology of soil microorganisms, whose population and community dynamics are important drivers of nutrient flow rates, primary productivity, and even biosphere-atmosphere exchange of trace gases, but constraints must be established, and these are two of the clearest used throughout this book. Rapid environmental change most certainly presents challenges to some species, in terms of individuals' ability to survive and reproduce, even as it presents distinct advantages to others, and no apparent benefit or disadvantage to yet others. Hence, the dynamics explored in this book include, in some cases, responses to climate change, and in other cases a lack of such response. But before we begin an accounting of these responses, from individual-level life history responses through population-level dynamics, to community-scale changes, to ecosystem dynamics, we must take a close look at climate change and the abiotic dynamics it encompasses.

CLIMATE CHANGE VERSUS GLOBAL WARMING

It is tempting to use the term *climate change* to refer to increases in Earth's temperatures or, similarly, to conflate the terms *climate change* and *global warming*. The term global warming is an oversimplification of the current changes in abiotic conditions on Earth because it implies that only changes in temperature are occurring, and that these temperature changes are unidirectional. Furthermore, the term has been applied most recently to connote solely human-driven temperature increases, which has promoted dismissal of the entire notion of climate change by those unconvinced by the evidence of a role for human beings in recent climate change. To draw an analogy that illustrates the importance of avoiding conflation of the terms climate change and global warming, we may recognize that evolution, like climate change, is a fact, whereas natural selection, like anthropogenic emissions of greenhouse gases, is the accepted theory explaining how this factual process occurs.

From an ecological perspective, climate change is most meaningfully considered as the suite of abiotic changes occurring across Earth coincident with the onset of the Industrial Revolution and progressing over the past 150 years, including rising temperatures, temperature variability, changes in precipitation and snow cover, and diminishing sea and land ice. All these changes can be linked to ecological dynamics, though it is probably fair to state that most research to date on the ecological consequences of climate change has focused on temperature changes.

TEMPERATURE CHANGES

The Earth's surface has warmed by approximately 0.4°C since 1850 (Brohan et al. 2006; Trenberth et al. 2007), and by approximately 0.2°C per decade over the most recent three decades (Hansen et al. 2006) (figure 1.2), largely as a result of increases in atmospheric CO_2 concentrations driven by the human use of fossil fuels (Hansen, Sato, et al. 2008). This increase hardly seems remarkable, and indeed, it would be difficult for you or me to detect from one day to the next. However, this single estimate, representing an average value calculated over the entire surface area of our planet and encompassing over a century and a half of data, masks a great deal of detail. The biological importance of this seemingly minute amount of change is evident in the widespread ecological dynamics it has elicited (Walther et al. 2002; Post, Forchhammer, et al. 2009). To understand the biological importance of climate change, we need to examine the abiotic data through the eyes of an ecologist.

From figure 1.2, two aspects of the recent temperature trend should become immediately apparent. First, there is considerable variation about the trend from year to year. The existence of this variation should signal to us as ecologists that not all organisms should be expected to respond similarly, in terms of rate and magnitude of change, to the recent and ongoing temperature trend. Short-lived or univoltine organisms, for example, may display life history and population dynamics that mirror interannual temperature variability as well as the overall temperature trend. Long-lived organisms, on the other hand, may display very gradual and at first almost imperceptible life history and population dynamical responses to the trend in temperature while displaying clearer responses to interannual variability about the trend itself.

The second aspect of the time series in figure 1.2 we may be particularly interested in as ecologists is the difference in the magnitude of temperature change between the Northern and Southern Hemispheres. Whereas global mean surface temperature has risen by 0.4°C over the past 150 years, the change in the Northern Hemisphere alone has been closer to 0.6°C (figure 1.2b), while that in the Southern Hemisphere has been about half that much, or 0.3°C (figure 1.2c). This indicates there has been considerable spatial heterogeneity across the globe in the magnitude and perhaps also rate of warming. When spatial resolution is added to the time series shown in figure 1.2, this variability across Earth's surface becomes readily apparent (figure 1.3). These data reveal that trends in annual mean surface temperatures over the past century (figure 1.3a), and since the onset of the rapid warming trend beginning in the late 1970s (figure 1.3b), vary considerably across the surface of the Earth (Trenberth et al. 2007). During the twentieth century, warming trends were most pronounced in Low Arctic central Canada, the southwestern United States, southeastern South America, and central and northern Asia (Smith and Reynolds 2005) (figure 1.3a). This pattern was largely reproduced during the last three decades of the twentieth century, although with greater apparent surface warming in northwestern South America, southern Africa, and southern Greenland (figure 1.3b) (Smith and Reynolds 2005). The challenge such spatial variation in the strength of local and regional warming poses to the science of ecology lies in the difficulty of predicting the response of any organism with a cosmopolitan or quasi-cosmopolitan distribution to ongoing and future climate change. This difficulty is compounded many times

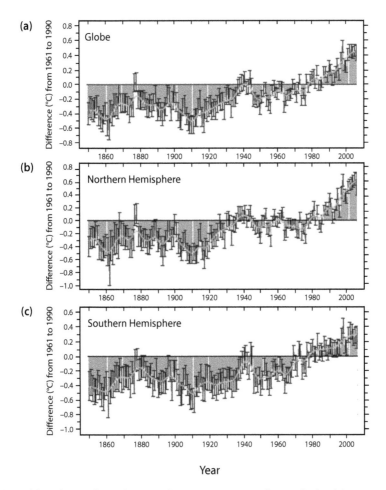

Figure 1.2. Estimates of annual mean surface temperature anomalies over land and the oceans for (a) the entire globe and (b, c) the Northern and Southern Hemispheres, respectively. Anomalies are expressed in relation to the mean for the period 1961–90. The data reveal that surface warming in the Northern Hemisphere has been approximately double that in the Southern Hemisphere. *Figure adapted from Brohan et al. (2006) and Trenberth et al. (2007).*

over if our interest lies in predicting how entire communities of organisms will respond to climate change over continental to global scales. Detecting a response to climate change in a single population or even, in some cases, multiple populations of a given species may not necessarily indicate how the species as a whole will respond to climate change throughout its distribution, especially if this distribution encompasses locales that have undergone widely different local temperature changes. Similarly, detecting responses to climate

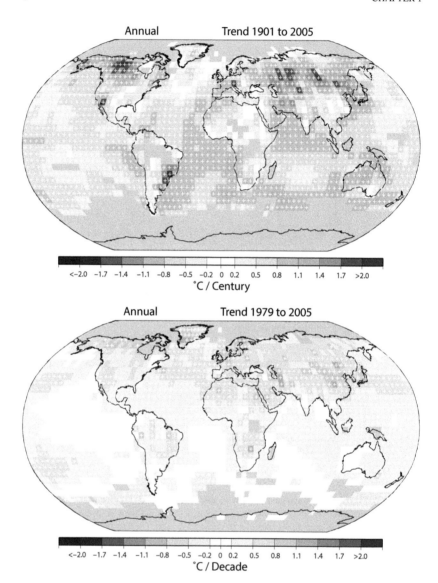

Figure 1.3. Spatial variation across the surface of Earth in the direction and magnitude of surface temperature trends (top) over the entire twentieth century and (bottom) during the period of most rapid warming, since the onset of the Northern Hemisphere warming trend in the late 1970s. Relative shading in both panels indicates the linear trend in annual mean temperature over the period indicated in each panel, and blank cells or gray areas indicate insufficient data for estimating a trend, which required a minimum of 66 and 18 years of data for the top and bottom panels, respectively. *Reproduced from Trenberth et al. (2007) and derived from data in Smith and Reynolds (2005).*

change in the species constituting a community of interest may indicate nothing at all about how the entire community will respond to future increases in temperature. The challenges inherent in such evaluations are explored in subsequent chapters.

Comparing recent temperature estimates with proxies of global mean surface temperatures derived from tree rings, sediment layers, and ice cores reveals that temperatures during the past few decades have exceeded those over the past two millennia (Mann et al. 1999; Mann and Jones 2003). Examining the recent temperature trend through a time-windowing approach reveals further interesting details that are of ecological significance. When the data on average global temperatures are displayed as decadal time series, it becomes evident that the rate of warming has not been constant over the past 150 years (figure 1.4) (Mann and Jones 2003; Trenberth et al. 2007). In each successively more recent multidecadal period plotted in figure 1.4, the slope of temperature change over time increases. The rate of change over the past twenty-five years is approximately four times greater than the rate of change over the 150-year period as a whole. This would seem to indicate that the rate of warming is increasing. From an ecologist's perspective, this should make us wonder whether organisms will be able to keep pace with the rate of temperature change if it continues to increase. A more refined perspective on this question would try to identify how responses to increasing rates of environmental change might vary among taxonomic groups with different life history strategies and generation times. For instance, certain species with highly plastic life history strategies or short generation times might be better able to match the pace of temperature change than might organisms with more highly conserved life history strategies or longer generation times.

Finally, it is worth noting as well that changes in interannual variability in temperature may prove just as important to the ecology of some species as the temperature trend itself. Here I refer not simply to heat waves or cold spells and the frequency of their occurrence but also to the magnitude of negative autocorrelation in temperatures between successive years. As we will examine in chapter 4 on population dynamics, abiotic fluctuations, when of sufficient magnitude, may influence the stability of population dynamics. Furthermore, the magnitude of serial (temporal) and spatial autocorrelation in temperatures may change as warming continues. Both of these parameters have the potential to influence ecological processes. For instance, increasing temporal autocorrelation in climatic conditions may lead to increases in population size in some types of systems that inevitably lead to population crashes (Wilmers et al. 2007b). Similarly, increasing spatial autocorrelation in temperatures over

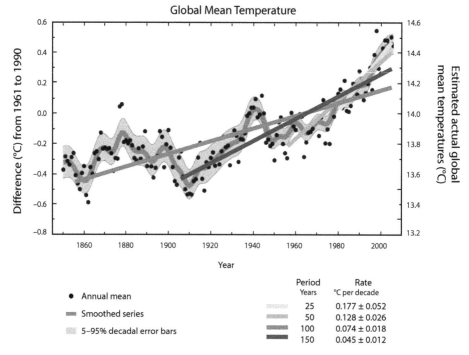

Figure 1.4. Observed annual global mean surface temperatures, based on station data, from 1850 to 2005, shown as solid black circles. The left-side *y*-axis is scaled to depict the data as annual anomalies from the mean for the period 1961–90, and the right-side *y*-axis shows the actual temperature values expressed as means calculated across measurements recorded at local stations. Linear trends are estimated for the entire period and for the most recent 100 years, 50 years, and 25 years, with rates for the respective periods shown as the slope of the linear regression for each. *Reproduced from Trenberth et al. (2007).*

large (i.e., subcontinental to continental) scales may increase the spatial synchrony of populations whose dynamics are environmentally entrained, thereby increasing extinction risk among those populations. For instance, temperatures recorded at weather stations along the west coast of Greenland became more spatially autocorrelated as Northern Hemisphere mean annual temperature increased, as did the population dynamics of caribou in the same area (Post and Forchhammer 2004). If increasing short-term—that is, annual to subannual— variability in abiotic conditions is a likely outcome of future climate change, this may have the potential to alter the stability and thereby persistence of some populations of organisms. Whether and how this might affect the stability properties of entire communities are questions examined in subsequent chapters.

PRECIPITATION CHANGES

Just as important in the ecology of some species as temperature changes, if not more so, have been changes in precipitation and aridification (Schlesinger et al. 1990). Here again, when we examine the abiotic data as ecologists, several features of these changes should capture our attention. Although more difficult to discern than the temperature trends described above, there appears to have been an overall decline across the globe in total annual precipitation over land since approximately 1950 (Mitchell and Jones 2005), although precipitation has begun to increase globally since approximately 1993 (figure 1.5) (Wentz et al. 2007). Moreover, it appears that variability among years in total precipitation over land has increased since approximately 1970. Considerable variation

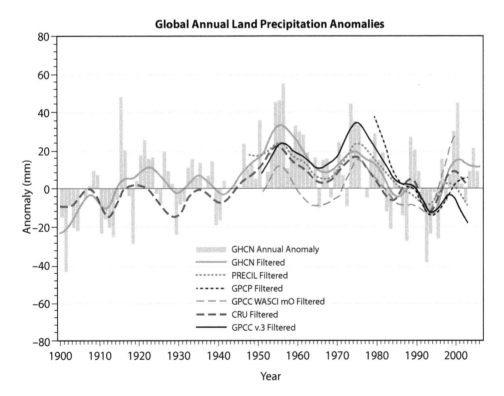

Figure 1.5. Global terrestrial annual precipitation anomalies relative to mean annual precipitation for the period 1981–2000. Source abbreviations are Global Historical Climate Network (GHCN), Precipitation Reconstruction over Land (PREC/L), Global Precipitation Climatology Project (GPCP), Global Precipitation Climatology Centre (GPCC), and Climate Research Unit (CRU). *Adapted from Mitchell and Jones (2005) and Trenberth et al. (2007).*

across the globe in trends in streamflow and surface runoff is expected to develop over the next half century, with increases in runoff projected for northern North America and Eurasia and decreases projected for western North America, southern Europe, southern Africa, and the Middle East (Milly et al. 2005). Both trends and variability in precipitation and water availability have an enormous potential to alter global primary productivity. Over the decade from 2000 to 2009, for instance, global terrestrial net primary productivity declined by approximately 0.55 petagrams of carbon annually, driven primarily by a pronounced Southern Hemisphere drying trend (Zhao and Running 2010).

To derive ecological context from such data requires, however, an indication of just how much of a decrease in annual precipitation over land has the potential to constitute a biological constraint on primary productivity if evaporation over land continues to increase with temperature. The Palmer Drought Severity Index (PDSI) provides this quantification as an indication of water stress (Alley 1984; Cook et al. 1999). The PDSI time series reveals worsening and persistent drought-like or drought-potential conditions since 1980 (figure 1.6) (Dai et al. 2004). Not only does this have the potential to alter primary productivity from global down to local scales, it also has the potential to drive shifts in

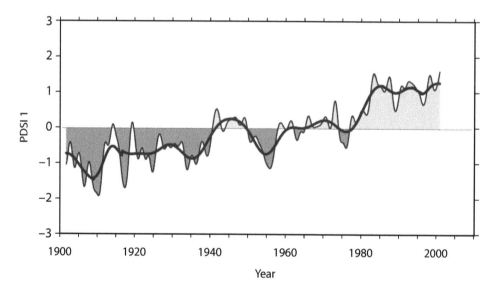

Figure 1.6. Time series for the Palmer Drought Severity Index (Alley 1984), reconstructed for the period 1901–2002. The index quantifies the amount of moisture taken up by the atmosphere relative to that released as precipitation, averaged annually over the surface of the globe. Increasingly positive values indicate increasing drought or drought-like conditions. The black trend line is the decadal average PDSI value. *Figure adapted from Dai et al. (2004).*

biome-wide plant community composition, transforming the vegetation types characteristic of the biomes we know today.

Aside from the recent trends in precipitation over land during the past several decades, there has also been considerable spatial heterogeneity across the globe in the magnitude and direction of changes in moisture balance. Whereas some regions, such as north temperate and south temperate zones, have received greater precipitation on average since the start of the twentieth century, other regions, most notably the Tropics, have become drier (Trenberth et al. 2007). Changes in moisture balance reflect the influence of temperature increases on the return of moisture to the atmosphere, and this may, over some regions, reflect increasing aridity despite increasing total annual precipitation. Spatial variation in the strength of the PDSI reflects this relationship, with increasing drought severity in tropical Africa, Central America, Malaysia, and the Amazon basin (Dai et al. 2004). Migratory species that travel along north-south gradients between breeding and winter ranges must contend with widely divergent precipitation trends and the consequences of these for resource availability at different stages during their annual life history cycles. As well, the spatial variation in the PDSI trend over the twentieth century suggests that primary productivity may respond differently to temperature trends where precipitation has increased, such as in Scandinavia, the U.S. Midwest, western Russia, and southeastern South America, compared to where it has declined along with rising temperatures, such as in equatorial and subequatorial Latin America and Africa.

CHANGES IN SNOW AND ICE COVER

Unlike temperature and rainfall, snow and ice are not simply abiotic parameters; rather, they should be considered in an ecological context as habitat for many species that are either snow or ice dependent or even obligate. Over the past three decades, Arctic sea ice has disappeared at the rate of 56,000 km^2 per year, precipitating Arctic-wide changes in tundra vegetation (Bhatt et al. 2010), while snow cover in the Northern Hemisphere has declined at a rate of approximately 100,000 km^2 per year (figure 1.7) (Post, Forchhammer, et al. 2009; Brown et al. 2010). To put the rate of loss of Arctic sea ice into perspective, deforestation in the Brazilian Amazon basin since the late 1970s has resulted in the loss of approximately 18,000 km^2 of rain forest per year, or one-third the rate of Arctic sea ice loss. There is likely little doubt in the minds of most people that deforestation in the Amazon basin threatens the persistence of animal species endemic to that region, and yet most people likely do not regard

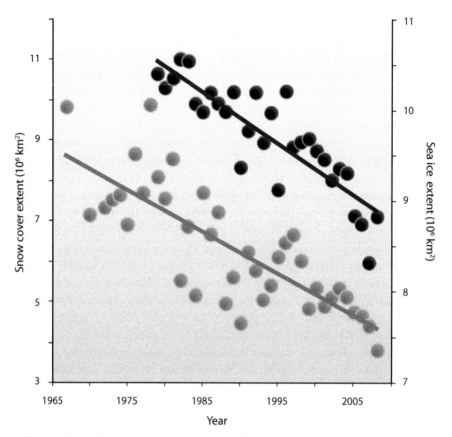

Figure 1.7. Annual totals for snow cover extent (gray) and sea ice extent (black), and their linear declines, over the Northern Hemisphere since 1966 and 1979, respectively, in millions of square kilometers. Both sea ice and snow provide crucial habitat and refugia for many snow- and ice-obligate and snow- and ice-associated species of animals in the Northern Hemisphere. *Modified from Post et al. (2009b).*

melting sea ice and vanishing snow cover from a similar perspective as to the difficulties such losses present to species that are dependent on ice and snow as habitat (Ray et al. 2008). Snow cover in particular is important to many species as cover during the period of offspring vulnerability, as a thermal refuge for periods of torpor or temporary inactivity during extreme cold, as an avenue of escape from predators, or as protection for delicate tissues in some plant species from wind and ice crystal abrasion during winter. Indirectly, trends toward earlier annual timing of snowmelt may influence the reproductive ecology of many plant species in alpine and arctic environments, where emergence and flowering are sensitive to the persistence of snow cover. As we will see in chapter 3, the decline of some species of alpine forbs in the U.S. state of Colorado may be related to increasingly early snowmelt.

What matters to ice-dependent and ice-associated species is not simply the total area covered by ice each year but also the quality of this ice. To draw another analogy between deforestation and diminishing sea ice as loss of habitat for ice-dependent and ice-associated species, we may compare multi-annual ice to old-growth forest. Not only is the total amount of sea ice declining across the Arctic, there is evidence that the amount of multi-annual ice is also declining (Barber and Hanesiak 2004). This phenomenon is comparable to the loss of old-growth forest. Ice that reforms every year and melts away later in the same year is annual ice; such ice is less stable than multi-annual ice and likely supports less life than old ice, in much the same way that the monospecific stands of second- or third-generation forest do not compare to old-growth forest in the diversity and abundance of species they support. Species that may be at particular risk of suffering declines as a result of decreases in the extent of snow cover and sea ice include wolverines (Schwartz et al. 2009; Brodie and Post 2010) and polar bears (Regehr et al. 2007; Molnár et al. 2008, 2010), respectively, but more subtle changes to the abundance and dynamics of flora and fauna may derive from the indirect effects of advances in the timing of snow and ice melt in northern environments, as chapter 3 explores in greater detail.

EL NIÑO–SOUTHERN OSCILLATION

The El Niño–Southern Oscillation (ENSO) phenomenon is characterized by periodic fluctuations in sea-surface temperatures in the equatorial Pacific Ocean that influence the seasonal, annual, and decadal magnitude and spatial patterns of precipitation and evaporation over land (Rasmusson and Carpenter 1982; Ropelewski and Halpert 1987; Trenberth and Hurrell 1994). During El Niño events, trade winds either stall or blow eastward from the western Pacific Ocean, elevating ocean surface temperatures across the equatorial Pacific and reducing cold-water upwelling along the Pacific coast of South America (Ropelewski and Halpert 1987). On land, El Niño conditions are associated with elevated temperatures and increased aridity in southeastern North America, northeastern South America, and southeastern Africa (Trenberth and Hurrell 1994).

Biological responses to El Niño events are well documented, and their recognition represents some of the earliest documentation of ecological consequences of contemporary climate change (Barber and Chavez 1983). Suppression of cold-water upwelling along the Pacific coast of South America during El Niño events reduces the productivity of near-shore waters because it drives phyto- and zooplankton populations into deeper, offshore waters as they track thermohaline zones that are favorable for their survival and reproduction (Barber and Chavez 1983). In turn, a suite of ecological consequences cascades throughout marine ecosystems as planktonic productivity declines. These consequences include the collapse of sardine fisheries, reduced recruitment in seabird colonies, die-offs of entire cohorts of sea lion pups as adults are forced to forage farther and farther from shore, and failed reproduction in Galápagos finches (Barber and Chavez 1983; Grant and Grant 2002). During the 1997–98 El Niño event, carbon retention by the equatorial Pacific Ocean increased as oceanic CO_2 efflux declined (Chavez et al. 1999). In Borneo, droughts accompanying El Niño events appear to trigger flowering and fruit production in dipterocarps (Brearley et al. 2007), but forest fires associated with El Niño events have led to 80–95 percent reductions in the densities of some species of fruit trees Malayan sun bears depend on for food (Fredriksson et al. 2007).

Projections of ENSO behavior and the expected frequency of El Niño events as Earth's climate continues to warm do not provide a clear picture of what to expect. The magnitude of El Niño events appears to be strengthening (figure 1.8). Expressed as the annual deviation from long-term mean sea-surface

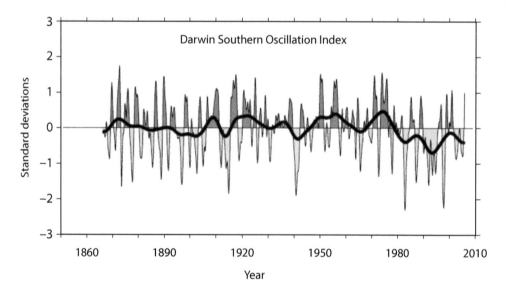

Figure 1.8. Time series of annual values of the Darwin Southern Oscillation Index, an index of the El Niño–Southern Oscillation, calculated as the difference between the normalized sea surface pressure at the Tahiti and Darwin stations between 1866 and 2005. Strongly negative values of the index are indicative of El Niño years. *Adapted from Trenberth et al. (2007).*

temperature, a time series of annual values of the Darwin Southern Oscillation Index suggests increasingly strong negative anomalies, indicative of El Niño conditions, between 1866 and 2005 (figure 1.8). Reports that the frequency of El Niño events may increase with future warming, and in fact may already be increasing, have been challenged. However, even if their frequency remains constant, increasingly strong El Niño events would certainly have the potential to disrupt ecosystems throughout the Tropics and subtropics.

PALEOCLIMATIC VARIATION

A crucial consideration as we examine the ecological consequences of current and future climate change is how contemporary climate change compares with what has occurred in the past. I refer to this consideration as crucial because I believe there is a great deal to be learned about the consequences of current climate change by looking at what happened to Earth's biota during previous warming episodes. I also believe there are some critical unknowns to be unveiled by considering the differences between contemporary climate change and paleoclimatic change.

The paleoclimatic record from the Cenozoic through the Pleistocene, which brings us to the present era, the Holocene, readily reveals several features of interest. First, Earth has been much warmer on many occasions in the deep past than it is today (Zachos et al. 2001). Second, climatic changes in the deep past appear to have occurred over much longer time scales than in the recent past (Zachos et al. 2001). This observation may, however, be an artifact—to some extent—of the greater resolution of more recent temperature proxies. Finally, there has been a close association between atmospheric temperatures and CO_2 concentrations over the past 60 million years (Pearson and Palmer 2000), obviously well before the Industrial Revolution, which heralded the most recent increase in atmospheric CO_2 and global temperatures. With respect to the first observation, however, we may be tempted to ask why it matters, in an ecological context, that Earth is warming now, given that it has been much warmer in the past. The most tempting reply may be that widespread extinctions have also occurred in the past, but the rebuttal to this would, of course, be that not all extinctions can be linked to past climatic changes, and, perhaps more important, that many taxonomic groups have persisted despite dramatic climatic changes in the past.

The next chapter is devoted to the most recent period of warming in Earth's history, the Late Pleistocene, but an introduction is warranted here. The Late Pleistocene may constitute a very apt model of the ecological consequences of warming that is ongoing today because it may have been comparable in rate to current warming, the Pleistocene-Holocene transitional warming having occurred over the course of decades (Alley et al. 2003). Important distinctions must be drawn, however, between current climate change and paleoclimatic changes, especially the most recent episode. First, the baseline temperature from which warming is proceeding today far exceeds that of the colder, drier Pleistocene. Second, human population size and growth rate, as well as the extent of human modification of the environment, outweigh those of early hominids or even of anatomically modern humans during the Pleistocene. Humans currently exert greater pressure on species in all of Earth's biomes than they have at any time during their existence, either through habitat destruction or through direct exploitation. Thus, taxonomic groups that persisted through previous climatic changes may not find the strategies that allowed them to do so then similarly viable today simply because of the difference our presence makes this time around. Indeed, understanding and foreseeing the outcomes in terms of species persistence of the nexus of climate change and human exploitation presents a particularly difficult challenge (Post and Brodie in review). This may be especially, and perhaps uniquely, the case in the Tropics, where human pressure on rare and endemic species is highest and where research on the ecological consequences of climate change is made difficult by the complexity of species interactions (Brodie, Post, and Laurance 2012).

STUDYING THE ECOLOGICAL
EFFECTS OF CLIMATE CHANGE

Conducting research on the ecological effects of climate change is an inherently multidisciplinary undertaking. It requires an understanding of pattern and process, of abiotic and biological dynamics, and it relies on an appreciation for the nuances of scale characteristic of ecological dynamics (Levin 1992), as well as diverse disciplinary approaches (Schlesinger 2010). That said, approaches to studying ecology in a climate change context fall principally into three categories of research: observational, experimental, and mathematical.

Observational studies generally make use of long-term data to draw relationships between records of abundance or occurrence of species, or springtime events, and climate change. A prototypical example of such studies is the analysis of a two-century record of phenological observations known as the Marsham data (Sparks and Carey 1995). This remarkable record of spring indicators was begun in 1736 by members of the Marsham family on their estate north of Norfolk, England (Marsham 1789; Margary 1926) and continued through five generations of Marshams until records ceased in 1947 (Sparks and Carey 1995). The data consist of observations indicative of the annual arrival of spring, including first flowering records for several species of plants; "leafing" dates of multiple species of trees, which are the dates on which leaves reached a specified size; and arrival or appearance dates of migratory birds, butterflies, and unspecified anurans (Sparks and Carey 1995). The utility of such a data set for investigating ecological responses to climate change is immediately obvious. First, the record spans two centuries, a healthy length of time by the standards of ecology, though with gaps in the records. Second, it quantifies variation in life history events of multiple species representing several useful categories of organisms for comparative analyses, including plants and animals, vertebrates and invertebrates, migrants and residents, forbs and trees, and homeotherms and ecototherms. Third, it quantifies the timing of events during two important periods, before and after the Industrial Revolution, or before and after humans markedly influenced Earth's climate. An additional remarkable feature of this data set is the fairly high degree of consistency in the quality and manner of recording of observations throughout the two-century period it covers. For such long-term data sets, this is probably a unique characteristic.

Sparks and Carey (1995) undertook an analysis of the entire Marsham phenological record that is fairly typical of those used with observational data. Their analysis focused on identifying correlations between phenological events in the data set and monthly weather data from a nearby station, and on fitting regression models that included time to account for temporal trends in the data and multiple candidate predictor variables from the list of weather variables displaying correlation with the phenological variables (Sparks and Carey 1995). This approach is useful for identifying trends in the ecological data and potential abiotic drivers of those trends. An obvious limitation of this approach is that other factors that have also changed over this period but are not included in the regression models might have contributed just as much or more to the observed ecological dynamics. Nonetheless, it is difficult to turn away from the opportunities afforded by such unique data sets in the interest of examining relationships between temperature or precipitation trends and ecological dynamics. Another limitation of the use of long-term ecological data sets in studying the effects of climate change is that many such collections of data were begun for purposes other than understanding ecological responses to climate change.

The Marsham record is not immune to this criticism, although it is probably closer in character to meeting this requirement than are many other similar long-term records because it was initiated with an interest in indicators of the arrival of spring. The long-term ecological monitoring project at the High

Arctic research station Zackenberg, in northeastern Greenland, is an example of a comprehensive, standardized program developed specifically to collect multi-annual data for the purpose of studying ecological responses to observed climate change (Forchhammer et al. 2008; Meltofte and Rasch 2008; Meltofte et al. 2008). Long-term observational data may be more readily available in Europe than in the United States, in part because there is a longer tradition of estate record keeping in Europe (Lauscher and Lauscher 1990). Nonetheless, the historical records of natural history observations made by American naturalists such as Aldo Leopold and Henry David Thoreau have come into use recently in analyses comparing the timing of events recorded earlier in the twentieth century with the timing of events at the same locations in the late twentieth century or early twenty-first century (Bradley et al. 1999; Miller-Rushing and Primack 2008).

There is some disagreement between advocates of experimental and observational approaches to the study of the ecological consequences of climate change (Agrawal et al. 2007b; Sagarin 2007; Wolkovich et al. 2012), and it is my personal impression that the experimental approach to this problem may be stronger in North America and the observational approach may be stronger in Europe. Experimental investigation of ecological responses to climate change most typically involves increasing near-surface temperatures on experimental plots. This can be achieved using elevated infrared heaters (Harte et al. 1995), warming coils buried under the soil surface (Bokhorst et al. 2008), or passive warming devices called open-top chambers (OTCs) that elevate near-surface temperatures by hindering the transverse movement of air over the ground within the enclosed chamber (Henry and Molau 1997; Marion et al. 1997; Sullivan and Welker 2005). Each of these approaches has inherent drawbacks. Overhead infrared heaters and warming coils both require a power source, and all three methods can alter the near-surface moisture regime or lead to soil drying. Additionally, OTCs, although designed to minimize unwanted side effects such as interference with evaporation and precipitation, may nonetheless create a rain shadow under the upper edge of the cone. In the Arctic, OTCs have also been employed in combination with snow fences to manipulate snow depth and, thereby, dates of snowmelt, to investigate their influences on plant community dynamics (Wahren et al. 2005) and plant ecophysiology (Welker et al. 2005). The timing of snowmelt is an important driver of the timing of life history events in a wide array of terrestrial species (Høye et al. 2007). In addition to temperature manipulations, some investigations of ecological responses to climate change have employed soil nutrient additions (Chapin and Shaver 1985; Hobbie and Chapin 1998; Shaver and Jonasson 1999; Arens et al. 2008). This approach has enjoyed particular favor in high-latitude studies.

In the Arctic, for instance, plant growth and ecosystem function are limited by the availability and turnover of soil nutrients, especially nitrogen (Shaver and Chapin 1991; Chapin et al. 1992a, 1992b; Shaver and Kummerow 1992; Chapin et al. 1995). The addition of fertilizer to experimental plots is intended, in this approach, to simulate the increased availability of soil nutrients in response to increased soil microbial activity and organic matter decomposition under warmer conditions.

Under the umbrella of mathematical approaches to studying ecological effects of climate change I would include statistical modeling, simulation modeling, and data-driven environmental niche modeling. The first form, statistical modeling, obviously goes hand-in-hand with long-term data collection. This approach has been widely used to quantify relations between recent climate change and population dynamics or life history variation. In fact, the study of ecological response to climate change has engendered the development of statistical modeling approaches specific to this task (Forchhammer, Stenseth, et al. 1998; Sæther et al. 2000, 2003; Post and Forchhammer 2001; Post et al. 2001; Ellis and Post 2004), based on preceding developments in the statistical analysis of time series data (Tong 1990; Bjørnstad et al. 1995; Framstad et al. 1997; Leirs et al. 1997). The main limitations of statistical analysis of time series data in a climate change context have to do with the problem of relating pattern to process, and to the frequent use in such models of harvest data rather than actual counts of individuals (Stenseth et al. 1999).

Simulation modeling and environmental niche modeling are closely related attempts to project the responses of species, communities, and ecosystems to expected future climate change. Ecosystem process models and gap models are forms of simulation modeling that attempt to quantify the carbon uptake potential of an ecosystem or biome in response to elevated atmospheric CO_2 and temperature. Ecosystem process models, such as the Terrestrial Ecosystem Model (TEM), work by—in a very simplistic interpretation—scaling up whole-plant ecophysiology of the dominant species in a biome to the scale of that biome (Melillo et al. 1993). Gap models, such as the Ecosystem Demography model, are demographic models that simulate the birth, reproduction, and death of individual plants of the dominant species in a community and scale up these processes to estimate carbon uptake by the entire biome (Moorcroft et al. 2001). Environmental niche modeling, on the other hand, is used to project the distribution and abundance of species under climate change scenarios (Jeschke and Strayer 2008). In this approach, the abiotic correlates of the presence and abundance of a species throughout its current distribution are used to predict the presence or absence of that species in accordance with changes in those abiotic correlates in time and space (Peterson et al. 2002). The main criticism leveled at environmental niche modeling is that abiotic environmental variables do not necessarily capture important biological interactions that also limit the distribution and abundance of organisms (Jeschke and Strayer 2008); this criticism is addressed in greater detail in chapter 5.

No single approach to the study of ecological consequences of climate change is adequate on its own, or without its particular set of limitations, some of which have already been acknowledged. Observational studies in the best cases are powered by long-term data, spatial replication of those data, and large-scale coverage, but suffer primarily from an inability to determine causality in the relationships they uncover (Agrawal et al. 2007b). Experiments, when properly designed, have the capacity to assign causality, but often suffer from small scales of study in both space and time. Even long-term, large-scale experiments suffer limitations arising from logistical and practical constraints, such as insufficient replication (Krebs et al. 1995). As well, incorporating biological realism into experiments and determining the magnitude of the treatment applied in them are far from trivial undertakings. For instance, microcosm experiments in the laboratory are useful for controlling unwanted background processes (Petchey et al. 1999) but may not translate readily into an understanding of ecological dynamics in nature. Similarly, detection of a significant response to resource manipulation may be an artifact of the level of resource manipulation applied rather than a biologically meaningful identification of the magnitude of resource limitation in nature. Proponents of time series analysis steadfastly defend the capacity of this approach to identify biological interactions in the structure and behavior of time series data (Turchin 2003), but withering criticisms of this view of the world are regularly leveled by experimentalists (Krebs 1998, 2003) and even, in some cases, by time series analysts themselves (Berteaux et al. 2006).

The argument for the utility of time series analysis in detecting the influence of interactions within and among species relates to what is referred to as the dimensionality of the time series data under inspection (Royama 1992). The dimensionality, or order, of the time series refers to the number of steps (n) back in time between which there is a relationship between an observation in year t and year $t - n$ (Royama 1992). First-order dynamics are assumed to indicate density dependence, second-order dynamics indicate a role of reproduction or trophic interactions in dynamics, and third-order dynamics indicate both (Framstad et al. 1997; Forchhammer, Stenseth, et al. 1998; Bjørnstad and Grenfell 2001). Similarly, as will be demonstrated in chapter 4, the order or dimensionality of climate terms in statistical models of population dynamics indicates direct or delayed effects of climate, the latter of which are presumed to operate through reproduction or trophic interactions (Forchhammer, Stenseth, et al. 1998; Post and Stenseth 1999; Post and Forchhammer 2001). Somewhat ironically, the best support to date that I am aware of for the capacity of time series analyses to capture biological interactions in determining the dimensionality of time series data derives from a laboratory experiment (Bjørnstad et al. 2001).

THE STUDY SITE AT KANGERLUSSUAQ, GREENLAND

Throughout this book, I refer occasionally to the results of my own research at my long-term study site in Low Arctic West Greenland, where I began obser-vational fieldwork in 1993 and where I implemented an ongoing, uninterrupted experimental complement to my observational work in 2002. My approach to the study of the ecological consequences of climate change thus employs ob-servation, experimentation, and modeling. I believe this constitutes a powerful combination of approaches, each one of which in isolation suffers from its own limitations but which in concert provide a potent toolkit. I began fieldwork at what would develop into a long-term study site near Kangerlussuaq, Green-land, in 1993. During that initial field season, my research focused on quanti-fying the timing of the caribou calving season in relation to the timing of the onset and progression of the plant growing season. Over the period for which station data are available, from 1974 to the present, mean annual temperature in Kangerlussuaq has increased at a rate of 0.68°C per decade (figure 1.9a), with the rate of warming since the initiation of my fieldwork there exceeding this figure by a factor of three (figure 1.9b).

The research protocol at my study site has been described elsewhere (Post 1995; Post et al. 2003, 2008a, 2008b; Post and Pedersen 2008) but will be reviewed briefly here. On a daily or near daily basis throughout the plant grow-ing season, permanently marked plots distributed over an approximately 6 km^2 area within the study site are monitored for plant phenology. These observa-tions include a tabulation of species present and their phenological states on all plots. The number of plots has increased from twelve in 1993–2008 to twenty-seven since 2009. Concurrently with observations of plant phenology, the entire study site, including an adjacent core calving area that has been in use by caribou since approximately 4,000 years years before the present (YBP) (Thing 1984; Meldgaard 1986), is monitored daily for the presence of caribou and muskoxen and their calves. All adults, yearlings, and calves are counted and recorded and their locations noted. These data are used to quantify the tim-ing of the caribou calving season and to monitor both herbivore species' use of the study site in relation to an ongoing herbivore exclosure and tundra warming experiment. This experiment consists of excluding both species of herbivore

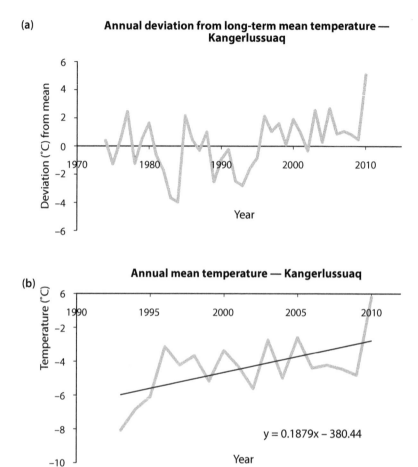

Figure 1.9. (a) Annual temperature anomalies derived from monthly temperatures at Kangerlussuaq Airport, Greenland, expressed as annual deviations from the mean for the period 1974–90. (b) Annual mean temperature in Kangerlussuaq since the onset of my research there in 1993. *Data obtained from the Danish Meteorological Institute (www.dmi.dk).*

from six 800 m² circular, fenced-in areas adjacent to control sites of the same size. Inside and outside these exclosures we have applied a passive warming treatment using OTCs that elevate surface temperatures by 1.5–3°C (Post et al. 2008a). Chambers are put in place on permanently marked plots in May of each year and removed again in August. Plots that receive this warming treatment are paired with adjacent control plots that remain at ambient temperature. The biomass of all standing vegetation, both live and dead, is quantified in late July or early August of each year using a nondestructive point-frame method that relates the numbers of hits by a lowered pin to the biomass on each plot. Data recorded in this manner also allow us to identify and monitor the number of plant species, and changes in these, on each plot. Over the course of this study we have been able to monitor changes in the timing and progression of the plant growing season, plant community dynamics in relation to warming and herbivory, offspring production by caribou and muskoxen, and the population dynamics of both species. Recently we have begun monitoring gas flux on the experimental plots to measure the effects of herbivory and experimental warming on ecosystem CO_2 dynamics. Many of the results of this study will be detailed in subsequent chapters where relevant.

Pleistocene Warming and Extinctions

The previous chapter focused on the past 150 years of climate change, which I referred to as recent and rapid climate change. Foreseeing the ecological consequences of the expected climatic changes that are projected to occur over the next century is an exceedingly complex undertaking. As suggested elsewhere (Post and Forchhammer 2006), however, it may be possible to garner insights into the potential consequences of future rapid climate change by considering the dynamics of plant and animal species and species assemblages during the Earth's most recent period of rapid warming.

THE PLEISTOCENE ENVIRONMENT
AS INDICATED BY ITS FAUNA

Examination of the faunal assemblages of the Pleistocene reveals details about their foraging strategies that indicate the environmental conditions of the Pleistocene, and how such conditions differed from contemporary conditions in the same regions. One of the most thoroughly studied of the Pleistocene biomes is the mammoth steppe of Beringia, the Pleistocene land mass that extended eastward from Siberia into interior Alaska. The accumulation of ice over land and coastal boundaries during the Pleistocene reduced sea levels sufficiently to expose land under the shallow Bering and Chukchi Seas (Sher et al. 2005). The resulting land bridge served as an important corridor for migration between Asia and North America. The term "mammoth steppe" was coined to describe the vegetation of this region because it was typified by arid, cold, windswept conditions that harbored a wide array of graminoids and forbs favorable as habitat for large herbivorous mammals (Guthrie 1982, 2001; Sher et al. 2005).

The picture of Beringia that emerges from the fossil evidence is one of a faunistically more diverse biome than any found in northern latitudes today (Guthrie 1968, 1982). Among the steppe species to survive the Pleistocene-Holocene transition and that remain extant in the tundra today are caribou, elk, muskoxen, Dall's sheep, moose, and wolves. Contemporary with these species were many more that were driven to extinction from the Beringian biome between the Late Pleistocene and its termination, including horses, steppe bison, helmeted muskoxen, dire wolves, lions, asses, camels, moose-stags, short-faced bears, sabre-toothed cats, badgers, and, of course, mammoths (Guthrie 1968, 1992; Matheus et al. 2003; Harington 2011). A special exception to these two lists is the saiga antelope, which survived beyond the end of the Pleistocene through range contraction to the Mongolian steppe, a low-latitude, high-elevation analogue to the mammoth steppe.

The three most abundant groups of large mammals in North America during the Pleistocene were horses, bison, and mammoths. Horses were of particular importance in terms of their contribution to Pleistocene faunal diversity and abundance, constituting one of the most geographically widespread species groups of Late Pleistocene North America (Guthrie 2003). All three of these groups were grazers, subsisting on a diet primarily if not exclusively comprised of grasses or grasslike sedges (Guthrie 1982). Their prevalence among, and indeed dominance of, the Pleistocene megafaunal species assemblage argues strongly for a biome vastly different from the boreal forest and tundra of the far north today (Guthrie 1968, 2006). For instance, of the nineteen species of large mammals recovered in Pleistocene fossil assemblages collected at four sites in interior Alaska, fourteen were ungulates and five were carnivores (Guthrie 1968). The mean live body mass among Pleistocene ungulates occurring at these sites was estimated at approximately 600 kg (Guthrie 1968). Today, only four of those species survive in the same region, with an average live body mass of approximately 180 kg (Guthrie 1968). Among the Pleistocene ungulates represented in the fossil assemblages at these sites, more than half, or eight out of fourteen species, were predominantly grazers, whereas among those species surviving there today, only one, Dall's sheep, is considered primarily a grazer (Guthrie 1968). Hence, if interior Alaska was representative of environmental conditions in Beringia in general, or indeed of the Arctic in general during the Pleistocene, we may surmise that it was much more amenable to the habitat requirements of grazers subsisting primarily if not exclusively on a diet of grasses and grasslike sedges then than it is today (Guthrie 2001).

We may also, after considering such numbers as those given above, wonder why Pleistocene mammalian diversity in the Arctic and far north was so much greater than that which is characteristic of the same region today. To support such a diversity and abundance of mammalian herbivores, the environment must have been highly productive (Sher et al. 2005). Even though it was colder and more arid, Beringia was likely characterized by a greater productivity and abundance of graminoids and forbs than the region is today (Blinnikov et al. 2011), and its plant communities were likely dominated by a higher diversity of plant species, belonging primarily to the Poaceae, Cyperaceae, and Asteraceae (e.g., *Artemesia*) families (Guthrie 2006). The floral diversity and productivity likely were not the result of climatic conditions alone but a product of interactions between climatic conditions and grazing pressure by large and small herbivores (Blinnikov et al. 2011), in accordance with the facilitation/promotion hypothesis outlined in the preface. By contrast, contemporary plant communities in Alaska and Siberia are dominated by relatively species-poor, unproductive, and in some cases chemically defended woody plants and shrubs, communities that do not support large numbers of herbivores and that are almost completely devoid of graminoids, which specialist grazers depend on. This difference indicates a much broader niche diversity during the Pleistocene than today, analogous to a checkerboard mosaic of environmental heterogeneity during the Pleistocene that contrasts with the simpler "striped" mosaic today (Guthrie 1989) and may hold clues as to what precipitated the widespread extinctions that occurred as the Pleistocene warmed into the Holocene.

The Pleistocene epoch, spanning approximately 1.3 million to 11,000 YBP, was generally colder, drier, and windier than the Holocene epoch we now live in. Dramatic shifts in climate during the Pleistocene are evident in paleoclimatic records. For instance, ice core data from Greenland that have been used to reconstruct a high-latitude climate record of the past 110,000 years indicate multiple sudden shifts in climate of 8–16°C warming in the span of a decade (Alley et al. 2003; Barker et al. 2012). The temperature over interior Greenland, for instance, rose abruptly by approximately 10°C several times during the Late Pleistocene, and cooled again gradually after each increase (figure 2.1). Alternatively, the Younger Dryas event near the end of the Pleistocene represented a rapid cooling of climate in the northern latitudes, followed by a warming that was also rapid, yet more gradual than the preceding cooling—coincident with rising global methane concentrations—that led to the termination of the Pleistocene and the onset of the Holocene (figure 2.1). Such shifts, or abrupt climatic changes, can occur in response to gradual but persistent forcing of the climate system that result in threshold changes in which climate suddenly switches to a new regime, akin to leaning over the side of a canoe until it unexpectedly tips (Alley et al. 2003). The relevance to contemporary climate change could not be clearer: even though it is possible to assign certainty to future gradual changes in climate based on relationships discerned through retrospective analyses of climate time series, predicting threshold changes is difficult, to say the least.

The popular conception that equates the Pleistocene epoch with the Ice Age, or even with a period of four major glacial cycles, is, however, mistaken. In reality, the Pleistocene was a period of multiple glacial and interglacial phases,

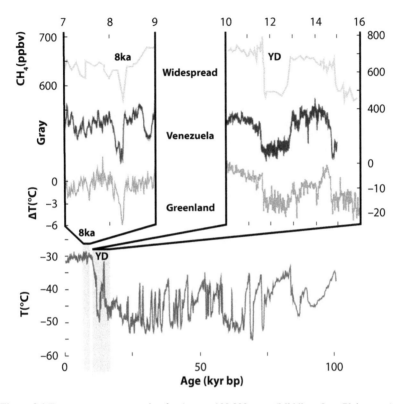

Figure 2.1 Temperature reconstruction for the past 100,000 years (Middle to Late Pleistocene) based on the GISP2 ice core extracted from the summit of the Greenland ice cap. The temperature estimates were derived from the ratio of ^{18}O to ^{16}O ($\delta^{18}O$) and from fractionation of the gas mixtures trapped in the ice (Severinghaus et al. 1998). The time scale on the x-axis proceeds from the deeper past on the right toward the present at the left. Temperature dynamics during the Younger Dryas (YD) and the most recent 9,000 years are exploded to illustrate the multiple occurrences of rapid warming. *Adapted from Alley et al. (2003).*

now referred to as oxygen-isotope stages, during which cooling phases were prolonged and warming phases were of much shorter duration but much more rapid (Shackleton et al. 1990), and, in the case of the terminal warming event at the Pleistocene-Holocene transition, occurred on the order of decades (Taylor et al. 1997; Alley et al. 2003). Glacial-interglacial dynamics were especially pronounced during the Late Pleistocene, from 70,000–60,000 YBP and onward. This period was characterized by multiple shorter-term temperature fluctuations within glacial phases, referred to, in the case of cooling events, as stadials, and in the case of warming events within glacial phases as interstadials. An even finer scale of climatic fluctuation, or flickering (Graham 2005), is

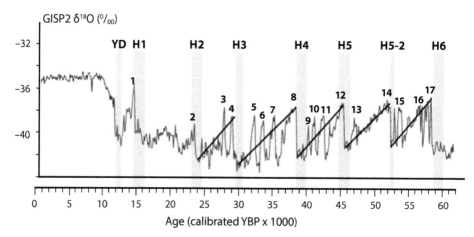

Figure 2.2. Shorter-term temperature cycles over the past 60,000 years, during the Late Pleistocene, reconstructed from the $\delta^{18}O$ signature in the GISP2 ice core from Greenland. The time series in this figure illustrates the occurrence of Heinrich events, when massive blocks of ice are released into the oceans by glaciers and ice caps, labeled H$_i$, and millennial-scale Bond cycles that terminate with Dansgaard-Oeschger events. *Reproduced from Graham (2005).*

evident during the Late Pleistocene within stadials, in which Heinrich events, characterized by the fracturing of ice shelves and the release of icebergs into the oceans, occur periodically and mark the endpoints of short-term Bond cycle cooling phases every 7,000–15,000 years that terminate with Dansgaard-Oeschger warming events every 2,000–3,000 years (figure 2.2) (Elliot et al. 1998). In contrast, the climate of the Holocene, the last 10,000 years of Earth's history, has been warm and stable. From the perspective of climate change ecology, therefore, the Late Pleistocene and the Pleistocene-Holocene transition are relevant because they represent the end of a prolonged period of climatic fluctuation on multiple temporal scales followed by rapid warming. By examining the consequences of this variability and trend for biome shifts, species losses or persistence, and community changes, we may gain insight into the ecological changes that the current warming period could portend. Not only did Earth's major biomes undergo extensive compositional changes during the late Quaternary and near the termination of the Pleistocene epoch (Jackson and Overpeck 2000; Jackson et al. 2000), they also underwent geographically large-scale redistributions (Williams et al. 2004), and did so rapidly, in some cases on the scale of decades (Williams et al. 2002; Post 2003a). Such rapid biome shifts must have posed consequences for the persistence of animal species whose abundance depended on forage availability or seasonal quality and whose niche parameters were at least partly defined by the distributions of plant species and their interactions with abiotic conditions.

BIOGEOGRAPHY AND MAGNITUDE OF PLEISTOCENE EXTINCTIONS AND CLIMATE CHANGE

The Quaternary Megafaunal Extinction was the most recent mass extinction event on Earth, and is so called because of the concentration of extinctions among mammalian fauna larger than 44 kg in body mass (Barnosky et al. 2003, 2004; Barnosky 2008). These extinctions commenced approximately 50,000 YBP, lasted until approximately 8,000–3,000 YBP, were diachronically distributed across the continents, and coincided in some cases with both major climatic shifts and the arrival or expansion of anatomically modern humans, *Homo sapiens sapiens*, especially those employing the Clovis point hunting technology (Barnosky et al. 2004). Broadly, these extinctions occurred in two waves, the first coinciding with the extinction of warm-adapted species during a period of rapid cooling between 50,000 YBP and 21,000 YBP, the Last Glacial Maximum (Barnosky et al. 2004; Nogués-Bravo et al. 2010). The second wave began approximately 21,000 YBP and lasted until approximately 8,000–3,000 YBP, spanning the period of rapid warming that characterized the transition from the Pleistocene to the Holocene (Alley et al. 2003; Barnosky et al. 2004; Guthrie 2006; Koch and Barnosky 2006; Nogués-Bravo et al. 2010). The temporal disaggregation of Pleistocene megafuanal extinctions is further evident in the timeline over which they unfolded on each of the continents. Extinctions occurred earliest in Australia, from 50,000 to 32,000 YBP, and were overlain considerably by slightly later-onset extinctions during the first wave to sweep through Eurasia, from approximately 48,000 to 23,000 YBP (figure 2.3) (Barnosky et al. 2004; Barnosky 2008). These multimillennial extinction waves followed the arrival of humans in Australia and do not, at first glance, appear to have coincided with any major climatic shift there, while those in Eurasia commenced before the appearance of *H. sapiens sapiens* but may have been precipitated by the onset of the climatic shift associated with the onset of the Last Glacial Maximum (Barnosky et al. 2004).

The period of rapid warming that followed the Last Glacial Maximum, beginning approximately 21,000 YBP, was characterized by a pulse of three additional extinctions. The first of these commenced in central North America approximately 15,600 YBP and lasted until approximately 11,500 YBP; the next was the second Eurasian extinction wave, spanning 14,000–10,000 YBP; and the most recent occurred in South America from approximately 12,000 to

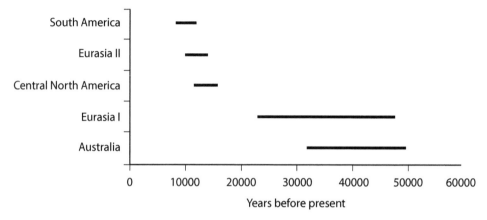

Figure 2.3. Timing and duration of the major megafaunal extinction events during the Pleistocene. The first two extinction waves commenced during the Middle Pleistocene on the Australian continent and in Eurasia, and persisted over several millennia. The second wave of extinctions began during the Late Pleistocene, was of much shorter duration, and was confined to central North America, Eurasia, and South America. As in figures 2.1 and 2.2, time proceeds from right to left along the *x*-axis. *Data adapted from Barnosky (2008).*

8,000 YBP (figure 2.3) (Barnosky 2008). All three of these events coincided with major climatic shifts toward generally warmer, wetter conditions and commenced generally in synchrony with or soon after the arrival of humans, in the case of North America, or in concert with their spread, in the case of Eurasia (Barnosky et al. 2004; Koch and Barnosky 2006). The South American extinctions are less temporally resolved but coincided grossly with the apparent arrival of humans approximately 12,900–12,500 YBP and with climatic changes (Barnosky et al. 2004).

The biodiversity loss associated with these extinctions is staggering. The mammalian megafauna of Australia suffered the greatest losses, with fourteen of sixteen genera, or 87.5 percent, disappearing in the 18,000-year interval over which extinctions unfolded on the continent (figure 2.4) (Koch and Barnosky 2006; Barnosky 2008). South America suffered a loss of 83 percent of its mammalian megafauna, or fifty of sixty genera, while North America experienced the extinction of 72 percent, or thirty-four of forty-seven, of its megafaunal genera (Barnosky 2008; Faith and Surovell 2009). Eurasia and sub-Sahelian Africa experienced comparatively milder extinctions, with the loss of 35 percent of its megafauna, or nine of twenty-six genera, in the former case and 21 percent, or ten of forty-eight genera, in the latter case (figure 2.4) (Barnosky 2008). These numbers do little, however, to convey the importance to ecosystem function and dynamics of the loss of species that were

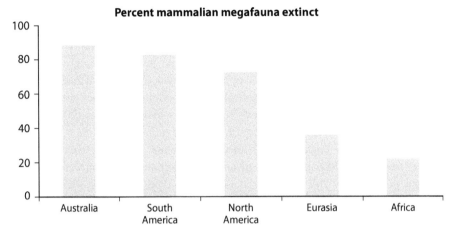

Figure 2.4. Magnitude of the megafaunal extinctions in terms of the percentage loss of megafaunal taxa, arranged in decreasing order according to the continents on which they occurred. *Data adapted from Barnosky (2008).*

in many instances likely of considerable keystone importance. The extinctions enumerated above included, for example, species such as caballoid and hemionid horses, the great Irish stag, mastodons, giant ground sloths, helmeted muskoxen, steppe bison, short-faced bears, hippopotamuses, straight-tusked elephants, woolly rhinoceroses, dire wolves, and the iconic woolly mammoth (Guthrie 1968, 2001, 2006; Matheus et al. 2003; Nogués-Bravo et al. 2010). As we will see in chapter 8, such large-bodied mammals, especially herbivores, may exert considerable influence on ecosystem function and the dynamics of ecosystem response to climate change. An additional point is worth making regarding the distinction between the two waves of extinction captured in these examples. As mentioned, the first coincided with a shift toward colder and drier conditions during the Last Glacial Maximum, whereas the second coincided with rapid warming and, perhaps less clearly, increasingly moist conditions during the Late Pleistocene. Although the first wave of extinctions appears to have been biased against warm-adapted species (Barnosky et al. 2004) and the latter against cold-adapted species (Guthrie 1968), where extinction risk is concerned, the magnitude of climate change may be just as important as the direction of change.

Assigning primary cause to the late Quaternary megafaunal extinctions has been an irresistible pursuit in the literature for decades, and one that has engendered persistent disagreement. It has been argued, for example, that the widespread, temporally concentrated, and synchronous nature of megafaunal extinctions throughout North America favors the intersection of several processes, including overhunting and climatic disruption of habitat (Faith and Surovell 2009). In contrast, the asynchronous timing of the disappearance of giant ground sloths from the South American mainland by approximately 11,000 YBP and from adjacent islands by approximately 4,400 YBP has been interpreted as evidence for a primary role of human exploitation in their extinction (Steadman et al. 2005). The reviews by Barnosky and colleagues (2004) and Koch and Barnosky (2006) suggest the concerted influences of human exploitation and climate change in most of the megafaunal extinction events since the Last Glacial Maximum, and of human exploitation itself on the extinctions in Australia. This seems to be a favored perspective on the Pleistocene megafaunal extinctions of Australia (Prideaux et al. 2007; Turney et al. 2008), where, for instance, the climate-related disappearance of browsing habitat for giant kangaroos, which were browse specialists, is considered secondary to the influence of human hunting in their extinction (Prideaux et al. 2009). Nonetheless, evidence from at least one Australian site, Cuddie Springs, has been interpreted as equivocal in relation to the question of whether the association of megafaunal remains with a human presence there is supportive of a primary role of human hunting in their extinction (Fillios et al. 2010). A recent analysis (Nogués-Bravo et al. 2010) employing environmental niche modeling or, as it is occasionally referred to, climate envelope modeling, suggests a clearer role for climate change in extinction events on all of the aforementioned continents. A more thorough exposition of the discipline of environmental niche modeling is given in chapter 5, on niche dynamics, but a brief introduction is warranted here. This approach involves quantifying the climate envelope of a species according to abiotic conditions that overlap its distribution during one climatic regime, and then using these parameters to estimate the amount of suitable habitat available to that species during another climatic regime with differing mean temperature, precipitation, or seasonality. Nogués-Bravo and colleagues (2010) used this approach to estimate the climate change footprints of the continents experiencing megafaunal extinctions represented in figure 2.4, and then to relate these coefficients to the magnitude of megafaunal extinction on those continents.

Nogués-Bravo and colleagues (2010) established as the basis for their analyses two hypotheses concerning the role of climate change in megafaunal extinctions during the late Quaternary. First, if climate change were to be implicated, we should expect to see taxonomic losses increasing with the magnitude of loss of the original climate envelopes on each continent (Nogués-Bravo et al. 2010). In other words, the climate change footprint, or total reduction in area of a certain climatic character, should be diagnostic of the magnitude of extinction if climate change were a driver of faunal turnover during the late Quaternary. The second hypothesis concerns indirectly the challenges to successful dispersal and redistribution posed by habitat loss resulting from rapid climate change during the Pleistocene-Holocene transition. According to this hypothesis, the continents experiencing the greatest magnitude of extinction should also have been characterized by the greatest distances of migration necessary to track similar climatic conditions during the period of climate change (Nogués-Bravo et al. 2010). For their analyses, Nogués-Bravo and colleagues (2010) focused on the terminal period of the Late Pleistocene, or that period from 21,000 YBP onward, when climate change was characterized by rapid warming and increasing precipitation.

In terms of the disappearance of total habitat area representative of cold and dry climates, Australia ranks first, losing 3–4 million km^2 (figure 2.5), with climate zones exhibiting these conditions retreating a total of 500 km at a rate of 17 meters per year (Nogués-Bravo et al. 2010). North America and Eurasia experienced similar levels of disappearance of areas characterized by cold, dry climates, both losing 1–4 million km^2 (figure 2.5), and undergoing northward shifts in these conditions by 1,500 km at a rate of 70 meters per year (Nogués-Bravo et al. 2010). Within these regions, however, the Arctic experienced a shift in cold, dry conditions northward at an astonishing rate of

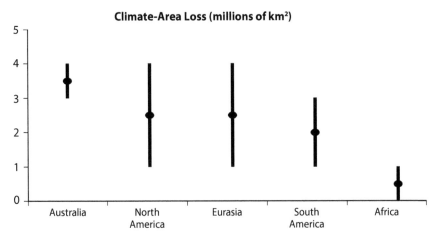

Figure 2.5. Estimated loss of climate-area in millions of square kilometers on each of the continents that experienced megafaunal extinctions during the Pleistocene. Although Australia is frequently regarded as the clearest example of human hunting–induced megafaunal extinctions, it ranks highest in total loss of climate-area during the Pleistocene. Climate-area loss is quantified based on the disappearance of climatic conditions that were present during the Pleistocene on each continent. *Data adapted from Nogués-Bravo et al. (2010).*

140 meters per year over a distance of 3,000 km (Nogués-Bravo et al. 2010). South America and sub-Sahelian Africa underwent milder losses of 1–3 million km^2 and 0–1 million km^2, respectively, both at a rate of 17 meters per year, for a total shift of 500 km (figure 2.5) (Nogués-Bravo et al. 2010). Using, conservatively, the minima of these climate-area losses together with the magnitude of megafaunal extinction on each continent, expressed as the percentage of genera going extinct on each continent according to Barnosky (2008), extinction loss across the continents scales roughly positively with the total loss of area representing cold, dry conditions owing to climate change at the end of the Last Glacial Maximum (figure 2.6). Similarly, the rank order of megafaunal extinctions among the continents scales with the rank order of their climate change footprints, with the exception of South America (figure 2.7a). Of note, the extinction rate of smaller mammals, those less than the 44 kg body mass threshold definitive of megafauna, also scales with the rank of the climate change footprint of the continents, though again with the exception of South America (figure 2.7b). Although the outlying nature of South American extinctions, which exceed what would be expected solely on the basis of the magnitude of Late Pleistocene climate change experienced there, may to some extent relate to the comparatively poorly temporally resolved faunal

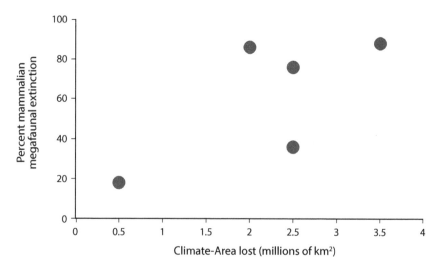

Figure 2.6. Scatter plot of the percent megafaunal loss during Pleistocene extinctions graphed against the total climate-area lost on each of the continents. The estimates for climate-area lost were minimum estimates, and both sets of data were adapted from Nogués-Bravo et al. (2010). In general, the extinction loss increases with loss of climate-area, suggesting an indirect role of Pleistocene climate change through habitat loss in the megafaunal extinctions.

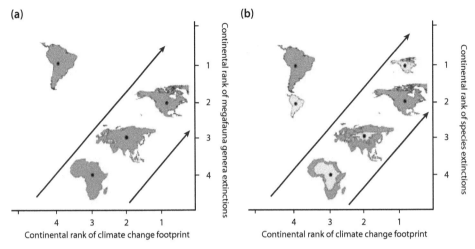

Figure 2.7. (a) Rank order of the continents experiencing Pleistocene mammalian megafaunal extinctions according to their climate change footprint, based on climate-area loss, and total number of megafaunal genera lost to extinction during the Pleistocene. (b) Plot depicting the same relationship as that in panel (a), but for small mammals. In both cases, South America experienced a disproportionately large extinction loss in relation to its climate change footprint. One possible explanation for this anomaly is that South America afforded less physical landscape space for poleward migration as temperatures rose during the Pleistocene, unlike North America and Eurasia. *Figure adapted from Nogués-Bravo et al. (2010).*

turnover data from that continent (Barnosky et al. 2004; Koch and Barnosky 2006; Nogués-Bravo et al. 2010), the possibility remains that South America acted nearly as an island continent owing to the isthmus of Panama potentially restricting range contractions or redistributions. Moreover, South America is inherently constricted toward southern high latitudes, which must also have limited poleward movement by cold-adapted species on that continent as Late Pleistocene temperatures rose.

CASE STUDIES OF PLEISTOCENE MEGAFAUNAL EXTINCTIONS

The assessment thus far has been largely anonymous with respect to the actual species involved in the megafaunal extinctions near the end of the Pleistocene. Here I focus on three taxa that indicate general differences in the manner in which extinction may result from climate change: horses, steppe bison, and mammoths. Horses were formerly one of the most abundant and diverse

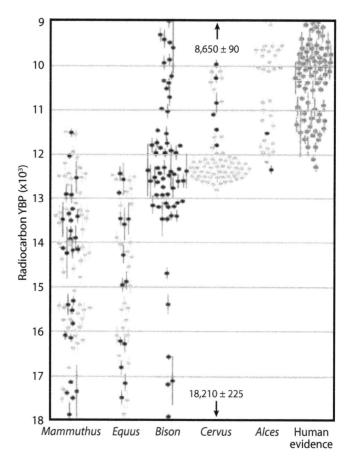

Figure 2.8. Presence of six megafaunal mammals, including humans, in the Alaska-Yukon region during the Late Pleistocene, based on fossil and archaeological evidence. Time proceeds vertically along the *y*-axis from past to present. Mammoths and horses are seen to have disappeared shortly after or preceding the arrival of humans into the region, respectively, implying a potential role of human hunting in the ultimate demise of mammoths but not of horses. *Figure adapted from Guthrie (2006).*

groups of large mammals in North America and Beringia (Guthrie 1982). Fossil evidence from Alaska and the Yukon Territory indicates they disappeared from Beringia approximately 12,500–12,000 YBP (figure 2.8) (Guthrie 2006). Prior to their extinction, their body size, as determined from metacarpal length in fossils, had undergone a rapid and perhaps accelerating decline from approximately 37,000 YBP until they disappeared from the fossil record 12,500 YBP (Guthrie 2003). This decline suggests environmental conditions had become unfavorable for growth to larger adult body sizes as the climate warmed. Similar declines in the adult body size of extant populations of other large herbivores, including red deer (Post et al. 1997) and, more recently, sheep (Ozgul et al. 2009), have been associated with current warming. As well, a rapid decline in the body size of red deer on the island of Jersey occurred during the last interglacial warm phase (Lister 1989). Guthrie (2006) has hypothesized that the climatically driven biome shift from productive graminoid- and sage-dominated swards to chemically defended dwarf shrub tundra drove the decline in body size of Pleistocene horses and their eventual extinction. Ruling out a role for human hunting in the extinction of horses in Beringia, Guthrie (2006) notes that the earliest remains indicative of human presence in Alaska date to approximately 12,200 YBP, or roughly three centuries after the disappearance of horses there (figure 2.8). A more recent assessment of extinction in horses, however, concluded that Late Pleistocene climate change was likely not the main driver of their extinction in Eurasia, where human extirpation was the more probable cause (Lorenzen et al. 2011).

Another of the most abundant species inhabiting northern latitudes during the Pleistocene was the steppe bison, which became completely extinct by the end of the Pleistocene. Beringian steppe bison taxonomy is complicated by the periodic mixing and isolation of populations throughout the stadial-interstadial cycles that characterized the Middle to Late Pleistocene in North America, when gene flow among subpopulations in episodes of east-west and north-south exchange was facilitated by exposed land bridges during the Early and Middle Pleistocene and inhibited by the presence of the Laurentide and Cordilleran ice sheets (figure 2.9) (Shapiro et al. 2004). Nonetheless, Pleistocene steppe bison originally appeared in Beringia during the Middle Pleistocene between 300,000 and 130,000 YBP, and disappeared from habitat remnants of Beringia by approximately 8,000 YBP (MacPhee et al. 2002), having eventually been replaced by woods and plains bison subspecies in northern and central North America (Roe 1970). Rather than relying on the use of fossils to reconstruct the distribution of steppe bison through the Late Pleistocene, Shapiro and colleagues (2004) used ancient DNA recovered from several hundred bison fossils and associated radiocarbon dates to reconstruct the effective population size of steppe bison and their dynamics in North America from 125,000 YBP onward. This reconstruction reveals three phases in steppe bison dynamics spanning the Middle to Late Pleistocene: an increase phase from 125,000 YBP until approximately 37,000 YBP, when the population entered a transition phase with stabilizing population growth, followed by a decline phase that persisted into the Pleistocene-Holocene transition (figure 2.10) (Shapiro et al. 2004).

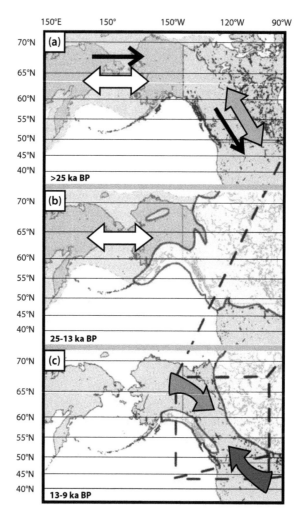

Figure 2.9. Inferred patterns of steppe bison gene flow (large arrows) in Beringia and colonization (small arrows) of North America during the Late Pleistocene. The colonization of North
America from Siberia, and gene flow between the two regions, was facilitated by the presence of
the Beringian land bridge until the onset of warming approximately 13,000 years ago. *Adapted
from Shapiro and colleagues (2004).*

The timing of these phases of population growth, stabilization, and decline
is critical to inferring the importance of human exploitation versus climate
change in the eventual demise of the steppe bison in North America. According to Shapiro and colleagues (2004), the onset of the steppe bison decline in
figure 2.10 cannot be attributed to the onset of the Last Glacial Maximum, the

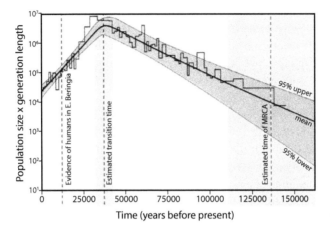

Figure 2.10. Skyline plot of reconstructed population size of bison in North America, derived from ancient DNA, from the Middle Pleistocene onward, with time progressing along the *x*-axis from right to left. The timing of the onset of the population decline coincides with a warming event called Marine Oxygen-Isotope Stage 3, and precedes the arrival of humans on the continent by several millennia. *Figure adapted from Shapiro et al. (2004).*

onset of rapid warming at the Pleistocene-Holocene transition, or the arrival of human beings because it predates all three events. Instead, the transition from population increase to decline appears to coincide with the warmest period of Marine Oxygen-Isotope Stage 3, the warmest part of the Late Pleistocene before its transition into the Holocene (Shapiro et al. 2004). Hence, the decline in Beringian steppe bison in North America may have been precipitated by an expansion of forests into the graminoid-dominated steppe tundra with which this species was closely associated throughout the Pleistocene (Shapiro et al. 2004), although this scenario does not preclude an ultimate influence of human exploitation in their eventual demise.

Woolly mammoths were one of the most iconic species of the late Quaternary, a species with morphological adaptations to the cold, dry climate and one whose presence was diagnostic of the mammoth steppe vegetation community representative of far northern landscapes during the Pleistocene (Guthrie 1968, 1982, 2001). The grazing-specialist dentition of this species, together with its hind-gut fermenting physiological morphology, is indicative of adaptation to high-volume throughput of low-quality forage dominated by graminoids (Guthrie 1982, 2006). Such habitat specialization may have left the woolly mammoth vulnerable to extinction owing to habitat loss associated with climate change, and there is good evidence that mammoth distributions waxed and waned in association with shifts between steppe-dominated and tree-dominated landscapes over the last 21,000 years in northern Eurasia (Stuart 2005). Nonetheless, the association of mammoth remains with early human settlements, the appearance of mammoths in cave art, and the spatio-temporal overlap between mammoth and human distributions and the seeming correlation with mammoth declines in the Late Pleistocene have been inter-preted as evidence for a role of humans in mammoth declines and extinction (Martin 1967, 1984; Surovell et al. 2005). As well, the survival of mammoths in some parts of northern Eurasia as recently as 3,000–4,000 years ago further complicates the implication of Pleistocene hunters in their extinction there and in North America (Stuart et al. 2004; Nikolskiy et al. 2011), in addition to com-plicating the implication of Late Pleistocene climate change in their demise (Lorenzen et al. 2011).

As the chronology compiled by Guthrie (2006) for interior Alaska and the Yukon Territory indicates, humans and mammoths may have overlapped there for 1,000 years before the demise of mammoths, arguing against the blitz-krieg, or sudden overkill, hypothesis (Martin 1984; Nikolskiy et al. 2011) for their extinction. Further evidence for a prolonged overlap between humans and mammoths and horses in interior Alaska during the Late Pleistocene comes from a recent analysis of ancient DNA collected from soils, which places the last appearance dates for these species there at approximately 10,500 YBP (Haile et al. 2009), considerably later than Guthrie's (2006) estimates, which were based on macrofossils. The interval during the Pleistocene-Holocene transition in the Alaska–Yukon Territory region that was of critical importance to mammoths and their habitat occurred between 13,500 and 11,500 YBP. It was during this period that the mammoth steppe, comprising cold- and dry-adapted plant species with low stature, such as graminoids and sages, experi-enced a climatic transition toward warmer, wetter conditions, which ushered in dwarf shrubs such as willow and birch, still common there today (Guthrie 2006). Following this transitional period, mammoths disappeared from the fos-sil record and were replaced by more generalist feeders such as elk, or wapiti, and moose (figure 2.11) (Guthrie 2006). Nogués-Bravo and colleagues (2008) used climate envelope models and reconstructed population models to infer the effect of habitat losses resulting from climate change during the Pleistocene-Holocene transition on mammoth extinctions in northern Eurasia. In this ap-proach, Nogués-Bravo and colleagues (2008) reconstructed the climate niche envelope of woolly mammoths based on concordance between locations and dates of finds of mammoth remains in Eurasia and paleoclimate reconstruc-tions for those locations. The paleoclimate reconstructions were used to es-timate the total area of suitable habitat for mammoths in Eurasia during five periods, 126,000 YBP, 42,000 YBP, 30,000 YPB, 21,000 YBP, and 6,000 YBP,

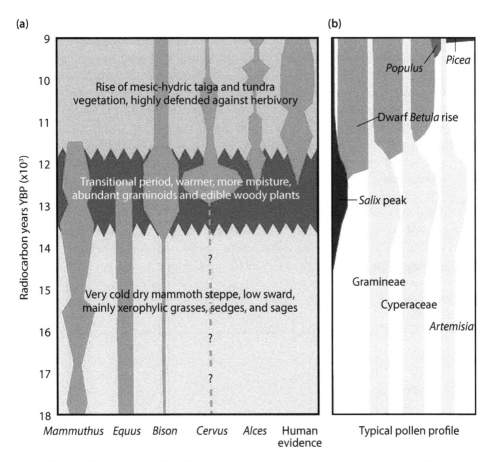

(a)

(b)

Radiocarbon years YBP (x10³)

9

10

11

12

13

14

15

16

17

18

Rise of mesic-hydric taiga and tundra
vegetation, highly defended against herbivory

Transitional period, warmer, more moisture,
abundant graminoids and edible woody plants

?

Very cold dry mammoth steppe, low sward,
mainly xerophylic grasses, sedges, and sages

?

?

?

Mammuthus   Equus   Bison   Cervus   Alces   Human
                                              evidence

Populus   Picea

Dwarf *Betula* rise

Salix peak

Gramineae

Cyperaceae

Artemisia

Typical pollen profile

Figure 2.11. (a) Reconstruction of the temporal patterns of presence, extinction, or arrival of the six mammalian megafaunal taxa depicted in figure 2.8, in relation to the timing of a major biome shift from a productive, graminoid-dominated herb sward to chemically defended shrub tundra, shown in panel (b) to the right. The biome shift is inferred on the basis of reconstruction of plant community structure and dynamics from ancient pollen. According to this view, the extinction of Pleistocene grazer-specialists coincided with the disappearance of graminoid-dominated communities in association with increasing temperature and precipitation. *Adapted from Guthrie (2006).*

to provide data encompassing a period that predates the earliest fossil remains of mammoths in Eurasia and extends through their demise in that region (Nogués-Bravo et al. 2008).

The total area of habitat suitable for woolly mammoths in Eurasia increased from virtually none 126,000 YBP to a maximum of approximately 10 million km² by 42,000 YBP, and then declined steadily through the terminal period of 6,000 YBP (figure 2.12a) (Nogués-Bravo et al. 2008). Both periods of minimum most suitable habitat for mammoths, 126,000 YBP and 6,000 YBP, coincided with warming transitions during interglacial periods, while those with maximum availability of most suitable mammoth habitat, at 42,000 YBP and 30,000 YBP, coincided with peaks in cold, dry conditions (Nogués-Bravo et al. 2008). Coincident with favorable climatic conditions during these latter two periods, the distribution of mammoths increased and expanded southward throughout northern Eurasia (figure 2.12b) (Nogués-Bravo et al. 2008). Coincident with the expansion of mammoths southward as the total area of climatically suitable habitat increased, anatomically modern humans appeared in Eurasia at approximately 40,000 YBP and began gradually expanding northward (figure 2.12b) (Nogués-Bravo et al. 2008), although notably, the zone of overlap between the two species appears to have been exceedingly narrow until humans reached far northern latitudes, around 6,000 YBP, by which time mammoths were nearly extinct. The reconstruction of woolly mammoth population size by Nogués-Bravo and colleagues (2008) suggests that the species increased

(a)

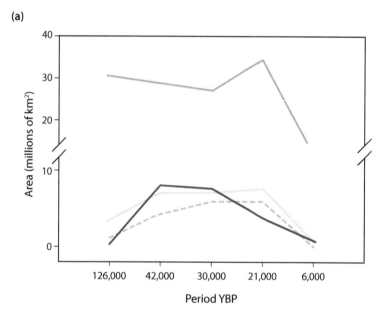

Figure 2.12. (a) Reconstructed area of suitable mammoth habitat in Eurasia from the Middle to Late Pleistocene, with time proceeding from left to right along the *x*-axis. The different lines depict quartiles of inferred suitable habitat for mammoths, with black representing total area most climatically suitable for mammoths and lighter shades depicting progressively less suitable habitat for mammoths. Periods with the least suitable area for mammoths coincide with warm interglacial periods at 126,000 and 8,000 years before the present. *Adapted from Nogués-Bravo et al. (2008).*   *(continued)*

in Eurasia from 126,000 YBP to a maximum by 21,000 YPB, after which its decline was precipitous.

The conclusion deriving from this analysis attributes the decline in the Eurasian population of woolly mammoths to a reduction in suitable habitat resulting from climate change, with the ultimate demise of the species a consequence of human hunting (Nogués-Bravo et al. 2008). Hence, in this case, it appears that the effect of climate change was to drive the abundance of the species downward to a population threshold at which it was vulnerable to exploitation by humans. Similarly, the temporally protracted nature of megafaunal extinctions in general across Eurasia has been interpreted as unsupportive of a primary role for humans except in the case of steppe bison (*Bison priscus*) and instead favors a role for climate-induced biome shifts precipitating declines, with humans contributing to extinction in declining species (Pushkina and Raia 2008). The important lessons to be drawn from these examples, fraught with qualifying assumptions as they are, are two: first, the effects of climate change on animal species may be indirect, operating through subtleties such as habitat

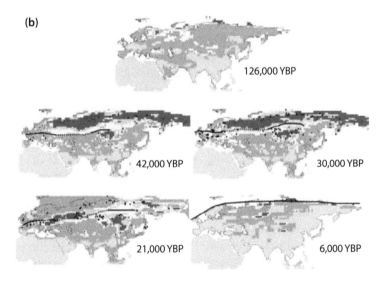

Figure 2.12. (*continued*) (b) Changes in the distribution of mammoths, depicted as black dots representing mammoth remains, coincident with the climatically driven expansion and then contraction of suitable mammoth habitat from the Middle to Late Pleistocene in Eurasia. The total area of climatically suitable mammoth habitat is depicted in each panel in shading, with black indicating the extent of most suitable habitat and lighter shades depicting the extent of progressively less suitable habitat. The black line in each panel beginning at 42,000 years before the present indicates the northward extent of evidence of human presence. *Adapted from Nogués-Bravo et al. (2008).*

loss; and second, human hunting pressure may not, in and of itself, be sufficient to drive species to extinction in some cases but may be capable of delivering the final blow to species whose persistence is already imperiled by climate change. These are lessons that, as explored further in chapter 7 in the discussion of the biodiversity implications of climate change, may well be of relevance to the conservation of threatened species today. In this case, however, the implication of critical importance to many species of conservation concern today may be that direct human exploitation, together with human alteration of habitat availability and connectivity, has pushed these species to population thresholds at which they are now especially vulnerable to contemporary climate change.

In the examples given above, it was inherently difficult to segregate the influences of human exploitation from those of climate change in the demise of Pleistocene megafauna. This has been a problem of nagging persistence in the megafaunal extinction literature because of the coincidence in time and space of human arrival or expansion into megafaunal habitats, evidence of human exploitation of Pleistocene megafauna, especially with the Clovis hunting technology, and the simultaneous onset of rapid climate change following the Last Glacial Maximum (Lorenzen et al. 2011). The inclusion of model assessments of small mammal extinctions by Nogués-Bravo and colleagues (2010) provides an essential perspective to the debate over the role of climate change in the late Quaternary extinctions because this taxonomic group is less likely to have been a focus of early human hunters. As shown in figure 2.7b, the rank order of continents according to their climate change footprint matches fairly closely their rank order according to the magnitude of small mammal extinctions, with, as noted, South America as an exception for possible reasons already outlined above. A closer examination of extinctions among smaller mammals during the Late Pleistocene has the potential to reveal more about the role of climate change exclusive of any influence of human exploitation in these extinctions. Just as important, examination of the spatial redistribution and local extinctions of small mammals during the Pleistocene affords important insights into the individualistic response of species to changes in biotic and abiotic environments that, as we will see in chapter 6, are of substantial relevance to understanding community-level response to ongoing and future climate change, as well as to understanding the concept of the community itself.

PLEISTOCENE MICROFAUNAL EXTINCTIONS
AND SPECIES REDISTRIBUTIONS

There is extensive evidence from North American cave deposits for species turnover and redistribution among small mammals during the Late Pleistocene

(Betancourt et al. 1990; Smith et al. 1995; Graham et al. 1996; Graham 2005, 2007). One estimate places the diversity of Late Pleistocene micromammalian fauna from the southeastern United States at approximately double the number of species currently extant there (Semken et al. 2010). Global or local extinction of small mammals is intuitively less readily ascribed to direct human exploitation than, for example, the extinction of larger-bodied mammals, which ostensibly would have provided greater reward per unit of effort expended locating, subduing, and rendering their products. Even today, human exploitation of wildlife prey species is biased toward the removal of larger species first (Jackson et al. 2001), and, within species, of larger individuals (Solberg et al. 2000).

Recent excavation of the Popcorn Dome site in the U.S. state of California reveals considerable biodiversity turnover among small mammals in association with rapid changes in climate during the Late Pleistocene (Blois et al. 2010). Blois and colleagues (2010) employed two metrics of diversity, evenness and species richness, to track variation in the composition of the small mammal community at their site through approximately the last 20,000 years of climatic change at the end of the Pleistocene. Evenness was used because it provides an assessment of diversity that is also sensitive to large variation in abundance. Hence, it accounts not only for the number of species present in a profile but also for the relative abundance of each species among the samples. This analysis detected two major faunal turnover events during the Pleistocene. The first event occurred between 15,000 and 14,000 YBP and was soon followed by a considerably more substantial turnover event between 11,000 and 7,500 YBP (figure 2.13) (Blois et al. 2010). Whereas the first event can be ascribed to shifts in relative abundance of small mammal species constituting the assemblage at the site, the second, greater turnover event was due to both changes in abundance and local loss of species (Blois et al. 2010). Both events coincided with episodes of rapid warming (figure 2.13a). Beginning approximately 15,000 YBP, but accelerating during the end-Pleistocene warming phase between 12,000 and 11,000 YBP, both evenness and species richness of the small mammal assemblage declined (figures 2.13b and c), with richness undergoing a 32 percent reduction (Blois et al. 2010). Reductions in evenness and richness both displayed strong associations with Late Pleistocene climate change. Species richness and the distribution of abundance among species in the small mammal assemblages were both greater during the colder Late Pleistocene than during the warmer Early Holocene (figure 2.14) (Blois et al. 2010). Blois and colleagues (2010) attribute this rapid and pronounced decline in small mammal diversity to the relatively short-term climatic fluctuations that characterized the Late Pleistocene, during which species differences in the ability to track environmental variability were key to determining

(a)

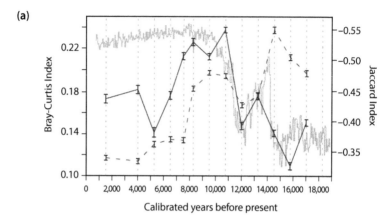

Calibrated years before present

Figure 2.13. (a) Dynamics of small mammal faunal diversity at Popcorn Dome in the U.S. state of California from the Late Pleistocene onward, with time progressing from right to left along the *x*-axis. Both the Jacard (solid black line) and Bray-Curtis (dashed line) indices quantify similarity between adjacent sample sets, and thereby, in this example, turnover in richness and diversity of communities as differences from one sample to the next, with greater values indicating higher turnover. Index estimates are superimposed on a temperature index (gray line) derived from the North Greenland Ice Core Project (NGRIP) ice core data to illustrate changes in faunal diversity in association with warming that commenced approximately 12,000 years before the present. *Adapted from Blois et al. (2010).*   *(continued)*

which species survived the transition, which species established at the site, and which species became extinct locally. This interpretation accords with the climate-equability model, according to which the distributions of species are determined to a greater extent by extremes in climatic seasonality than by climatic means (Graham and Grimm 1990). As well, it supports the notion that increasing climatic variability, especially over relatively short timescales, may represent a substantial disadvantage to species independently of the trend in average climatic conditions.

SPATIAL, TEMPORAL, AND TAXONOMIC HETEROGENEITY IN PLEISTOCENE REDISTRIBUTIONS: LESSONS TO BE LEARNED

Although extinctions have been the major focus of this chapter, there is also a great deal to be learned about the ecological consequences of rapid climate change for species assemblages by examining episodic and gradual redistributions of species throughout the Pleistocene epoch, especially during the Late

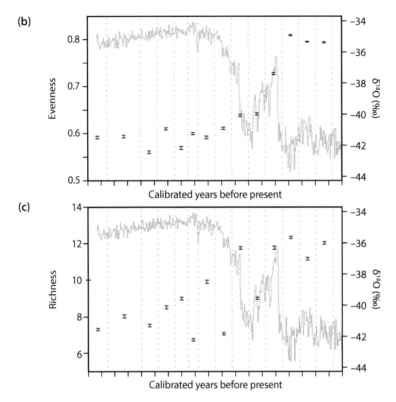

Figure 2.13 (*continued*) Changes in species evenness (b) and species richness (c) in the small mammal community during the Late Pleistocene at the Popcorn Dome site in the U.S. state of California, superimposed on temperature change (gray line) derived from the NGRIP ice core data. Both indicate a decline in small mammal diversity at the site in association with warming. *Adapted from Blois et al. (2010).*

Pleistocene. For instance, the late Quaternary megafaunal extinctions and associated collateral biodiversity losses among megafaana and smaller mammals resulting from biogeographic range adjustments may be indicative of the consequences of future warming (Barnosky et al. 2011). The spatiotemporal dynamics of tree distributions throughout the Late Pleistocene as based on pollen maps consistently reveal species-specific responses to climatic fluctuation during glacial-interglacial cycles (Jackson and Overpeck 2000; Williams et al. 2004) and, in many cases, the co-occurrence of species that do not occur together today (Jackson et al. 2000). Such sympatric-allopatric dynamics are evident in, for example, the changes in distributional patterns of hemlock and beech trees in North America. These species currently overlap but displayed disparate ranges at the end of the Last Glacial Maximum, 12,000 YBP, and

(a)   **(b)**

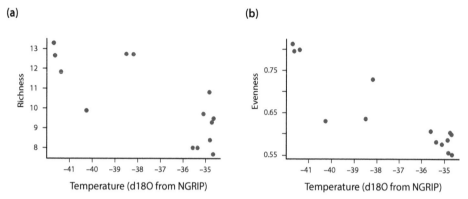

Figure 2.14. Association between species richness (a) and species evenness (b) of the small mammal community at the Popcorn Dome site, California, with temperature as derived from the NGRIP ice core data. *Derived from Blois et al. (2010)*.

assumed increasing range sympatry beginning about 6,000 YBP (figure 2.15) (Graham and Grimm 1990). Similarly, there is paleo-evidence from montane zones that plant species assemblages did not move in concert upslope in response to Pleistocene warming episodes or downslope in concert in response to cooling episodes; rather, species redistributed themselves individualistically and at differing rates along elevational gradients throughout the Pleistocene (Livingstone 1975; Betancourt et al. 1990; Graham and Grimm 1990).

The shifting faunal distributions that occurred throughout the Late Pleistocene in association with climatic fluctuation display even more marked species specificity than that which is evident from individualistic dynamics of plant species. The Pleistocene species assemblage that included the heather vole and red-backed vole and the bog lemming in North America dissolved as the northward advance of the red-backed vole lagged behind that of the heather vole and bog lemming (Graham et al. 1996; Graham 2005). As well, there appears to have been little directional congruence in the spatial dynamics of Pleistocene mammalian redistributions, as several species underwent east-west expansions and contractions while other species previously co-occurring with those underwent north-south expansions and contractions, and yet others went extinct entirely (Graham and Mead 1987; Graham 2005). The previously species-rich faunal assemblage of a diverse array of small mammal species that typified eastern North America during the Pleistocene also dissolved owing to the differing rates and directions of response by those species to changing climatic and environmental conditions (Graham et al. 1996), despite generality in directionality among some species (Lyons 2003).

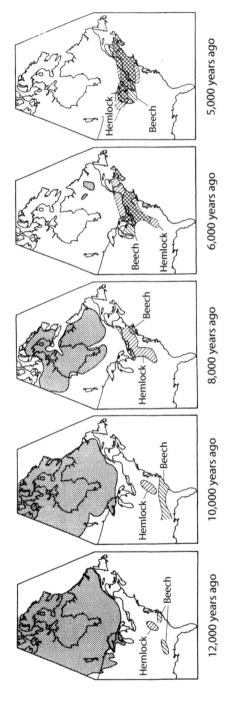

Figure 2.15. Changes in the distribution of hemlock and beech in the northeastern United States from the Late Pleistocene onward, in association with the retreat of the Laurentide ice sheet. This example illustrates differences in rates and direction of species-specific responses to climate change and how these can alter species composition by bringing previously spatially disjunct species into co-occurrence at some sites. *Adapted from Graham and Grimm (1990).*

Such individualistic responses of species to environmental change during the Pleistocene do not, however, necessitate randomness of response among species, or their ostensible assembly into communities, because the possibility of shared resource or niche requirements among some species may promote similarity in tracking of environmental gradients through time (Graham 2005). Additional examples of disparity in direction and rate of response to climate change among species whose distributions overlapped during the Pleistocene but do not do so today are explored in greater detail in chapter 6, on community dynamics, where the concept of no-analogue communities will be a major focus. However, the salient lesson emerging from this brief overview is that species may not move in predictable directions or at predictable rates based simply on their association with more predictably changing abiotic parameters or even on their association with other species. This, of course, raises the possibility that the notion of niche conservatism, that species' habitat and resource utilization patterns as well as relationships with other species remain fixed over time, may be a questionable basis for predicting changes to species distributions under scenarios of future climate change. This last point will be revisited in chapter 5, on niche dynamics.

RECONSIDERING THE MEGAFAUNAL
EXTINCTIONS: THE ZIMOV MODEL

An alternative scenario was presented by ecologists familiar with the ecosystem consequences of the foraging behavior of large herbivores (figure 2.16) (Zimov et al. 1995). Since the landmark studies of wildebeest and beaver foraging ecology by McNaughton (1976) and Naiman, Melillo, and Hobbie (1986), respectively, ecologists have developed a growing appreciation for top-down controls of ecosystem function. As we will see in chapter 8 on ecosystem dynamics, ecosystems ecology can most generally be divided into two subdisciplines, one that views ecosystem processes as driven mainly by abiotic dynamics, such as the availability of energy and nutrients in systems, the so-called bottom-up approach, and one that views ecosystem processes as driven mainly by the influences of animals on the rates and spatial distribution of nutrients and energy through the environment, the so-called top-down approach (Loreau 2010). In this latter subdiscipline, McNaughton demonstrated in a series of pioneering studies in the 1970s and 1980s the influence of wildebeest foraging activity and migrations on concentrations of nitrogen in Serengeti soils and the rate of nitrogen turnover in the system (McNaughton 1976, 1983, 1985; McNaughton et al. 1988). This body of work has shown clearly

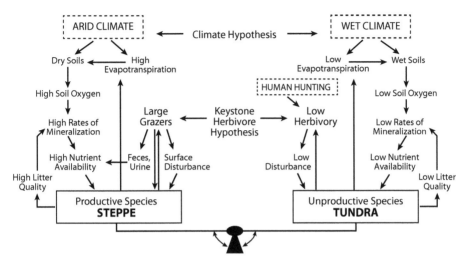

Figure 2.16. The Zimov model depicts contrasting hypotheses of drivers of the biome shift that occurred in the far north during the transition from the Pleistocene to the Holocene. According to the climate hypothesis, increasing temperatures and precipitation drove the shift from a productive, graminoid-dominated steppe community with abundant large mammalian herbivores to an unproductive, chemically defended shrub tundra, which ultimately led to the demise of grazing-adapted megafauna. In contrast, according to the keystone herbivore hypothesis, the extinction of Pleistocene megafaunal grazers by, for example, human hunting drove the biome shift by allowing or promoting the expansion of shrub tundra in association with warming. *Adpated from Zimov et al. (1995).*

that in grazing ecosystems inhabited by large mammals, primary productivity is stimulated by tissue removal by herbivores and the enhancement of nutrient turnover through urinary and fecal inputs by herbivores (McNaughton et al. 1988, 1989; Augustine and McNaughton 1998). Furthermore, tissue removal through herbivory may, in grazing ecosystems, facilitate the development of reciprocal dynamics that stimulate herbivory through increases in the nitrogen concentration of aboveground plant tissues (McNaughton et al. 1988; Post and Klein 1996).

Structural modification of ecosystems, with resultant alteration of nutrient dynamics over large spatial scales and long periods, may also occur as a consequence of the foraging behavior of smaller mammals. Naiman and colleagues have illustrated, for instance, the massive, ecosystem-wide effects exerted by beavers through damming behavior on nutrient turnover between and within aquatic and terrestrial systems, as well as their role in forest succession throughout much of North America (Naiman et al. 1986). In northern Minnesota, beavers converted 13 percent of the total forest area to wetlands—meadows and ponds—through damming activities spanning six decades, increasing soil carbon content and doubling plant-available nitrogen through the formation of pond sediments (Naiman et al. 1994). Across the entire United States, reductions in beaver populations through widespread hunting and trapping in the seventeenth through the early nineteenth centuries are estimated to have precipitated the conversion of 195,000–260,000 km^2 of wetlands to dry land (Naiman et al. 1988).

Applying the top-down perspectives of such landmark studies, Zimov and colleagues (1995) examined the relationships between the extinction of large herbivores at the end of the Pleistocene and biome-wide shifts in plant community composition from productive, graminoid-sage steppe to unproductive, shrubby tundra (*sensu* figure 2.8). They applied the cause-and-effect relationship in the opposite direction. In this scenario, climatic warming and increased precipitation during the Pleistocene-Holocene transition would have reduced soil oxygen content and nutrient cycling and availability, thereby promoting the development of unproductive tundra vegetation (Zimov et al. 1995). The outcome would have been indistinguishable from similar effects driven by the extirpation of keystone herbivores through human hunting (figure 2.16) (Zimov et al. 1995). The factors limiting and driving seasonal and interannual variation in soil oxygen concentrations and their effects on soil microbial processes, and thereby ecosystem function, are complex (Burgin et al. 2010). Nonetheless, the Zimov keystone herbivore hypothesis is relevant in the context of climate change because it suggests that grazers such as horses, bison, and mammoths would have exerted a reinforcing influence on the development and maintenance of productive graminoid-dominated steppe vegetation through the mechanical removal of leaf tissue, fertilization of soil through urine and feces deposition, and surface disturbance, all of which would have promoted high soil nutrient availability. The Zimov hypothesis forces us to consider, then, whether climate change hastened the demise of Pleistocene megafauna of the mammoth steppe by shifting plant community composition away from favorable habitat for such species or whether human elimination of one or a few keystone herbivores precipitated a biome shift that then precipitated the extinction of other species of large herbivores that depended on the mammoth steppe plant communities. We will return to the topic of interactions among climate change, keystone herbivores, and plant community dynamics in chapter 8.

RELEVANCE TO CONTEMPORARY CLIMATE CHANGE

If rapid warming during the Pleistocene-Holocene transition contributed to, or even acted as the main driver of, mass extinctions, such a scenario would seem

to suggest that contemporary climate change has a similar capacity to precipitate species losses. If so, then we may rightly ask, where is the evidence for
widespread population declines at the species level, or where is the evidence
for a role of climate in geographically widespread population change? I believe
the evidence is already before us. As we will see in chapter 7, on biodiversity,
many population declines have already been linked to recent climate change,
and some species losses or declines may be similarly associated with recent
warming trends and precipitation changes. Perhaps more illuminating, though,
is evidence for geographically large-scale synchrony in population fluctuations
of some species. The population dynamics of caribou and reindeer (both of the
genus *Rangifer*), for instance, display statistically significant correlated fluctuations over, in some cases, hundreds to thousands of kilometers (Post and
Forchhammer 2006), and caribou populations throughout the Arctic are currently in decline (Vors and Boyce 2009). Although this example falls short of
constituting definitive evidence for imperilment of the persistence of a species
as a result of contemporary climate change, it does suggest that climate change
has the capacity to induce population synchrony over large spatial scales, and
thereby to contribute to extinction risk, because, as we shall see in chapter 4,
synchronously fluctuating populations are more likely to go extinct simultaneously than those that fluctuate asynchronously (Heino et al. 1997).

Life History Variation and Phenology

An individual's life history comprises progressive stages in its development, maturation, reproduction, and eventual mortality. A suite of traits is associated with this progression. Traits typically of interest in a life history are the timing of emergence, hatching, or birth; size or length at emergence, hatching, or birth; timing of the annual period of reproduction; age or size at which the onset of reproduction occurs; age or size at which reproduction ceases; and age or size at the time of death. The study of the timing of life history events and the manner in which they vary in space and time constitutes phenology.

Despite its relation to individual-level phenomena, in the context of climate change phenology is studied most commonly at the population or site level, with a focus on such phenomena as annual first arrival or departure dates of migrating birds, first flowering dates of plants, leaf-out dates of trees, or green-up dates derived from satellite data. Very few studies have focused on individual-level phenological variation, perhaps because the expression of life history traits is easier to observe at the population level. Yet variation in the timing of the life history events that are the basis of phenological studies invariably reflects individual strategies aimed at maximizing offspring production and survival. In my view, it is important to consider individual-scale phenological variation in addition to the more commonly studied phenomena at the population level because processes evident in the former can be instrumental in explaining patterns observed in the latter. After all, natural selection operates on individual-level variation but results in patterns detectable at the population level. In this chapter, I review examples illustrating patterns in phenological responses to observed and experimental climate change before arguing for a more rigorous analytical approach to quantifying the interaction between climatic and biotic factors in such dynamics and presenting a framework for doing so.

GEOGRAPHIC AND TAXONOMIC VARIATION IN PHENOLOGICAL RESPONSE TO CLIMATE CHANGE

The most commonly observed phenological response to recent climate change is an advance in the timing of early life history events such as migration, plant emergence or flowering, amphibian breeding, or egg-laying dates in birds (Brown et al. 1999; Inouye et al. 2000; Gibbs and Breisch 2001). In theory, it should be in the interest of the individual to reproduce as early as possible, all other things being equal, to increase the probability of successful reproduction before death (Williams 1966). In practice, however, advancing life history events to the earliest possible annual timing may incur fitness costs. For instance, forbs in the Rocky Mountains in the United States respond to warmer springs and earlier snowmelt by advancing their flowering phenology, but appear to incur increased flower loss to frosts because the date of the last spring frost has not advanced with the date of snowmelt (Inouye 2008).

Phenological advances have been recorded for biomes across the Earth, though in differing magnitudes (Schwartz 2003). Several comprehensive reviews have been published over the past decade documenting and summarizing these advances (Walther et al. 2002; Parmesan and Yohe 2003; Root et al. 2003), some of the highlights of which are worth reviewing here. For instance, a meta-analysis of results from sixty-one studies examining phenological responses to warming in 694 species or functional groups over the past half century revealed an overwhelming tendency toward advancement of life history events with warming (figure 3.1) (Root et al. 2003). Overall, responses to warming varied from a delay of 6.3 days per decade documented for breeding

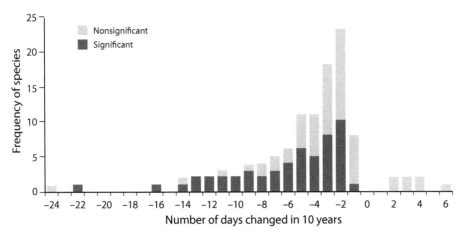

Figure 3.1. Frequency distribution of observations of the phenological advance per decade, since approximately 1950, by life history events in invertebrates, amphibians, birds, and plants in relation to temperature, according to a meta-analysis by Root et al. (2003). Negative associations between the timing of life history events and temperature far outnumber positive associations, none of which is statistically significant. Warming has thus advanced phenology from 1 to 22 days per decade when data from all taxa are combined, though most advances have been modest and on the order of 2–3 days per decade. This analysis was restricted to species displaying a phenological response to climate change. *Adapted from Root et al. (2003).*

dates of the North American Fowler's toad (*Bufo fowleri*) to an advance of 24 days per decade for breeding dates of the common murre (*Uria analge*). Notably, however, neither of these extremes reflected a statistically significant trend, but significant changes in phenology across taxonomic groups ranged from advances of 1 to 22 days per decade (Root et al. 2003). Phenological advancement in response to warming was strongest for birds and weakest for trees (figure 3.2). This analysis concluded that the mean phenological response to warming across all species and taxonomic groups examined was an advance of 5.1 days per decade, with responses at mid-latitudes (32° to 49.9° north) averaging an advance of 4.2 days per decade and responses at higher latitudes (50° to 72° degrees north) averaging an advance of 5.5 days per decade (Root et al. 2003). Hence, phenological responses to warming were larger at latitudes experiencing greater warming.

A similar meta-analysis published in the same year in the same journal produced slightly different results. This study assessed reports of phenological responses to warming in 172 species, including breeding dates of amphibians, breeding site arrival dates of migratory birds and butterflies, nesting dates of birds, and dates of first flowering in forbs and leaf bud burst in trees (Parmesan and Yohe 2003). In this analysis, the mean rate of advancement across taxonomic groups and regions was approximately half that reported in the study by Root and colleagues (2003), or 2.3 days per decade on average (Parmesan and Yohe 2003). Eighty-seven percent of the phenological changes were consistent in direction with expected responses to warming.

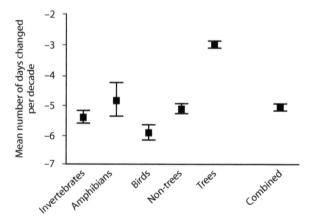

Figure 3.2. Mean number of days by which the timing of temperature-dependent life history events has advanced in invertebrates, amphibians, birds, and plants, and the overall mean advancement across all of these groups. The analysis was confined to species that displayed a significant phenological response to increasing temperature. *Adapted from Root et al. (2003).*

In light of the magnitude of difference between these two studies in the overall mean phenological response to warming, it is tempting to speculate that it owes in some part to the data used in the two studies, to methodology, or to both. A subsequent reassessment of the two meta-analyses concluded that their difference owed to methodology (Parmesan 2007). The primary difference between these two meta-analyses, according to Parmesan (2007), lies in the filter applied in selecting the data to be included in each. In the first study, the authors selected a subset of available data that included only results reporting responses to temperature because, as stated by the authors, the purpose of the paper was to quantify the magnitude and direction of phenological response to warming (Root et al. 2003). In the second study, the authors amassed the results of all studies that fit their criteria for length and quality of data and whose purpose was to investigate the potential for phenological response to warming (Parmesan and Yohe 2003). Hence, the first meta-analysis was designed to quantify mean response to warming, whereas the second was designed, first, to investigate whether there was a detectable response to warming across taxonomic groups, and second, if so, to quantify the magnitude of this response.

A more recent, and additionally more comprehensive, investigation of phenological responses to warming focused on joint attribution, or attribution of warming to anthropogenic forcing, followed by attribution of phenological dynamics to anthropogenically forced warming (Root et al. 2005). This was the first such attempt to link observed plant and animal phenological dynamics to human-induced warming, and as such it represents an important advance over previous meta-analyses. The approach taken by Root and colleagues (2003) was to apply temperature data produced by the HadCM3 (a coupled climate model, version 3, developed at the Bradley Centre, Bracknell, UK) general circulation model (GCM) (Gordon et al. 2000) in analyses of observed phenological dynamics in three scenarios: natural temperature variation only, anthropogenic temperature variation only, and combined natural and anthropogenic temperature variation. Comparing the strengths of the relationships between the timing of various life history events in an array of taxa and the three climate scenarios should, according to the authors, resolve whether the observed phenological dynamics are attributable to natural climatic variation, anthropogenically forced climatic variation, or their combination (Root et al. 2005). The justification for this approach is based on the conclusion in the Third Assessment Report of the Intergovernmental Panel on Climate Change that the observed global temperature trend and dynamics during the twentieth century matched most closely simulations from the HadCM3 GCM that included combined natural and anthropogenic contributions to climate (Houghton et al. 2001).

In their more recent meta-analysis, Root and colleagues (2005) utilized data on 145 species that displayed significant phenological trends and represented forbs, trees, birds, and invertebrates from a total of forty-two sites in Europe, North America, and Asia. Across all species, events, and sites, phenological advancement averaged 3.2 days per decade (figure 3.3). Among the regions represented in the data analyzed in this example, phenological advancement was greatest in Europe, averaging 5 days per decade, and was approximately twice the rate there as in North America (figure 3.3). Similarly, phenological advancement was nearly four times more rapid at high latitudes (4.4 days per decade on average) than at low latitudes (figure 3.3). Both birds and woody plants displayed a rate of phenological advancement of approximately 5 days per decade on average (figure 3.3). In their comparison of relations between observed phenological dynamics and natural, anthropogenic, and combined climatic dynamics, Root and colleagues (2005) detected no association using natural climatic forcing alone, but did detect similarly strong associations with anthropogenic climatic dynamics and combined natural and anthropogenic forcings, though somewhat stronger associations with the combined forcings. This finding strongly suggests that human-induced climate change is the driver of a broad array of phenological shifts throughout the Northern Hemisphere.

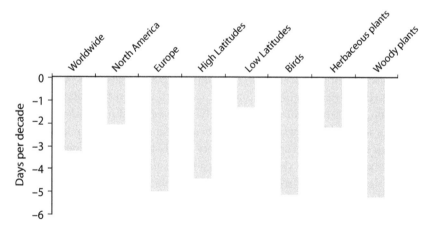

Figure 3.3. Average rates of phenological advancement in days per decade among species examined in a meta-analysis by Root et al. (2005). Estimates are shown for all species averaged across the Earth; North America and Europe individually; sites at high and low latitudes; and for birds, herbaceous plants, and woody plants. Phenological responses to recent warming appear most pronounced in Europe, at high-latitude sites, and among birds and trees. *Data summarized from Root et al. (2005).*

The rates of phenological advancement reported in these meta-analyses suggest advances in the range of 2–5 days per decade across taxa and between low-mid and mid-high latitudes. Only a single study to date has compared rates of phenological advancement among taxonomic groups at a high-latitude site (Høye et al. 2007). In contrast to high-latitude systems, the representation of low- to mid-latitude systems in phenological data sets has been widespread spatially and extensive temporally. For this reason, data from high-latitude systems are particularly valuable because they provide the basis for a point of comparison of rates of phenological advancement in response to warming between low and high latitudes, the latter of which have experienced the more rapid rate of warming.

This high-latitude study by Høye and colleagues (2007) drew on ten years of observational records from the High Arctic monitoring site at Zackenberg, Greenland, that was mentioned in chapter 1. Observations included dates of flowering in six species of vascular plants, emergence dates in twelve groups of arthropods, and dates of clutch initiation in three bird species (Høye et al. 2007). Among the records displaying significant trends over time, phenological advancement ranged from approximately 15 to 35 days per decade and averaged, across all records, 14.5 days per decade (figure 3.4). The primary driver of phenological dynamics at this study site appears to be the annual date of snowmelt, which had advanced over the same period covered by the data in figure 3.4 by 14.6 days (Høye et al. 2007), remarkably similar to the mean rate of phenological advancement across taxa at the site. This analysis suggests that rates of phenological advancement in response to warming have the potential at very high latitudes to outpace those at lower latitudes by nearly an order of magnitude.

PATTERN AND SCALE IN PHENOLOGICAL DYNAMICS

Patterns in satellite-derived images of primary productivity suggest a lengthening of the plant growing season in recent decades (Myneni et al. 1997), whereas data on plant phenological dynamics from studies conducted at plot and sub-landscape scales indicate shortened phenophases, or phenological events, in response to warming (Sherry et al. 2007; Steltzer and Post 2009; Miller-Rushing et al. 2010). The contrast may be resolved, however, by recognizing the difference between phenology in the context of individual life history strategies of disparate species and landscape-scale patterns of phenology, and by recognizing the difference between local, species-specific phenological dynamics and those occurring at the landscape scale.

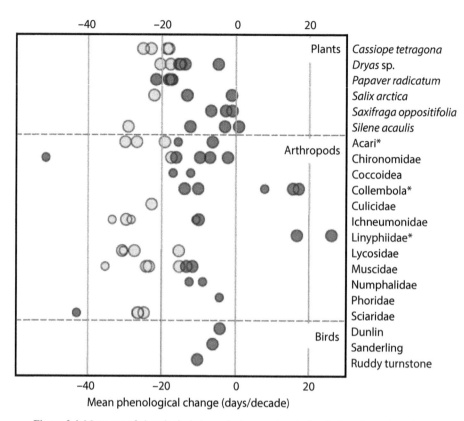

Figure 3.4. Mean rate of phenological change in days per decade for plants, arthropods, and birds observed from 1996 to 2006 at Zackenberg, northeastern Greenland. Light gray indicates statistically significant change. Although most taxa displayed advancing phenology over the period of observation, some arthropods displayed increasingly delayed emergence phenology, though none of those trends was significant. The rate of advance in the significant cases in this analysis far exceeded the rate of advance reported for other biomes in the meta-analyses of data from lower-latitude systems by Root et al. (2003, 2005). The timing of snowmelt is the ostensible driver of spring phenology at this study site. *Adapted from Høye et al. (2007).*

A comparison of phenological patterns at these different scales makes this apparent. As mentioned, satellite-derived data on the timing of primary productivity at large spatial scales, such as the scale of continents or subcontinents, suggest that the length of the plant growing season has increased in recent decades (Myneni et al. 1997; Karlsen et al. 2009), presumably in response to increasing temperatures. For instance, Normalized Difference Vegetation Index (NDVI) data from Fennoscandia reveal a trend toward earlier onset of the growing season by 0.27 days per year, and a delay in the termination of the growing season by 0.37 days per year, from 1982 to 2006 (Karlsen et al. 2009). There are a number of possible explanations for such patterns. The timing of initiation of plant growth may advance with warming even as the timing of cessation of plant growth may remain constant or even be delayed with warming (Steltzer and Post 2009). Alternatively, or additionally, there may be an increase in the spatial variability of the timing of initiation or cessation of plant growth, or both, in association with warming because of the interaction between the influences of landscape heterogeneity and climate on phenology. In the first case, we should expect observations and experimentation to support the patterns evident in analysis of satellite, or NDVI, data; namely, at the plot scale, plants initiate growth earlier but either do not advance the timing of cessation of growth or delay it in response to warming. In the second case, we should expect phenological data observed at landscape scales to display a pattern of increasing variability among sites in the timing of phenological events with warming.

There appears to be a wealth of empirical support for the predictions of the latter scenario but not for those of the former. An analysis of plot-based data on the timing of flowering by several species of forest floor herbaceous plants provides an example of this. These data were collected over a fifty-year span as part of a phenological monitoring program in Norway that covered several hundred square kilometers and four degrees of latitude (Lauscher and Lauscher 1990; Post and Stenseth 1999; Nordli et al. 2008). These data reveal a positive association between climatic warming and variation among sample sites in the timing of flowering (figure 3.5); in other words, warming appears to increase spatial variability in the timing of flowering and, presumably, commencement of the growing season among sites separated by hundreds of kilometers (Post and Stenseth 1999; Post, Stenseth, et al. 1999). Similarly, a comprehensive analysis of fifty-one years of plant phenological data from Germany indicates that variability among sampling sites in the timing of early phenological events increases with warming (Menzel et al. 2006). Indices of the onset of the plant growing season based on satellite data reveal the same pattern: increasing spatial variability in plant phenology with warming at the landscape scale (Mysterud et al. 2001; Pettorelli et al. 2005, 2006).

By contrast, however, experiments and observations at smaller scales, over meters to hundreds of meters, indicate that warming is associated with a compression in time and space of the plant growing season. My colleagues and I conducted an experiment at my field site in West Greenland to investigate whether the increase in spatial variability in the timing of plant phenological events with warming observed at the continental to subcontinental scales was repeatable at smaller spatial scales. In this experiment, we used open-top

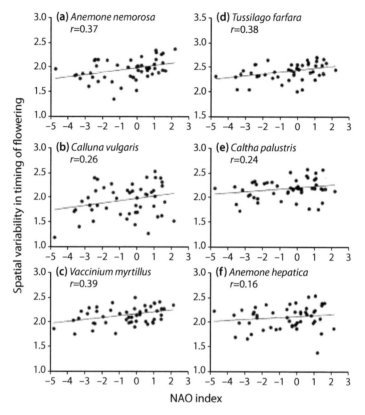

Figure 3.5. Increase in variation in the timing of first flowering by six species of forest floor forbs among sites across Norway over a 50-year period in the twentieth century in relation to the North Atlantic Oscillation (NAO) index. In Norway, local temperatures correlate positively with the NAO index. Such patterns indicate increasing spatial heterogeneity in phenological events with warming across hundreds of square kilometers, a pattern that is also evident in satellite-derived plant phenological data but is opposed by observations at smaller scales, on the order of hundreds of square kilometers. *Adapted from Post and Stenseth (1999)*.

chambers to passively warm experimental plots measuring 1.5 m in diameter distributed across the landscape over an area of approximately 3 km². This experiment revealed a reduction in variability in the timing of phenological events both within our plots and among them across the landscape (Post et al. 2008b). Interestingly, this result applied only to the earliest phenological events measured and was reversed for the latest phenological events observed (Post et al. 2008b). Moreover, another experiment we conducted on the same plots revealed that experimental warming compressed rather than extended the phenological sequence of multiple life history events in several plant species

(Post et al. 2008a). Multi-annual observational data on the timing of multiple plant phenological events on plots distributed across the study site also corroborated the results of our warming experiments (Post et al. 2008b).

A resolution of this apparent contrast of divergent phenological patterns at large and small spatial scales has been proposed recently (Steltzer and Post 2009). Satellite data are not capable of resolving individual species. Therefore, it is not currently possible, using satellite data, to determine whether a prolongation of the total number of days over which plant growth occurs annually actually reflects an increase in the growing season length for any individual species. Based on phenological studies that have tracked multiple events within species, it appears that, while different life history phases are differentially responsive to warming in some species (Post et al. 2008a), warming advances and accelerates the timing of some life history events, shortening their duration. Nonetheless, as multiple studies have shown, species are differentially responsive to warming, with some advancing their phenology considerably in response to rising temperatures and others responding less so or even delaying their phenology in response to warming (Høye et al. 2007; Sherry et al. 2007). Hence, divergence of species along a temporal axis with warming, with vernal species advancing their phenology and autumnal species displaying less or no advancement in their phenology (figure 3.6), would result in what appears to be a lengthening of the growing season as

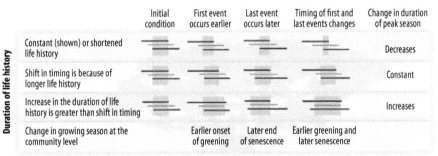

Shift in timing of life-history events

Figure 3.6. Possible scenarios of changes in the timing and duration of phenological events among species, and the consequences of these different scenarios for inferences from satellite-derived data concerning the response of growing season length to climatic warming. The growing season for the entire plant community, or as observed over a large region such as at the continental or regional scale, will be seen to increase in response to warming if species segregate temporally because of differential advancement and shortening of their individual growing seasons, as at upper right of the figure, or if all or most species initiate growth earlier and prolong growth. Current satellite technology does not permit distinguishing among species and therefore cannot resolve which mechanism is in operation at large spatial scales. *Adapted from Stelzter and Post (2009).*

ascertained from satellite data, even though no individual species increased the duration of its growth period.

Long-term data on the timing of flowering by multiple species of alpine forbs at the Rocky Mountain Biological Laboratory in the U.S. state of Colorado provide support for the notion of temporal segregation of species' life histories in response to warming. At this study site, Nevada pea (*Lathyrus lanszwertii* var. *leucanthus*) reaches peak flowering midway through the growing season, but it also exhibits a prolonged flowering that overlaps temporally with that of big-leaved lupine (*Lupinus polyphyllus*), with which it might compete for insect pollinators (Forrest et al. 2010). An analysis of thirty-three years of data on the timing of flowering of *L. leucanthus* in relation to other species within the community revealed that during late snowmelt years, the timing of flowering of *L. leucanthus* overlapped more with that of late-flowering species, but during early snowmelt years, *L. leucanthus* underwent disproportionately greater advances in its flowering phenology compared to other late-blooming species, reducing the temporal overlap with them (Forrest et al. 2010). Although this shift in its flowering phenology resulted in more overlap with early-flowering species in early snowmelt years, in general, the temporal overlap between the phenology of *L. leucanthus* and other species was reduced by advancing its timing of flowering in warm years (Forrest et al. 2010). Similarly, the flowering times of early- and late-flowering species of the genus *Mertensia* observed at the same study site are more divergent in warm years, when early-flowering species advance their phenology to a greater extent than do late-flowering species (Miller-Rushing and Inouye 2009).

PHENOLOGY AND THE AGGREGATE LIFE HISTORY RESPONSE TO CLIMATE CHANGE

Despite an early quantitative emphasis in the study of phenology on relations among the timing and duration of multiple phenophases in the seasonal growth of individual plant species (Mahall and Bormann 1978), phenology is commonly studied today as the response of the timing of single life history events to observed or experimental changes in climate. In reality, life history strategies comprise the timing and duration of multiple life history events that are linked to varying degrees with one another in succession. The sequence of phenological or life history events and their durations, leading up to and including reproduction, may be referred to as the aggregate life history of an organism (Post et al. 2008a). The relevance of the aggregate life history perspective to the study of phenological responses to climate change derives from Lewontin's (1965) classic treatment of the idealized fecundity schedule.

An organism's total lifetime investment in offspring production can be viewed in relation to the timing of the onset and duration of the key life history events determining the reproductive life span (Lewontin 1965). These key events might comprise, for instance, birth, the time of peak reproductive output, and death; the corresponding chronology describes the age at or timing of first reproduction, the age at or timing of peak reproduction, and the age at or timing of an organism's final reproductive event (figure 3.7) (Lewontin 1965). Lewontin (1965) described the inverse V-function of the relationship between fecundity and the sequence of life history events determining lifetime reproductive output (figure 3.7) as the idealized fecundity schedule and explained how shifts in one or all three of the key life history events in the schedule would affect population dynamics, with the greatest consequences seen with a rigid translation of the entire fecundity schedule (figure 3.7a). The different contributions of variation in each component of the idealized

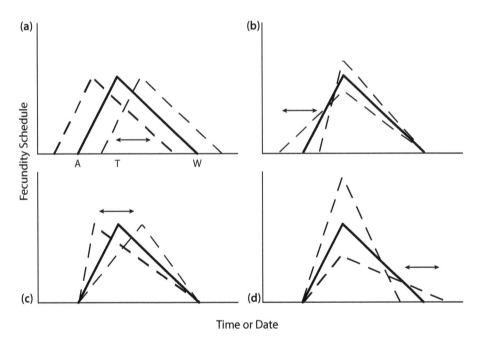

Figure 3.7. Idealized fecundity schedule and the manner in which changes in the timing of key life history events, such as the age at or timing of first reproduction (A), age at or timing of peak reproduction (T), and age at or timing of the last reproductive event in an organism's life cycle (W), alter the shape of the generalized inverse V-function of lifetime reproduction. The shift in panel (a) represents a rigid translation of the entire fecundity schedule, (b) represents a change in age at first reproduction, (c) represents a shift in age at peak reproduction, and (d) represents a shift in the termination of reproduction. Lewontin (1965) determined that the rigid translation in panel (a) has the greatest consequences for population dynamics of all of the possible scenarios.

fecundity schedule to offspring production and population dynamics suggest that as ecologists interested in life history responses to climate change, we would be well advised to incorporate Lewontin's (1965) insights into our study of phenology.

The idealized fecundity schedule can be adapted for the study of the response of phenological sequences to climate change by incorporating, for instance, the timing of individual life history events in a series of events leading up to and including reproduction. These events might consist of, in birds, the annual timing of arrival at breeding grounds, annual date of nesting, and annual date of egg laying, or, in plants, the annual timing of emergence, flower set, and blooming, as well as, in both cases, the duration of these events (Post et al. 2008a). The dates of discrete phenological events representing stages in the life history of an organism may then be represented in our adaptation of Lewontin's idealized fecundity schedule by plotting them along the x-axis, as in figure 3.7, along with the corresponding duration of each of these events on the y-axis. To fit an approximation of the inverse V-function used by Lewontin (1965), we may assume that the duration of any particular life history event (yi), or the total duration of all observed life history events (Y), will be related to the timing of initiation of that event (xi) for any sequence of life history events i to n, which in the bird example suggested above might comprise arrival, nesting, and laying, so that

$$Y = \begin{cases} a_i + b_i x_i & x \in [i : i + 1] \\ \cdots & \cdots \\ a_n + b_n x_n & x \in [n - 1 : n] \end{cases} = f(X). \tag{3.1}$$

Then the aggregate life history (L) of an organism may be quantified as

$$L = \int_i^n f(X)\,dx \tag{3.2}$$

over the range of life history events observed ($i \ldots n$). Equation (3.2) quantifies the area of the polygon delineated in parameter space by the points representing the duration of each phenological event, plotted against the dates on which those events occurred (as in figure 3.7). The area under this polygon will then represent an organism's aggregate life history (Post et al. 2008a).

To demonstrate the utility of this approach, a two-year warming experiment was conducted at the Kangerlussuaq study site to investigate the aggregate life history response of one forb species and two species of dwarf shrubs to warming. The objective of the experiment was to determine whether the aggregate life histories of each of the species was altered by warming, either expanding or contracting, and, if so, to identify which of the individual life history events constituting the aggregate life history was responsible for the aggregate response to warming. Our results indicated that, in one species, alpine chickweed, the dates of emergence and blooming were unaltered by experimental warming, whereas the date of flower set advanced in response to warming; overall, the aggregate life history of this species was unchanged by warming (Post et al. 2008a). In contrast, in both gray willow and dwarf birch, warming resulted in a reduction of the aggregate life history by advancing all three life history events observed in both species and by reducing the duration of one of them in gray willow and two of them in dwarf birch (Post et al. 2008a). Interestingly, recent work on parasite life cycle dynamics in the Arctic indicates that climatic warming has reduced the developmental period of the larvae of the parasitic lungworms that infect muskoxen (Kutz et al. 2005, 2009), further evidence for the responsiveness of not just the timing of phenological events but their duration as well to climate change. Moreover, there is mounting evidence for an increase in voltinism in many species of invertebrates in response to climatic warming (Altermatt 2010; Martin-Vertedor et al. 2010; Poyry et al. 2011). Further incorporation of an aggregate life history perspective into the study of phenological responses to climate change should improve our understanding of the extent to which individual life history events and their durations respond differently to climate change, as well as how such differential responses contribute to productivity and population dynamics. Such insights are not possible when individual life history events are studied in isolation.

TEMPORAL DEPENDENCE AND A MODEL
OF PHENOLOGICAL DYNAMICS

Almost universally, studies of phenological responses to climate change rely on comparing data on the timing of events and data on variation in temperature or precipitation. In some cases, trends in phenological time series alone are used to infer advances in the timing of life history events in response to presumed warming trends. Interannual variation is an important component of phenological response to climate change (Forchhammer, Post, et al. 1998), however, because of the obvious year-to-year variation in temperature and precipitation about the trends in each (see chapter 1).

Analyses focusing solely on relations between abiotic parameters and the timing of life history events such as emergence and flowering in plants, or migration, breeding, and offspring production in animals, overlook the potentially important contributions of other factors that may operate independently of climate change or, perhaps, in concert with it, such as density dependence and variation in resource availability, as suggested by the facilitation/promotion hypothesis described in the preface. One method of capturing the signal of these other factors in time series data on phenological dynamics is by quantifying temporal dependence in the data.

The timing of a given life history event can be viewed as a response to the influences of, and even interactions among, climate, conspecifics, and resources (figure 3.8) (Post et al. 2001). In this model, climate may act directly or indirectly on the life history event of interest at the scale of the individual: climatic conditions may influence resource availability, which in turn influences the timing of the life history event. The timing of this event has the potential to influence density or abundance the following year, such as by altering the timing of reproduction, thereby influencing offspring survival and recruitment to the population. Density in turn acts on resource availability, which, as mentioned above, may influence the timing of the focal life history event.

The model in figure 3.8 was developed as a framework for applying autoregressive time series analysis to phenological data, so in some sense it is focused on life history dynamics. As such, it is also fairly simplistic because it ignores, for instance, any effect of climate on density, which might in turn influence the timing of life history events, an interaction addressed later in this chapter. Nonetheless, the model is useful for taking into account statistically the relationship between the timing of the focal life history event in any given

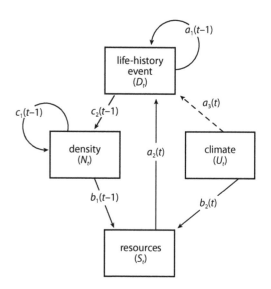

Figure 3.8. Heuristic model of life history dynamics in relation to climate, population density, and resources. In this model, climate is assumed to influence the timing of life history events of individuals in the focal species by determining resource availability, but may be adapted to include a direct influence of climate on the timing of the life history event. The relationships in this model form the basis for the development of a statistical model of phenological dynamics incorporating the influences of density dependence, resource availability, and climate on the timing of phenological events. *Adapted from Post et al. (2001)*.

year and its timing in successive years before attempting to quantify the influence of climate on the timing of the event itself. The coefficient a_1 in this model quantifies the relation between the timing of life history events in successive years and is referred to as *temporal dependence* (Post et al. 2001).

Temporal dependence is to the timing of life history events what density dependence is to population dynamics. The analogy can be developed from an analytical perspective. Imagine you were interested in analyzing the contribution of climatic variation to the population dynamics of a species of interest. The data in hand consist of annual observations of the number of individuals in the population and annual measurements of some relevant abiotic variable believed to limit abundance in this species, such as temperature or rainfall. A significant correlation between the population dynamics time series and the weather time series might suggest a relationship between the two, but it could do nothing more than offer this suggestion. A major limitation of this analysis would, of course, be the fact that it did not take into consideration any influence of density dependence, which is a powerful force in population dynamics. Statistical methods abound for incorporating density dependence into time series analyses of data on abundance through autoregression (Royama 1992), in which the coefficient quantifying the relationship between an observation in time step t and an observation in time step $t-1$ can be translated statistically as quantifying the strength of direct density dependence (Bjørnstad et al. 1995).

In a similar sense, a correlation between a phenological time series and a weather time series can only suggest a relationship between the two. Certainly, we must allow for the possibility that there is a biological constraint on the timing of a given life history event in one year and its timing in previous years. This can be represented heuristically as the process represented by the feedback loop a_1 in figure 3.8, and quantified statistically as temporal dependence. This can be formalized mathematically by specifying an equation characterizing dynamics in each of the boxes in the model in figure 3.8, and then linking these through substitution. An exposition of this model (Post et al. 2001) is worth reiterating here as a means of illustrating the statistical interpretation of temporal dependence.

We may begin with general functions describing the timing of a given life history event, d, resource availability, s, and density, n, in year t as follows:

$$d_t = f(d_{t-1}, s_t) \tag{3.3a}$$

$$s_t = g(n_{t-1}, u_t) \tag{3.3b}$$

$$n_t = n_{t-1} h(n_{t-1}, d_{t-1}) \tag{3.3c}$$

in which u_t represents some climatic or other abiotic variable in year t, and equation (3.3c) represents a simple density-dependent function describing first-order population dynamics.

The functions at each compartment in this heuristic model may take many forms, but in this case, for simplicity, we will assume that the life history event of interest bears some relation to its timing in the previous year and to resource availability in the current year; that resource availability is density dependent and related to climate in the current year; and that density is related to density in the previous year in multiplicative fashion and to the timing of some life history event that influences, for example, birth rates and thereby recruitment into the population. Approximating functions for each compartment as additive on a log scale produces the following equations:

$$D_t = a_0 + a_1 D_{t-1} + a_2 S_t \tag{3.4a}$$

$$S_t = b_0 + b_1 N_{t-1} + b_2 U_t \tag{3.4b}$$

$$N_t = c_0 + (1 + c_1)N_{t-1} + c_2 D_{t-1} \tag{3.4c}$$

Solving equation (3.4b) for N_{t-1} and substituting the resulting equation into (3.4c) produces an equation for density as a function of resource availability, climate, and the timing of the life history event of interest:

$$N_{t-1} = c_0 + (1+c_1)\frac{1}{b_1}(S_t - b_0 - b_2 U_t) + c_2 D_{t-1}. \tag{3.5}$$

Equation (3.5) can now be substituted back into equation (3.4b), yielding an equation of resource availability in relation to climate and a lagged influence of the timing of the life history event of interest:

$$S_t = b_0 + b_1 c_0 + (1+c_1)\left[\frac{1}{a_2}(D_{t-1} - a_0 - D_{t-2}) - b_0 - b_2 U_{t-1}\right] \\ + b_1 c_2 D_{t-2} + b_2 U_t. \tag{3.6}$$

This equation can now be substituted back into the equation for the timing of the life history event of interest, equation (3.4a), which, after gathering terms and simplifying, yields

$$D_t = (b_1 c_0 a_2 - a_0 c_1 - c_1 a_2 b_0) + (1 + a_1 + c_1)D_{t-1} \\ + (a_2 b_1 c_1 - a_1 - a_1 c_1)D_{t-2} + a_2 b_2 U_t + (-a_2 b_2 - c_1 a_2 b_2)U_{t-1} + \varepsilon_t \tag{3.7}$$

in which the term ε_t captures unexplained variance. We now have an autoregressive equation of life history dynamics in terms solely of delayed timing of the life history event of interest and climate that is amenable to statistical analysis. Referring momentarily back to the heuristic model in figure 3.8, we can observe that climate was assumed to exert only an indirect influence on the timing of life history events through its influence on resource availability (coefficients b_2 and a_2 in figure 3.8). The potentially direct influence of climate on the timing of life history event, coefficient a_3 in figure 3.8, was not taken into account in the derivation of our model in equation (3.7). This direct influence of climate is likely of key importance in early-season life history events of plants, such as the timing of emergence or bud burst, or the timing of emergence in invertebrates (Sparks and Carey 1995; Høye 2007; Høye et al. 2007; Post et al. 2008a, Post, Forchhammer, et al. 2009). Such an influence can readily be incorporated into our model, and a detailed exposition of this has been undertaken elsewhere (Post et al. 2001).

Note that the structure of this model allows for the potential influence of the timing of life history events in prior years on their timing in the current year (the coefficient a_1 in figure 3.8). This means, biologically, that the timing of a given life history event is not unconstrained by prior experience, and it allows for the possibility that, for instance, early flowering in one year may be followed by later flowering in the next year if this process is negatively autocorrelated, or that early breeding in one generation may promote even earlier breeding in the next generation through heritability in this trait, thereby generating positive autocorrelation. Statistically, the construction of our model in this fashion allows for decomposition of the coefficient of direct temporal dependence (the coefficient quantifying the influence of the timing of a life history event in the preceding year on its timing in the current year) into statistical density dependence and pure temporal dependence (Post et al. 2001). This is achieved by deriving the coefficient of direct temporal dependence for a time series on phenological dynamics using the full model in equation (3.7) in autoregressive analysis and then subtracting from this an estimate of coefficient a_1 in figure 3.8 derived from a simple linear regression of D_t against D_{t-1}. The resulting difference is an estimate of the signal of statistical density dependence from the time series on phenological dynamics.

Applying autoregressive time series analysis to phenological data in this manner allows us to account first for temporal dependence and then to quantify the contribution of climatic variation to the remaining variation in the phenological data. We can also analyze variation in the extent of temporal dependence among populations or among sites over which the phenological data have been collected. Finally, we can examine relationships between temporal dependence and general life history characteristics of the populations or species of interest, or relationships between temporal dependence and the strength of climatic influence on the timing of life history events.

Two examples of such analyses illustrate their utility. In the first, fifty years of data on first flowering dates for multiple species of forbs in several populations in Norway were analyzed using autoregressive time series analysis

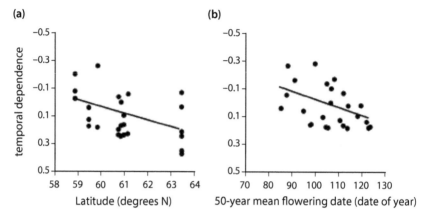

Figure 3.9. Relationships between the strength of statistical temporal dependence, a term that quantifies the relationship between the timing of life history events in successive time steps, and (a) latitude (°N, ranging from 58 to 64) and (b) the 50-year mean flowering date (ranging from day 70 to day 130) for three species of forest floor forbs in Norway. The coefficient of temporal dependence must be compared to 1 for statistical significance; hence, temporal dependence was strongest at southern latitudes and among early-blooming plants. *Adapted from Post et al. (2001).*

applied to the model in figure 3.8. This analysis revealed statistically strong relationships between latitude and the strength of temporal dependence displayed in these populations: temporal dependence was stronger in southern than in northern populations (figure 3.9a) (Post et al. 2001). As well, temporal dependence was stronger in earlier blooming than in later blooming populations and species (figure 3.9b). Finally, the strength of temporal dependence in flowering dates between successive years, and the strength of the statistical density dependence derived from phenological data, increased with the strength of climatic influence on the timing of flowering (figure 3.10). These relationships suggest that vernal species, that is, those that bloom early in general, may be more responsive to climatic warming, perhaps because they have been selected for life history strategies that capitalize on alleviation of environmental constraints on the timing of reproduction, which in turn may translate into a greater dependence in any given year on the timing of flowering in the preceding year.

In the second example, the model in figure 3.8 was applied to an autoregressive time series analysis of fifty years of data on arrival dates of six species of migratory birds in 135 populations breeding in Norway (Forchhammer, Post, and Stenseth 2002). This analysis revealed differences between long- and short-distance migrants in the strength of the relationship between spring

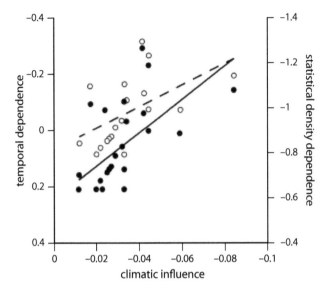

Figure 3.10. Relationships between statistical temporal dependence and statistical density dependence and the strength of the climatic influence on the timing of flowering for three species of forest floor forbs in Norway. Because of the manner in which the statistical model used to quantify these coefficients is derived from the heuristic model in figure 3.8, the coefficients increase in strength as they become more negative. Hence, both temporal and density dependence increased with the strength of the climatic influence on phenological dynamics in these species. *Adapted from Post et al. (2001).*

arrival dates on breeding grounds and warming associated with the North Atlantic Oscillation. The relationship between winter warming and earlier arrival on breeding grounds was stronger in short-distance migrants than in long-distance migrants (figure 3.11) (Forchhammer, Post, and Stenseth 2002). Interestingly, the strength of temporal dependence was nearly three times as great in long-distance migrants as in short-distance migrants, although this difference was not statistically significant (figure 3.11a). Finally, the strength of temporal dependence increased with the latitude at which populations bred in short-distance migrants but bore no such relationship to latitude in long-distance migrants (figure 3.11b).

Although the interpretation of statistical coefficients in this approach may seem at first glance far from intuitive, the advantage to applying a heuristic model such as that in figure 3.8 in a quantitative framework over the more common practice of analyzing phenological time series in linear, stepwise regressions is that it allows the identification of processes in patterns in the data. Analyses of long-term and interannual variation in egg-laying dates in

(a)   **(b)**

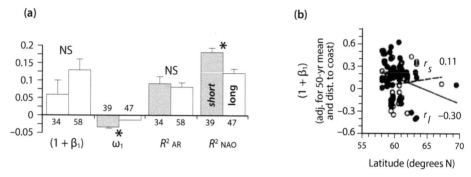

Figure 3.11. (a) Results from application of a model of life history dynamics such as that in figure 3.8 to long-term data on timing of arrival by short-distance (solid bars) and long-distance (open bars) migratory birds breeding in Norway, indicating variation in the strength of temporal dependence, climatic influence, and the contribution of the North Atlantic Oscillation index to variation in arrival dates, between both types of birds. The arrival time of short-distance migrants displayed stronger temporal dependence, climatic influence, and association with the NAO index than did that of long-distance migrants, whose timing may be more strongly influenced by changes in day length. (b) Variation in the strength of temporal dependence in spring arrival dates with latitude of breeding grounds in short-distance (open circles) and long-distance (solid circles) migrants breeding in Norway. NS indicates lack of statistical significance between column pairs. The x-axis labels in (a) indicate, respectively, the strength of direct density dependence [$(1 + \beta_1)$], the strength of climatic influence (ω_1), the coefficient of determination of the autoregressive component of the model (R^2_{AR}), and the coefficient of determination of the North Atlantic Oscillation component of the model (R^2_{NAO}). *Adapted from Forchhammer et al. (2002).*

British birds appear in hindsight to have been hampered by the inability to resolve the potential contributions of variation in resource availability to variation in the timing of life history events independently of climatic influences (McCleery and Perrins 1998). Feeding experiments in one passerine species that has figured prominently in studies of advancing egg-laying dates in response to climatic warming have demonstrated that birds can breed earlier in response to more abundant food (Nager et al. 1997). This is quite obviously the sort of relationship represented by coefficient a_2 in figure 3.8, yet a simple linear regression analysis is not capable of identifying this process in variation in the timing of egg laying when data on resource availability are unavailable. Moreover, inferring an influence of climate change on the timing of life history events on the basis of trends in phenological time series alone is inherently problematic, especially in instances in which statistical relationships with abiotic data vanish when temporal trends in both are accounted for (McCleery and Perrins 1998). More convincing analyses are those that demonstrate relationships between long-term trends in phenological time series and those in climatic variables in addition to significantly related interannual fluctuations in both (Forchhammer, Post, et al. 1998). Although such examples illustrate the utility of application of statistical models such as that derived above, more explicit treatment of the interaction between climatic and resource- or density-dependent factors in the timing of life history events can be developed from insights derived from a general theoretical model of the timing of life history events in individuals.

THE IWASA-LEVIN MODEL AND ITS RELEVANCE TO CLIMATE CHANGE

In seasonal environments, individuals are faced with a dual challenge in timing reproduction to ensure the viability of their offspring. First, individuals must contend with the risk of experiencing complete reproductive failure if environmental disturbance fatal to offspring occurs after reproduction. Avoiding this risk would be a simple matter if the latest possible date of disturbance were known in advance, in which case natural selection would favor those individuals that reproduced only after the last date of disturbance. Aside from such foreknowledge being exceedingly unlikely, if not impossible, in the natural world, individuals are faced with the challenge of competing with conspecifics for resources necessary for successful reproduction, including, for instance, space, nutrients, and mates, among many others. Such competition is likely to intensify as the season of reproduction progresses, thereby necessitating balancing the risk of reproducing too early in relation to environmental disturbance events and too late in relation to competitors. The Iwasa-Levin model presents a formalized treatment of the evolutionarily stable timing of reproduction in relation to these challenges (Iwasa and Levin 1995) and is of relevance here in understanding how climate change may influence individual life history strategies and the timing of life history events in relation to reproduction.

In the simplest scenario explored in the development of this model, competition is assumed to be minimal or absent, in which case individuals time reproduction solely in relation to the probability of experiencing environmental disturbance, which diminishes with the progression of the breeding season (Iwasa and Levin 1995). If x denotes the latest possible date of occurrence of some environmental disturbance, such as a late frost, snowstorm, or freezing rain, then $f(x)$ denotes the probability distribution of x, or the probability of environmental disturbance occurring on any given date throughout the breeding season. Then, according to Iwasa and Levin (1995), the reproductive success of an individual that commences breeding on any given day y is given by

$$\psi(y \mid x) = \begin{cases} e^{-by} & \text{if } y > x \\ 0 & \text{if } y < x \end{cases} \tag{3.8}$$

in which b quantifies the strength of the relationship between the timing of reproduction and reproductive success. Hence, according to equation (3.8), reproduction will fail if timed before the last possible date of environmental disturbance. In the absence of competition, then, reproductive success can be expressed as a function of the probability of avoiding environmental disturbance, as

$$\Phi(y) = \int_0^y f(x)\,dx\,e^{-by} \tag{3.9}$$

When competition among conspecifics for resources related to reproduction is included in the Iwasa-Levin model, successful offspring production depends on initiating reproduction before competitors do but after environmental disturbance is likely to occur. In this case, the reproductive success of an individual declines with the number of conspecifics that initiate reproduction earlier than that individual, and is given by

$$\psi(y \mid x) = \begin{cases} \exp\left[-a \int_x^y g(z)\,dz\right] & \text{if } y > x \\ 0 & \text{if } y < x \end{cases} \tag{3.10}$$

in which the coefficient a quantifies the intensity of competition (Iwasa and Levin 1995). In this case, as a increases, individuals must commence reproduction increasingly early following the date of last disturbance, or as soon after it as possible, to minimize competition and maximize the chance of successful reproduction. In contrast, when competition is minimal or absent, that is, when a is set to zero, there is no advantage to reproducing as early as possible after the last date of disturbance (Iwasa and Levin 1995). The dependence of reproductive success on the combined pressures of environmental disturbance and competitive intensity is then given by

$$\Phi(y) = \int_0^y \exp\left[-a \int_x^y g(z)\,dz\right] f(x)\,dx \tag{3.11}$$

To understand the relevance of the Iwasa-Levin model in the context of phenological dynamics and the timing of life history events in relation to climate change, we must explore the consequences of climate change for the time-distributed probability of fatal environmental disturbance and competition. I will do this by presenting a heuristic model that illustrates how climate change

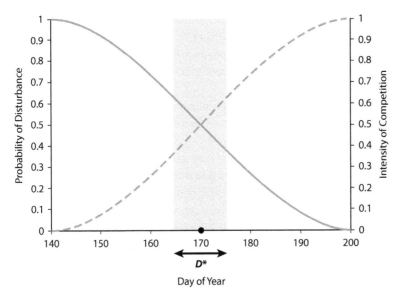

Figure 3.12. Conceptualization, deriving from the Iwasa-Levin model, of the relationship between the probability of environmental disturbance that is fatal to offspring (solid line) and the intensity of competition with conspecifics for resources necessary for successful reproduction as the breeding season progresses (dashed line). The probability of disturbance (solid line) declines throughout the breeding season from absolutely likely (probability = 1) to absolutely unlikely (probability = 0), while the obverse is the case for the intensity of competition (dashed line), which increases throughout the breeding season with the arrival or establishment of competitors at the breeding site. The optimal date of reproduction, D^*, is defined here as the date on which the probability of disturbance has declined to 0.50 and the intensity of competition has increased to 0.50. The window during which successful reproduction may occur, shown here in gray, is defined as the period between the earliest possible date of successful reproduction (the lower bound on the window, given as the date on which the probability of disturbance has declined to 0.60) and the latest possible date of reproduction (the upper bound on the window, given as the date on which the intensity of competition has increased to 0.60).

may interact with competition with conspecifics to alter the timing of, for example, reproduction, following from the predictions of the Iwasa-Levin model. First, let us consider the simple scenario in which the probability of environmental disturbance in the form of late frost, freezing rain, or severe drought, for example, declines from absolutely likely (probability of occurrence = 1) to absolutely unlikely (probability of occurrence = 0) as the reproductive season progresses (figure 3.12, solid line). At the same time, let us imagine that the intensity of competition for resources necessary for successful reproduction increases from near zero prior to the onset of the reproductive season to one, or maximum intensity, late in the reproductive season (figure 3.12, dashed line).

We may arbitrarily assume that the optimal timing of successful reproduction occurs at the date on which the probability of environmental disturbance has declined to 50 percent, but this value is assigned here purely for simplicity, and may more conservatively be assigned to a date with a lower associated probability of disturbance.

From this starting point, then, the individual must achieve a balance between reproducing too early with respect to the probability of environmental disturbance and too late with respect to time-dependent intensification of competition. Let us assume in this example that the lower bound defining the earliest possible date of successful reproduction is the date on which the probability of disturbance is 60 percent and the upper bound defining the latest possible date of successful reproduction is the date on which the intensity of competition has increased to 60 percent. These values are arbitrary and used here simply for illustration, and need not be equal. The window of reproduction is then centered on the optimal date of reproduction and is represented by the gray bar and associated arrow along the x-axis in figure 3.12. Individuals reproducing earlier than the optimal date, D^*, but within the window of optimal timing may still reproduce successfully but are exposed to a greater risk of suffering reproductive failure owing to environmental disturbance than are those individuals that reproduce on the optimal date. By comparison, individuals reproducing later than the optimal date but within the window of optimal timing may also reproduce successfully but are at greater risk of suffering reproductive failure because of more intense competition than that experienced by individuals reproducing on the optimal date. I make the assumption in this heuristic model that reproduction outside the optimal window results in reproductive failure and does not represent a viable strategy.

We can explore the consequences of climate change for the optimal timing of reproduction and the length of the window of viable reproductive dates by examining changes in D^* and this window in relation to changes in the time-distributed probability of disturbance and intensity of competition. First, we will examine the scenario of an abbreviated season of disturbance. This would be representative of, for example, a reduced probability of a fatal frost or late-season snowstorm as climate warms and will have the effect, in the model presented in figure 3.12, of increasing the slope of the relationship between the probability of disturbance and date during the breeding season. Subsequently we will examine the consequences of an extended season of likely disturbance. This would be representative of, for example, an increased or prolonged probability of fatal drought as climate warms and will have the effect, in the model in figure 3.12, of reducing the slope of the relationship between the probability of disturbance and time.

In the first alternative to the default scenario of figure 3.12, let us examine the case in which the period over which environmental disturbance can occur is abbreviated as a result of warming, while the period over which competition intensifies during the reproductive season remains identical to that in the default scenario. In this case, the last possible date of disturbance occurs earlier than in the default scenario, and the optimal date of reproduction, D^*, advances (figure 3.13a). As well, the lower bound on the window of reproductive opportunity advances, while the upper bound, determined by the intensity of competition for resources required for successful reproduction, remains unchanged. The net effect is a lengthening of the window of reproductive opportunity (figure 3.13a), which, at the population scale, should emerge as a decrease in synchrony in the timing of reproduction by individuals. Because the optimal date of reproduction has now moved to the left within the window of reproductive opportunity relative to the default scenario, we can expect a positively skewed distribution of reproductive dates within the population. If we combine an abbreviation of the period over which environmental disturbance can occur with delayed onset of competition relative to the default scenario, not only do the optimal date of reproduction and the lower bound of the window of reproductive opportunity advance, the upper bound on this window is simultaneously delayed (figure 3.13b). This results in further lengthening of the window of reproductive opportunity and an even more strongly positively skewed distribution of reproductive dates within the population (figure 3.13b). However, the optimal date of reproduction has not changed from that depicted in the scenario in figure 3.13a. In other words, the net effect of the interaction between the alleviation of the environmental constraint on the onset of reproduction by climatic warming and the alleviation of density-dependent competition has been to lengthen the duration of the reproductive season at the population level as a consequence of increased temporal dispersion of the timing of reproduction by individuals. This would suggest, for example, that the length of the growing season can increase in response to warming only if competition for reproductive resources is simultaneously reduced.

Next we shall examine the consequences of an advance in the onset of competition for reproductive resources. First, we consider the consequences of an advance in the progression of the intensity of competition for resources necessary for reproduction against the default scenario for the seasonal decline in the probability of environmental disturbance from figure 3.12. In this case, the optimal date of reproduction remains unchanged, but the upper and lower bounds on the window of reproductive opportunity converge on a single date (figure 3.13c). Because in this case the optimal date of reproduction falls after the date representing the collapsed window of reproductive opportunity, complete reproductive failure is the likely outcome. A similar if more pronounced result accrues when an advancement of the progression of competition coincides with a prolongation of the period over which environmental disturbance can occur. This is depicted in figure 3.13d, in which the latest possible date of environmental disturbance is extended later into the reproductive season while at the same time competition intensifies earlier in the season. In this case, the optimal date of reproduction is delayed considerably relative to the optimal date in the default scenario in figure 3.12. Moreover, the upper bound on the window of optimal reproduction, or the latest possible date for successful reproduction, now occurs earlier than the lower bound on the window of optimal reproduction, or the earliest possible date of reproduction (figure 3.13d). Here again, reproductive failure is likely to ensue.

The final scenario of advancement represents the case in which both the latest possible date of disturbance and the intensity of competition advance together. The result in this case is an advancement of the optimal date of reproduction similar in magnitude to that occurring in the scenarios represented in figures 3.13a and 3.13b. In this case, however, the added influence of the advancement of competitive intensity results in a narrowing of the window of optimal reproduction (figure 3.13e). Hence, in this case, alleviation of environmental constraints on the timing of reproduction by an effect of, for example, warming on the probability of a fatal frost occurring early in the reproductive season, combined with an intensification of competition for reproductive resources, will select for increased population-level synchrony in the timing of reproduction by individuals. In effect, what this means for individuals encountering this scenario is that, while earlier reproduction is possible, there is less room for error in missing the optimal date for successful reproduction.

Next, we will explore scenarios involving only delayed progression of either disturbance or competition, or both. In the first such scenario, a delay in the termination of environmental disturbance during the season of reproduction has the effect of prolonging the period over which a catastrophic disturbance may occur. In a climate change context, such a scenario may arise if rising temperatures increase the probability of occurrence of, for example, extreme drought or wildfires later in the reproductive season than would be the case under the default climate scenario. If such a prolongation of the period of disturbance occurs in the absence of any change in the progression of the intensity of competition from the default scenario, the result is a delay in the optimal date of reproduction (figure 3.13f). Because both the lower and upper bounds on the window of optimal reproduction converge on a single date that precedes the optimal date of reproduction (figure 3.13f), reproductive failure is predicted in this case. By contrast, if the probability of disturbance through time remains unchanged from the default scenario but the onset and progression of

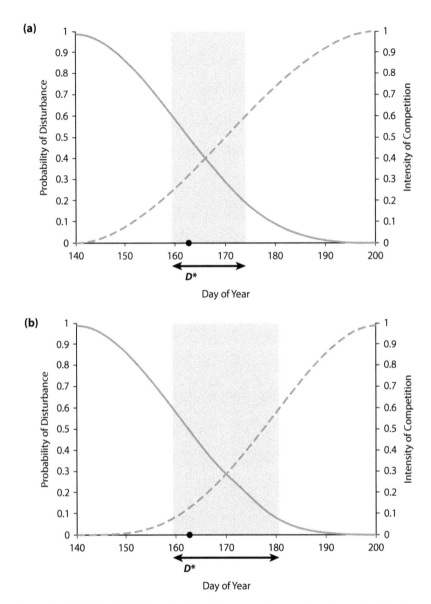

Figure 3.13. Solid and dashed lines in all panels represent the time-dependent probability of disturbance and intensity of competition, respectively, as in figure 3.12. In panel (a), the latest possible date of disturbance has advanced and now occurs earlier, while the progression of the intensity of competition remains unchanged from the default scenario in figure 3.12. In this case, the optimal date of reproduction has advanced, but the window of reproductive opportunity remains unchanged. In panel (b), the latest possible date of disturbance has advanced, while the onset of competition is delayed. In this case, the optimal date of reproduction has advanced, and the window of reproductive opportunity is extended.   (*continued*)

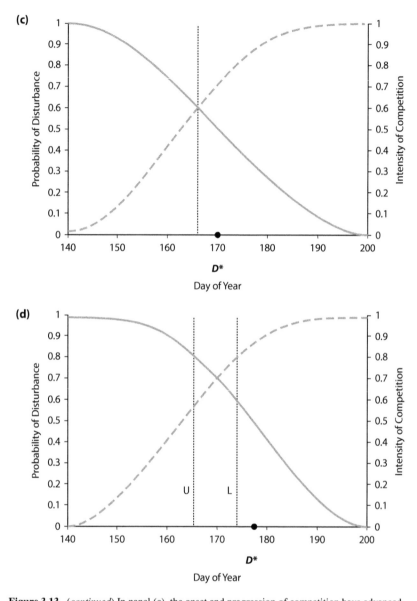

Figure 3.13. (*continued*) In panel (c), the onset and progression of competition have advanced and now occur earlier than in the default scenario in figure 3.12, while the probability of disturbance in relation to time remains unchanged from the default scenario. In this case, the lower and upper bounds defining the earliest and latest possible dates of successful reproduction occur on the same date, which precedes the optimal date of reproduction, D^*; reproductive failure will result. In panel (d), the onset and progression of competition have advanced while the probability of disturbance is delayed. In this case, the upper bound on the window of reproductive opportunity occurs before the lower bound, and both of them precede the optimal date of reproduction; reproductive failure will again result in this scenario.   (*continued*)

(e)

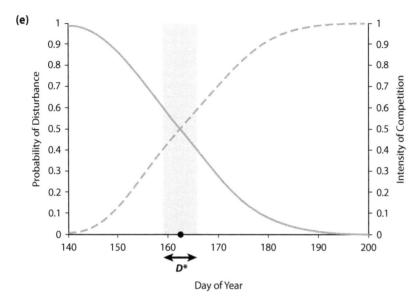

Figure 3.13. (*continued*) In the scenario in panel (e), the probability of disturbance remains unchanged from the default scenario in figure 3.12, but the onset and progression of competition have both advanced. Here the optimal date of reproduction has advanced, as has the window of reproductive opportunity, which is now compressed compared to that in the default scenario.

(*continued*)

competition are delayed, the optimal date of reproduction remains constant with respect to that predicted under the default climate scenario (figure 3.13g). As well, the onset of the season of optimal reproduction, defined by the lower bound on the window of reproduction, remains unchanged compared to the default scenario, but the length of the window of optimal reproduction is extended because the upper bound on it occurs later in the season than it does in the default scenario, resulting in a positively skewed distribution of reproductive dates (figure 3.13g). Finally, in the case in which the probability of environmental disturbance is prolonged by climate change and the onset and progression of competition are delayed, the optimal date of reproduction is delayed compared to the default scenario (figure 3.13h). In this case as well, the window of optimal reproduction is compressed because, while both the lower bound (earliest possible date) and upper bound (latest possible date) are delayed, the slopes of both the curves of disturbance and competition are increased under this scenario. The result at the individual level will be a delay in the timing of reproduction, while the result at the population level will be an increase in the synchrony of reproduction among individuals. Interestingly, if the timing of the onset of competition is advanced under this final scenario

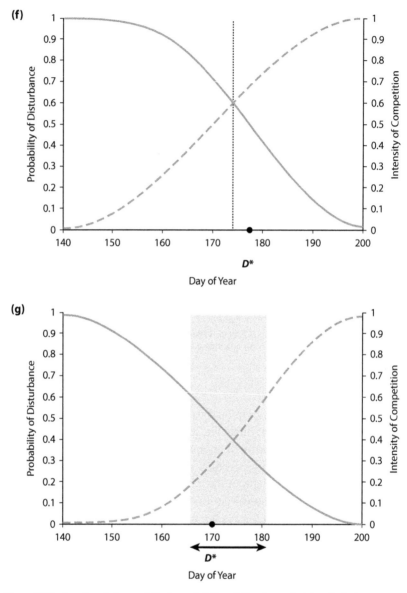

Figure 3.13. (*continued*) In panel (f), the probability of disturbance is delayed relative to the default scenario in figure 3.12, while the onset and progression of competition remain unchanged. In this case, as in the scenario illustrated in panel (c), the window of reproductive opportunity has collapsed to a single date, which precedes the optimal date of reproduction, and reproductive failure will result. In panel (g), the probability of disturbance is unchanged from the default scenario, but the onset and progression of competition are delayed. In this case the optimal date of reproduction remains unchanged from the default scenario, but the window of reproductive opportunity is longer, and extends later into the season.   (*continued*)

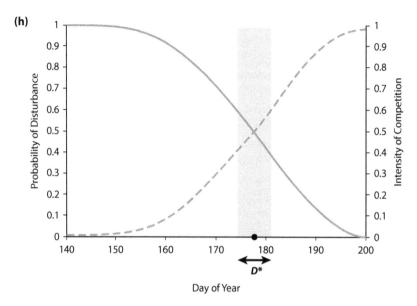

(h)

Day of Year

Figure 3.13. (*continued*) In the scenario in panel (f), the probability of environmental disturbance is prolonged, while the timing of onset of competition is delayed. The result is a delay in
the onset of—and compression of the length of—the window of optimal reproductive timing.
This scenario should promote delayed yet more synchronous timing of reproduction at the
population level.

rather than delayed, the result will be reproductive failure, for the optimal date
of reproduction will now fall outside the upper and lower bounds on the window of optimal reproductive timing because the upper bound (the date of 60
percent probability of competition) now occurs before the lower bound (the
date of 60 percent probability of disturbance).

This comparison among scenarios of change has revealed several key insights regarding the consequences of interactions between climatic and biotic
factors for the timing of life history events in response to climate change. Before delving into these, however, it is worth reiterating the assumptions under
which the heuristic model in figure 3.12 was developed, assumptions that derived from the Iwasa-Levin model. It was assumed that earliest possible reproduction is favorable but constrained by the probability of environmental
disturbance, which is fatal for offspring and diminishes in likelihood as the
reproductive season progresses. It was furthermore assumed that delayed reproduction is unfavorable because of the increasing intensity of competition
for reproductive resources as the reproductive season progresses. Under these
circumstances, the following insights emerge.

First, the timing of reproduction can be expected to advance under climate change only if the probability of environmental disturbance associated with climatic conditions diminishes earlier or at an advanced rate throughout the growing season. This is the case regardless of whether the intensity of competition for reproductive resources among conspecifics (i.e., density dependence) progresses at an advanced rate, at a delayed rate, or remains unchanged relative to baseline conditions throughout the growing season. Second, the length of the season of optimal reproduction can increase under climate change only if the probability of environmental disturbance diminishes earlier or at an advanced rate and competition is delayed or remains unchanged from baseline conditions. Third, an increase in the length of the season of optimal reproduction is accompanied by increasing positive skewness, but never negative skewness, in the distribution of reproductive dates within the population, a scenario under which it becomes increasingly difficult for individuals to reproduce before the optimal date of reproduction. Fourth, because an earlier and more rapid alleviation of environmental constraints on the timing of reproduction, combined with an advancement in the intensification of competition, translates into not only an advance in the optimal date of reproduction but also a shortening of the length of the season of reproduction (figure 3.13c), it is under these circumstances that individuals will face increased pressure to time reproduction optimally. Finally, reproductive failure will result under any scenario in which climate change prolongs the probability of environmental disturbance unless this prolongation is also accompanied by a delay in the onset and intensification of competition. A corollary to this conclusion is that any advance in the intensification of competition will similarly result in reproductive failure unless it is offset by an earlier reduction in the probability of environmental disturbance. Hence, delayed reproduction is a viable strategy only when competition is reduced or negligible (Iwasa and Levin 1995), and the interaction between influences of density dependence and climate change on the timing of life history events appears to be consistent with the facilitation/promotion hypothesis rather than with the tension hypothesis. This makes it unlikely that satellite-derived patterns of extended growing seasons reflect strategies of prolongation of the timing of reproduction by individuals, but rather increases in temporal heterogeneity among species.

MODELING THE CONTRIBUTION OF PHENOLOGY TO POPULATION DYNAMICS

In light of the importance of variation in the timing of life history events such as reproduction to offspring production and survival, we should expect phenological dynamics to have the potential to contribute to variation in abundance. The connections among climatic variation, phenology, and abundance or population dynamics are, however, difficult to demonstrate quantitatively, in part because phenological studies and studies of population dynamics have not typically been pursued in concert within the same study system (Miller-Rushing et al. 2010). Moreover, when such phenomena have been studied in concert in the same system, it is rarely the case that the studies were initiated with an interest in the implications of climate change for phenological variation and of phenological variation for population dynamics. Nonetheless, individual-based studies of the timing of offspring production conducted as part of longer-term population monitoring studies indicate the utility of such approaches for quantifying the role of climate change in phenological dynamics and subsequent effects on population dynamics. The longitudinal study of red deer on the Isle of Rhum, for instance, has demonstrated advancing timing of parturition in response to the simultaneous influences of climate, density, and advancing plant phenology (Post 2003c; Moyes et al. 2011).

In the more common circumstance in which concurrent data on interannual variation in the timing of life history events and on population dynamics are lacking, the contributions of climatic variation to phenological dynamics, and of phenological dynamics to variation in abundance, can be inferred through applying a modification of the phenology model in equation (3.7) to time series data on abundance. Returning to the heuristic model in figure 3.8, we can see that it implies that variation in density or, by extension, population size is a function of the timing of life history events as related to reproduction and mortality. Although density in this model is assumed to influence resource availability, which in turn influences the timing of life history events, the dependence of density on the timing of life history events provides an avenue of feedback from resource availability to density. Furthermore, climate is assumed in this model to influence density only indirectly, through, primarily, resource availability and its influence on the timing of life history events related to reproduction and mortality. While in the optimal situation we would test for the influences of resource availability and the timing of life history events such as reproduction on population dynamics using data on all three, it is very rarely the case that such a comprehensive, multi-annual data set would be at our disposal. We can, however, adapt the model in equations (3.4a–c) to this purpose, and we can do so with the alternative form of equation (3.4a) above that includes a direct influence of climate on the timing of life history events, indicated by the dashed arrow in figure 3.8. We will take this approach because it allows us to draw inferences about the interaction between the influences of climate and temporal dependence in the timing of life history events on population dynamics.

We begin with a set of equations equivalent to equations (3.4a–c), except that in this example the equation for the timing of life history events includes the influences of resources and climate:

$$D_t = a_0 + a_1 D_{t-1} + a_2 S_t + a_3 U_t \qquad (3.12a)$$

$$S_t = b_0 + b_1 N_{t-1} + b_2 U_t \qquad (3.12b)$$

$$N_t = c_0 + (1 + c_1)N_{t-1} + c_2 D_{t-1} \qquad (3.12c)$$

We proceed by solving equation (3.12c) for D_{t-1} and inserting the resulting equation into equation (3.12a), giving:

$$D_t = a_0 + \frac{a_1}{c_2}[N_t - c_0 - (1 + c_1)N_{t-1}] + a_2 S_t + a_3 U_t \qquad (3.13)$$

Equation (3.12b) can now be substituted into equation (3.13), and, after setting $t = t - 1$ in the resulting modification of equation (3.13), it can now be inserted back into equation (3.12c), giving:

$$N_t = c_0 + (1 + c_1)N_{t-1} + c_2 \left\{ a_0 + \frac{a_1}{c_2}[N_{t-1} - c_0 - (1 + c_1)N_{t-2}] \right.$$
$$\left. + a_2(b_0 + b_1 N_{t-2} + b_2 U_{t-1}) + a_3 U_{t-1} \right\}. \qquad (3.14)$$

After simplifying and gathering terms, we have an equation for abundance or density in the current year as a function of past densities and climate that incorporates the influences of the timing of life history events and resource availability and that is amenable to an autoregressive analysis of time series data on annual abundances:

$$N_t = (a_0 c_2 - a_1 c_0 + a_2 b_0 c_2 + c_0) + (1 + a_1 + c_1)N_{t-1}$$
$$+ (a_2 b_1 c_2 - a_1 - a_1 c_1)N_{t-2} + (a_2 b_2 c_2 + a_3 c_2)U_{t-1} \qquad (3.15)$$

Notably, according to equation (3.15), the additive effects of temporal dependence in the timing of life history events and of density dependence generate first-order, direct density dependence, as is evident in coefficient $(1 + a_1 + c_1)$. By contrast, the multiplicative interactive effects of temporal dependence and density generate second-order, delayed density dependence, as is evident in coefficient $(a_2 b_1 c_2 - a_1 - a_1 c_1)$. As well, equation (3.15) indicates that the direct effects of climate on life history dynamics and on resources should be expected to manifest as delayed (one-year lagged) effects of climate on density.

TRENDS AND STATISTICAL CONSIDERATIONS

Among the first pieces of high-profile evidence to emerge documenting the responses of wildlife to recent climate change were reports of earlier breeding dates in amphibians and birds (Beebee 1995; Crick et al. 1997; McCleery and Perrins 1998; Crick and Sparks 1999; Dunn and Winkler 1999). Trends in phenological data are inherently intuitive because they indicate in some cases an increasingly earlier occurrence of events that we innately identify as harbingers of spring. Yet trends can pose challenges to the statistical association of changes in the timing of life history events with changes in climate.

The edible frog (*R. esculenta*), the common toad (*Bufo bufo*), the natterjack toad (*Epidalea calamita* or *B. calamita*), and three species of salamanders of the genus *Triturus* all displayed trends toward earlier spawning in England between 1978 and 1994 (Beebee 1995). Spawning dates of the natterjack toad and edible frog had advanced by approximately two to three weeks by the end of this period compared to spawning dates at the beginning. Spawning dates of the three salamander species had advanced by five to seven weeks by 1994 compared to spawning dates at the end of the 1970s and early 1980s (Beebee 1995). Trends in the spawning dates of those species displaying significant trends correlated negatively with late winter–early spring temperatures. Similarly, egg-laying dates of twenty species of birds breeding in England (31 percent of the species for which data were analyzed) displayed significant advances by, on average, 8.8 days over the period from 1971 to 1995 (Crick et al. 1997).

These examples strongly implicate recent climatic warming in the advancement of breeding dates in amphibians and birds. However, the attribution of observed trends in phenological data to climate change is far from straightforward. A correlation between two time series that both display significant trends may represent a spurious relationship unless the correlation remains after the trend has been removed (Royama 1992). Moreover, phenological time series data may be significantly temporally autocorrelated, meaning that observations in successive years are not independent of each other. As we noted above, this process may arise through biological limitations on the timing of life history events imposed by their timing in previous years. Assigning relations between two autocorrelated time series requires, at a minimum, reducing the degrees of freedom associated with the significance test of their correlation coefficient (Post and Stenseth 1998) or, preferably, accounting for this autocorrelation using autoregressive analysis (Forchhammer, Post, et al. 1998), as described above.

This is an issue that is inherent to investigations of phenological response to climate change because abiotic factors such as temperature display trends due to climate change, and ecological time series data may, if they encapsulate biological processes related to climate, display trends as well. Moreover, other factors influencing phenological dynamics may have undergone temporal changes during the period over which phenological data were collected. Primary among these would be changes in population size or density and changes in habitat or resource availability. Interannual variation in data such as first arrival dates or first egg-laying dates may be vulnerable to changes in population size from year to year in two ways. First, there may be a sampling effect associated with population size: larger populations may display greater variation among individuals in the timing of life history events, especially those at the tails of the distributions, such as first arrival or departure dates. Second, there may be an actual density-dependent influence on the timing of life history events such as initiation of migration or breeding owing to the influence of intraspecific competition on resource availability, including space.

An especially comprehensive analysis of long-term variation in arrival and departure dates of migratory birds breeding in the United Kingdom attempted to take into account the influence of population density (Sparks et al. 2007). The analysis was hampered by use of an index of population status rather than of population size or density, but it detected a marginally significant relation between spring arrival dates and this index, suggesting that factors other than changes in climate alone do indeed play a role in variation in migration timing (Sparks et al. 2007). This analysis also revealed only a weak correlation between the average spring arrival date of ten common species of migrants and spring temperature on their breeding grounds but a much stronger correlation with spring temperature on their departure grounds, suggesting an influence of proximal environmental cues in the timing of departure for breeding (Sparks et al. 2007). In this case, Sparks and colleagues (2007) suggested that the species displaying the greatest decline in abundance over the period covered by the data also displayed more minimal advances in spring arrival dates. Similarly, Tryjanowski and colleagues (2005) demonstrated an effect of population status on the magnitude of the trend in spring arrival dates for thirty species of birds breeding in Poland on the basis of twenty years of data. Species in populations displaying an increasing trend over this twenty-year period also displayed the strongest trends toward earlier arrival on breeding grounds (figure 3.14). This association was attributed to the greater probability of early detection in species increasing in abundance and a lower probability of early spring detection in declining species (Tryjanowski et al. 2005).

Additionally, trends in resource availability and quality may influence the timing of life history events independently of or in addition to the effects of density. For instance, as mentioned above, experiments have demonstrated that proffering supplemental food to captive birds advances their egg-laying dates. It seems plausible that warming could influence the resource base for migratory and resident birds, thereby influencing their timing of initiation of egg laying. Obviously, this is more difficult to envision for amphibians, whose development and life cycles are more closely linked directly to abiotic factors such as temperature and hydroperiod. Nonetheless, analytical models derived from linked equations expressing dynamics of reproductive dates, resources,

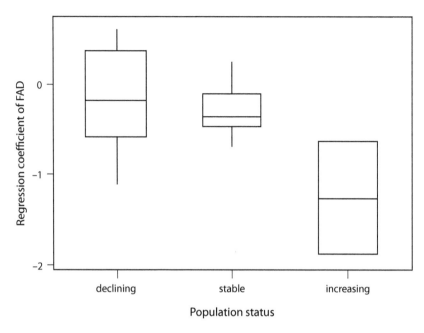

Figure 3.14. Trends in first arrival dates (FAD) of 30 species of migratory passerine birds breeding in Poland grouped according to their population status. Increasing populations are more likely to display a negative trend in arrival dates, which might be an indication of a tendency toward disproportionate observance of early migrants in such populations. *Adatped from Tryjanowski et al. (2005).*

and density, such as equations (3.4) and their associated models, should be applied more widely to improve our understanding of the factors contributing to life history responses to climate change.

EMPIRICAL EXAMPLES LINKING CLIMATE, PHENOLOGY, AND ABUNDANCE

Direct effects of climate-change driven variation in phenology on abundance may be more likely in species whose life history dynamics are closely temperature dependent. As we have already seen, the timing of life history events such as emergence and flowering in many plant species is highly responsive to variation in temperature and temperature-related events such as the annual timing of snowmelt. At the Rocky Mountain Biological Laboratory (RMBL) in the U.S. state of Colorado, the timing of snowmelt is, as mentioned, a strong determinant of the timing of plant growth and phenology, but the interaction among temperature, snowpack, and the timing of spring snowmelt is complex because warm years may also be characterized by deep snowpack, which melts later than shallow snowpack (Inouye et al. 2000). The timing of flowering of multiple species of alpine and subalpine forbs has been tracked at the RMBL for three decades, with data on some species consisting of daily or twice daily counts of numbers of flowers produced on each of several 4 m² plots, as well as, in some cases, the number of flowers per rosette (Inouye et al. 2003). These data indicate that some vernal species, such as *Delphinium barbeyi*, *Erigeron speciosus*, and *Helianthella quinquenervis*, bloom earlier in warm years, which might be expected to present individuals of this species with an advantage in terms of numbers of flowers, fruits, and seeds produced (Inouye 2008). Despite increasing mean monthly temperatures at the RMBL, however, the date of the last annual spring frost has not advanced; as a consequence, early-blooming flowers are exposed to a greater risk of frost damage, which in the case of *H. quinquenervis* has resulted in a doubling of the proportion of flowers lost to frost damage (Inouye 2008). Such losses are likely to reduce considerably recruitment in these species (Inouye 2008). Indeed, the abundance of *Mertensia ciliata* has declined at the RMBL, with this species having completely disappeared from some of the study plots, presumably because the species experiences loss of flowers in warm years, either because of earlier blooming, which leaves flowers susceptible to frost damage, or because warm years that induce early flowering also result in drought stress for this species (Miller-Rushing and Inouye 2009).

MORE COMPLEX AND SUBTLE FORMS
OF PHENOLOGICAL VARIATION

The preceding examples of phenological variation focused primarily on interannual variation in the timing of life history events. More complex and subtle variations are possible, and in operation. As we will see in greater detail in chapter 6, variation in the phenology of resources may have implications for consumer dynamics, and in some cases these consumer-resource dynamics may occur independently of the implications of variation in resource phenology for resource dynamics themselves. For instance, long-term data from the ecological monitoring station at Zackenberg, Greenland, indicate that snow cover and the time of snowmelt have both advanced annually in association with a 2°C rise in June temperature since 1996, eliciting an advance in the flowering phenology of several forb species and a dwarf willow of, on average, 2.5 days per year (Forchhammer et al. 2012). In contrast to the patterns observed at the RMBL in Colorado, however, this advance in phenology has not translated into changes in the growth or flower production of those plants; it has, though, been accompanied by an increase in offspring production by muskoxen that consume those plants (Forchhammer et al. in review). Hence, the relevant metric for quantifying demographic consequences of trends in phenology induced by climate change depends not only on the species undergoing a phenological response to climate change but also on the species with which it interacts (Visser and Both 2005; Forchhammer et al. 2012).

Variation in abundance related to phenological dynamics of resources may be linked to a wider array of annual events than those related directly to reproduction. An example of this is seen in the daily counts of the number of caribou and muskoxen present at my study site near Kangerlussuaq, Greenland, recorded annually. These data reveal a pattern of waxing and waning in the abundance of both large herbivores at the site that has a strong seasonal component (figures 3.15a and b). These patterns reflect the seasonal movement into and out of the study site by both species. In the case of caribou, the phenological dynamics represented by their movement into the study site reflect the annual migration to the calving grounds they have used for the past several thousand years (Thing 1984; Meldgaard 1986). In the case of muskoxen, however, there is no annual migration into and out of the study site because the local population is resident in the area around the site. Rather, muskoxen here, as elsewhere, likely exhibit a pattern of movement among favored foraging areas more indicative of small-scale rotational migrations unrelated to the timing of offspring production (Forchhammer and Boertmann 1993; Forchhammer and Boomsma 1995; Forchhammer et al. 2005). In both cases, however, we might expect the timing of peak abundance of these herbivores in the study site to coincide with the phenology of forage plants, which also display interannual variation in the timing of peak emergence. There is, however, considerable variation among years in the magnitude and sign of the correlation between plant phenology and the timing of herbivore presence at the study site (figure 3.16). Moreover, this variation is not consistent between herbivore species, which may have implications for vegetation response to climate change. For instance, the average daily abundance of both caribou and muskoxen at the study site influences the biomass response of vegetation to warming (Post and Pedersen 2008). Whereas muskoxen appear to influence the biomass response to warming of the entire plant community, and deciduous shrubs in particular, caribou appear to exert an influence primarily on graminoid response to warming (Post and Pedersen 2008). Hence, the timing of herbivore pressure on vegetation may be an important component of plant community dynamics and ecosystem function that is more subtle than the more obvious contributions of plant phenological dynamics.

(a)

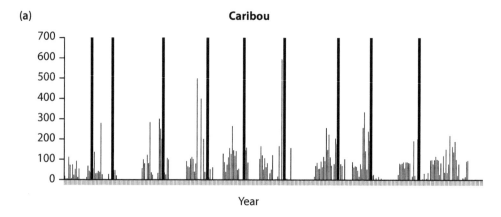

Caribou

(b)

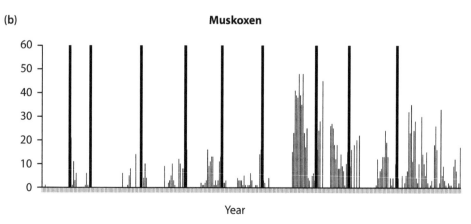

Muskoxen

Figure 3.15. The numbers of adult caribou (a) and muskoxen (b) at the study site near Kangerlus-suaq, Greenland, vary seasonally and interannually. The former scale of variation is related to the phenology of herbivore movement into and out of the study site during each summer of observation, and may be related to the dynamics of the plant growing season. The latter scale of variation is represented by changes in the peak numbers of individuals of each species observed annually and encompasses population dynamics, integrating the processes of immigration, emigration, birth, and death.

Finally, the interaction between phenotypic plasticity and the predictability of environmental conditions influencing, for example, the annual timing of key life history events may also be of fundamental importance to the ability of species to adapt to ongoing and future changes. The ability of individuals to adjust the annual timing of life history events such as the onset of repro-duction to match variation in environmental cues represents a special form of phenotypic plasticity that we might term phenological plasticity. Recent

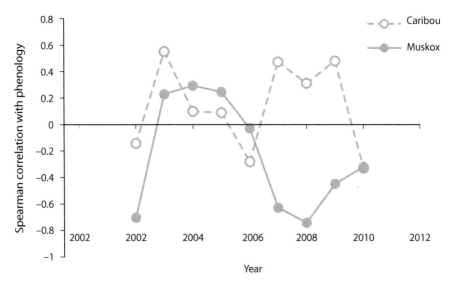

Figure 3.16. Variation across years in the magnitude and sign of the correlation between numbers of caribou and muskoxen observed at the study site near Kangerlussuaq, Greenland, on a daily basis and the proportion of plant forage species emergent on those dates each year.

modeling suggests that fitness is buffered against environmental variability by phenological plasticity only when the environmental cue and the phenological response are tightly coupled (Reed et al. 2010). Under conditions in which the predictability of the environmental cue eroded, phenological plasticity in response to environmental variability precipitated population declines and enhanced extinction risk (Reed et al. 2010). Hence, the extent to which variation in phenology in response to climatic variation and change enhances the reproductive success of individuals, and thereby population persistence, appears to depend on the extent to which the environmental cue eliciting the phenological response is actually indicative of optimal conditions for reproduction.

Population Dynamics and Stability

As a starting point in what will develop into a rather lengthy treatment of the related subjects of population dynamics and stability in relation to climate change, let us consider a deceptively simple illustration of the notion of stable point equilibrium in population dynamics (figure 4.1a). This figure derives from the now classic treatment of stability in deterministic and fluctuating environments by May (1973b). In it, we observe that the densities of two populations, N_1 and N_2, have settled into equilibrium with one another and remain locked in this arrangement because of the strongly stabilizing influence of the compression factor, represented by the symbol λ, which tends to nudge the densities of each population toward this stable point equilibrium (May 1973b). Although May (1973b) originally developed this treatment to describe stability between interacting populations, we may also view N_1 and N_2 as consecutive densities within the same population viewed through time because these, of course, may also interact in the specific sense that density in time step one may influence density in time step two, just as density in time step two may depend on density in time step one. In this case, what is depicted in figure 4.1a is the end result of a tendency of the population to settle into damped oscillations, leading to a constant density through time.

The important feature of this illustration, as I explore in much greater detail below, is that this stable point equilibrium is possible only in strictly deterministic environments. As soon as we introduce environmental fluctuation or variation, changes in population size or density from one time step to the next become less predictable and, by extension, less likely to be characterized by point stability. This is represented in figure 4.1b, in which the densities of the interacting populations, or of a single population through time, are represented as a cloud of possible densities, the distribution of which about the previous stable point equilibrium is determined by the magnitude of the dispersion factor, represented by the symbol σ^2 (May 1973b). The effect that inclusion of this dispersion factor, which accounts for the influence of environmental variation on population density, has on stability in this conceptualization of population dynamics is to reduce the likelihood of achieving stability, in effect increasing the total area of instability relative to that of stability within the parameter space represented by the model describing such dynamics (figure 4.1c) (May 1973b). For our purposes, which relate to understanding the implications of climate change for population dynamics and stability, we should take immediate note of the fact that the dispersion factor exerts its influence without regard to direction or trend in environmental variation. In other words, climate change may have the potential to influence population stability, and thereby persistence, not simply because of the direction of population response to increasing temperature or trends in any other abiotic parameter but rather by the sheer fact that it may increase environmental contributions to population dynamics. Hence, we may discover that the magnitude of population response to climate change matters at least as much as, or perhaps even more than, the direction of population response to climate change.

ESTABLISHING THE FRAMEWORK FOR ADDRESSING POPULATION RESPONSE TO CLIMATE CHANGE

Understanding the contribution of abiotic variation to population dynamics will prove central to our efforts to predict the persistence of populations in a changing climate. Understanding persistence, in turn, is of fundamental importance to efforts to predict the fate of species in a changing climate. And understanding the fate of species in a changing climate is of essential importance to understanding the implications of climate change for species interactions and community dynamics and persistence. This chapter establishes a foundation for what unfolds in the rest of the book as we examine, in subsequent chapters, community dynamics and stability, biodiversity, and ecosystem function.

Population dynamics, or the variation in abundance of a population through time, can be decomposed into two components, density-dependent and density-independent processes. Density-dependent processes are those involving competitive interactions among members of the same species within the same population that influence survival and reproduction. Density-independent processes are those that do not involve interactions with other members of the same species in the same population but rather owe to external factors such as environmental variation. It is this latter set of processes that has relevance to climate change, though density dependence certainly has a role to play in the response of populations to climate change. Density dependence may, for instance, operate in tension with climatic influences on dynamics, in which case density dependence will tend to limit abundance to densities below which climatic effects on changes in population size are minimal. Alternatively,

(a)

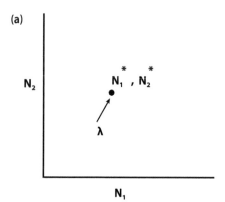

N$_t$ through time

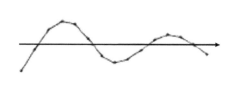

(b)

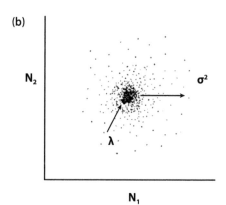

(c)

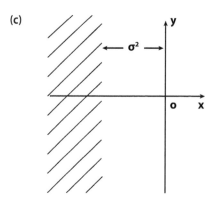

density dependence may facilitate or promote climatic influences on dynamics, in which case density dependence will act together with climate to limit population size.

Classically, density dependence is considered stabilizing in population dynamics (May 1973a; Maynard Smith 1974), in which case strong density independence is necessary to destabilize population dynamics, a clear instance of the tension hypothesis outlined in the preface. As discussed later in this chapter, however, density dependence is not always stabilizing. Nonetheless, the stabilizing properties of density dependence derive from its simultaneous, or in some cases separate, influences on mortality and fecundity. As population density rises, so might the numbers of individuals dying. However, this does not necessarily indicate a density-dependent increase in mortality because the more individuals there are in the population at any given time, the more there are to die. What matters in this case is the per capita rate of mortality. If this rate increases with density, the increase is considered density dependent. Similarly, if the birth rate, or the number of offspring per female in the population, declines with rising density, this is considered a density-dependent reduction in fecundity.

In a strictly density-dependent, deterministic scenario, rates of mortality and fecundity determine whether the population growth is positive or negative. At low densities, fecundity (birth rate) should exceed mortality (death rate), in which case the population will increase. At high densities, mortality should exceed fecundity, in which case the population will decline. The intersection between fecundity and mortality, or the point at which the lines describing their relationships with density cross, is considered a stable equilibrium density. However, it is not necessary for both fecundity and mortality to be density dependent for their lines to intersect, resulting in an equilibrium density; this can also occur when only one of them is density dependent. This point

Figure 4.1. A stable point equilibrium (N_1^*, N_2^*) is possible in deterministic environments if the eigenvalues (λ) of the interaction between two populations (N_1 and N_2) have negative real parts, as represented in panel (a). Although originally presented in terms of two interacting species in a simple community, this concept may be adapted to describe the interaction between densities of a single population in successive time steps, in which case the stabilizing factor, λ, acts as a compaction force driving the population toward a stable point equilibrium through monotonically damping population dynamics. In stochastic environments, the dispersion factor, σ^2, acts to drive the dynamics of the interacting populations away from a stable point equilibrium, as in panel (b). In this case of dynamics in a single population, environmental stochasticity acts to move the population toward destabilizing, potentially chaotic, dynamics. This scenario is visualized in the parameter space in panel (c), where the dispersion factor, σ^2, moves the stability region, which is shaded, further to the left along the x-axis of the stability plane than would be the case in a deterministic environment, where stability lies simply in the negative region. *Adapted from May (1973b).*

of intersection is an equilibrium density because all densities will exhibit a tendency to approach this density, and it is stable because at this population density, fecundity and mortality are equal, so there will be no change in density from this point. If such an equilibrium exists, then the interesting question becomes one of why populations change in size through time.

The obvious answer to this question is that population dynamics do not behave deterministically but rather stochastically. This stochasticity, or nondeterministic behavior in population dynamics, can arise as a consequence of demographic stochasticity or environmental stochasticity. The former represents unpredictable variation in demographic parameters, such as birth and death rates, while the latter represents unpredictable variation in environmental parameters (Caswell 2000; Boyce et al. 2006). In part, environmental stochasticity derives from density-independent influences on fecundity and mortality (Leirs et al. 1997). If we are interested in looking for a signal of climate change in population dynamics, then a good starting point would be to look for climatic influences on fecundity and mortality. The simplest conceptualization of population dynamics in discrete time, however, relates changes in abundance from one time step to the next to four processes: birth, immigration, death, and emigration, the classic BIDE parameter space describing population dynamics. Density independence may, of course, act on any of these processes, so we should also consider looking for signals of climate change in immigration and emigration.

A less obvious answer to the question posed above, however, is that far from being stabilizing, density dependence may, under certain conditions, actually become destabilizing. Even purely deterministic models of population growth are capable of generating a wide array of dynamics, from highly stable to chaotic (May 1973b, 1974, 1975a, 1975b; May et al. 1974). For instance, comparisons of empirically derived estimates of density dependence and the intrinsic rate of population increase in single-species models for invertebrates, using deterministic models of population growth, have demonstrated that a combination of either or both strong density dependence and large rates of increase can produce chaotic dynamics (Hassell 1974; Hassell et al. 1976). Whether this applied as well to vertebrates, which typically display much longer return times in dynamics and are less prone to near-instantaneous irruptions, seemed unlikely until an analysis of nonlinearity in the dynamics of a population of Soay sheep on an island off the coast of Scotland revealed that highly overcompensating density dependence in rates of mortality and fecundity exerted a destabilizing influence on dynamics (Grenfell et al. 1992). In this case, nonlinearity in the strength of density dependence contributed to its destabilizing influence. In other words, population lows were followed by overcompensating offspring production, leading the population to overshoot carrying capacity and precipitating overcompensating mortality (Grenfell et al. 1992). It is also possible for nonlinearity in density dependence to exert a stabilizing influence on dynamics, as has been demonstrated using statistical capture-recapture models for African rodents (Leirs et al. 1997). In this instance, dynamics were best reproduced using a model incorporating nonlinear density dependence combined with deterministic density independence and stochastic environmental seasonality (Leirs et al. 1997).

The utility in examining both population dynamics and stability in relation to climate change relates primarily to extinction risk. We may be able to identify a signature of climate change in population dynamics if we are able to demonstrate analytically an influence of climate on the dynamics of a given population. This may tell us very little, however, about the contribution of climate change to extinction risk in that population. What we really need to know, in this context, is what this climatic influence does to the observed dynamics. We might infer, for example, that a population of a given species may benefit from climate change because a retrospective analysis of its dynamics identified a positive influence of temperature or some similar proxy on changes in abundance. From a stability standpoint, however, what really matters is how climate, or climate change, influences *variation* in abundance. Even a positive climatic signal in dynamics might destabilize population dynamics and precipitate a population crash or local extinction if, for instance, it increases the variation in abundance or population size from one time step to the next (Ellis and Post 2004).

Climate change might also contribute in surprising ways to extinction risk through interactions with density dependence. Synergy of this type is evident in the classic study on the crash of reindeer on St. Matthew Island, part of the U.S. state of Alaska (Klein 1968). In this example, reindeer were introduced to the island, where none had existed before, and therefore found themselves exposed to an abundance of forage in a predator-free environment. Over the course of several climatically favorable years, the population irrupted until intraspecific competition had increased to such an extent that fecundity began declining. During an exceptionally harsh winter, extreme mortality set in, and nearly the entire population died off. Hence, in this case, biotic interactions— namely, intraspecific competition—facilitated the catastrophic population response to a harsh winter. Less intuitive, however, is the interaction between minimal competition during the early phase of population growth and more favorable climatic conditions that facilitated the rapid population increase, which ultimately left the population vulnerable to a crash. In theory, populations exposed to increasingly autocorrelated environments, in which favorable climatic conditions occur in runs of several consecutive years, may increase in density to the point at which a single unfavorable year precipitates a catastrophic crash (Wilmers et al. 2007b). Hence, short-term studies indicating positive associations between temperature or precipitation increases (or both) and population size may overlook the possibility that density dependence can interact with such abiotic trends in an ultimately devastating fashion.

It is also essential to recognize, as alluded to above, that changes in the variability of environmental conditions from one time step to the next may be just as important to population dynamics, particularly stability, as the influence of trends in environmental conditions themselves. Stochastic climatic events such as droughts can induce complete reproductive failure and loss of entire cohorts. Populations of finches on the Galápagos Islands, for instance, experience periodic drought that obliterates reproductive efforts in such years, producing gaps in the age structure and cohorts composing the populations (Grant et al. 2000).

CLASSIC TREATMENTS OF POPULATION STABILITY VIEWED AFRESH THROUGH THE LENS OF CLIMATE CHANGE

In any discussion of stability, it is essential to distinguish between ecological stability and mathematical stability. Maynard Smith (1974) made this distinction in typically eloquent fashion. To paraphrase here, persistence in an ecological context is relative, whereas mathematical stability is an absolute property. Hence, ecological stability should be considered in the most relative terms possible, while mathematical stability can be characterized quantitatively through numerical solution. Although some species, communities, or even ecosystems may give the impression of stability as a result of their long-term persistence, "no ecosystem persists forever, but some are more persistent than others" (Maynard Smith 1974).

The stability of populations may be reduced by many factors, the most counterintuitive of which may be density dependence. Classically, we are taught that density dependence is stabilizing. As was described above, this notion derives from the fact that as populations increase toward carrying capacity, birth rates begin to decline with increasing density, while death rates begin to rise with increasing density. Hence, in populations below carrying capacity, birth rates exceed death rates, and there should be a positive net contribution to population growth. In contrast, in populations above carrying capacity, death rates exceed birth rates, and there should be a negative net contribution to population growth. As a consequence of these opposing forces, populations should constantly "search out" K, or carrying capacity, which represents a stability point, for two reasons: first, at K, birth rates equal death rates, so there are no additions to or subtractions from the population (i.e., there is zero net growth), and second, densities below K should increase toward it, while those above K should decline toward it.

Many types of population dynamics can, however, arise from deterministic, density-dependent models, including chaotic dynamics, which are highly unstable (May 1972, 1974). Chaotic dynamics are unstable because, as fluctuations in density become more random and pronounced, there is an increasing probability of the population density—and hence population size—reaching zero. An empirical analysis using data from studies of the population dynamics of invertebrates illustrated that population dynamics could move through a parameter plane from highly stable, monotonic growth at low rates of increase and/or weak density dependence, through increasingly dynamic fluctuations such as monotonic damping, through limit cycles, to, finally, chaos, with increases in either or both the intrinsic rate of increase and density dependence (Hassell et al. 1976). More recently, Grenfell and colleagues (1998) performed stability analyses on long-term data on the population dynamics of Soay sheep inhabiting an island off the coast of Scotland and found that highly overcompensating density dependence generated long-term instability. In this case, birth rates during favorable years were so high that they pushed the population above a stable threshold, beyond which density-dependent mortality was so excessive that it precipitated population crashes.

The early studies by May and Hassell were of seminal importance in asserting that many types of population dynamics could be derived from fairly simple models of deterministic, density-dependent population growth. The more recent study by Grenfell and colleagues was important because, in corroborating the observations of May and Hassell, it demonstrated that density dependence, far from being solely stabilizing, could in fact become destabilizing if it was overcompensatory. All these studies reflected the early and persistent influence in the field of population dynamics of the foundational papers that focused the attention of population ecologists for decades on the role of density dependence in population limitation and regulation.

Deterministic models of population growth are mathematically convenient. But the main limitation of deterministic models of population growth is immediately apparent: they ignore environmental variation in time and space (Maynard Smith 1974). Similarly, the classic treatments of population dynamics largely ignore individual variation in birth and death rates, which both contribute to the stability of dynamics at the population level (Lominicki 1988). Nonetheless, for very large initial population sizes, deterministic models can approximate the dynamics of populations in stochastic environments. The problem arises when population sizes are small, in which case stochastic models may predict extinction, whereas deterministic models may predict only the lack of an equilibrium (May 1973a).

The mathematical context of stability in population dynamics was developed earlier as the concept of neighborhood stability (Lewontin 1969) and was soon followed by the first mathematical formalization of factors, including climate, that might erode stability in population dynamics (Levin 1970). Although Levin's (1970) interest lay in developing a framework for investigating the role of environmental stochasticity in competitive exclusion and coexistence of species at the community level, the insights gained from his treatment of the dynamics of the populations composing the community are of relevance here. Primary among these was the observation that environmental stochasticity is capable of generating dynamics that deviate from point stability but that still may achieve dynamic equilibrium (Levin 1970).

A similar conclusion was reached in a more extensive mathematical treatment of population dynamics and stability in stochastic environments by May (1973b). May's treatment provided a seemingly trivial modification of the simplest deterministic model of population growth:

$$\frac{dN(t)}{dt} = f[N(t)], \tag{4.1}$$

in which N represents population size at time t and $dN(t)/dt$ is the intrinsic rate of population increase. For such a purely deterministic model, an equilibrium is achieved by setting the growth rate of the population equal to zero:

$$0 = f(N^*), \tag{4.2}$$

in which N^* indicates a stable point equilibrium.

This stable equilibrium point can be likened, in topological terms, to a cup in a landscape in which the population rests at some stable density (May 1973b). Displacing the population from this stable point density requires a disturbance of some kind, such as environmental variability through time, which might occur as a result of climate change. The effect environmental variation has on this system of equations is to render consideration of an equilibrium population size or density meaningless. Instead of N^* we must now think in terms of the probability distribution $f^*(n)$ (May 1973b). That is, rather than a stable point equilibrium, the population experiencing random environmental variation now describes a cloud of points representing the probability distribution of population densities, the scatter of which, σ^2, scales with the degree of environmental variability. Revisiting the topological analogy, we can see that the stable landscape of the deterministic model, in which the population rested at point stability in a cup, is now, in May's words, "heaving up and down like the floor of a fun house" (May 1973b).

These opposing contributions of stabilizing density dependence and de-stabilizing environmental stochasticity can be represented quantitatively at the equilibrium probability distribution, where net population growth is set to zero, as

$$0 = -\frac{d}{dn}[M(n)f^*(n)] + \frac{1}{2}\frac{d^2}{dn^2}[V(n)f^*(n)] \qquad (4.3a)$$

or

$$\frac{d}{dn}[M(n)f^*(n)] = \frac{1}{2}\frac{d^2}{dn^2}[V(n)f^*(n)], \qquad (4.3b)$$

in which the left-hand side of equation (4.3b) represents the friction term condensing the probability distribution of population density and the right-hand side of equation (4.3b) represents the diffusion term driven by environmental stochasticity dispersing the probability distribution of population density (May 1973b). Hence, stability in stochastic environments is possible only in the condition in which the friction and diffusion processes are equivalent. Any trend in climate combined with increasing climatic variability would therefore require a commensurate strengthening of density dependence, though not overcompensatory in nature (*sensu* Grenfell et al. 1992), to maintain stability. Maintenance of population stability under a climate change scenario of increasing interannual variability in climatic conditions requires, therefore, tension between the strengths of density-dependent and density-independent influences on population growth (Post 2005).

Environmental stochasticity such as that represented by interannual variation in climatic conditions may have the effect of jostling the population out of point stability through a series of densities that approximate point stability, depending on the degree of environmental variability. The stabilizing influence of density dependence and the potentially destabilizing influence of environmental stochasticity, which may be represented by climatic variation (Levin 1970), are thus in opposition with each other, the former acting to compact the probability cloud of population densities and the latter acting to disperse it (May 1973b). If density dependence is strong in comparison to density independence, we should expect stability; in contrast, if it is weak in comparison to density independence, we should expect instability and, possibly, local extinction. The effect a systematic change in climate, that is, a trend in temperature or precipitation in addition to variability in either or both, would have on this system is analogous to a tilting of the landscape in addition to its up- and downward heaving, rendering the probability of the population settling on a stable density negligible to null. As we will see later in this chapter, however, with reference to a case study of the population dynamics of Svalbard reindeer, multiple equilibria may be possible in this scenario.

INCORPORATION OF CLIMATE
INTO TIME SERIES MODELS

In a strict sense, the movement of climatic effects from some unspecified error term in models of population dynamics into a term incorporating their influence on changes in abundance or density from one time step to the next was first made explicit in an analysis of the dynamics of southern pine beetles in relation to weather and density dependence (Turchin et al. 1991). Turchin and colleagues (1991) simply split the error term in their model into two components, one representing weather and the other representing the residual variance unexplained by the other variables in their model. This development was important in setting the stage for the transition from phenomenological models of population dynamics to statistical models of population dynamics. Moran (1953a, 1953b), as we will see later in this chapter, can probably be credited with putting a weather term into a simple statistical model of population dynamics, but he provided no conceptual basis explaining how to bridge the gap between a conceptual, phenomenological, or heuristic model of dynamics to his statistical model. Despite Moran's early contribution to the discussion of weather in population dynamics and synchrony, time series models of population dynamics, which are really statistical models of population dynamics, remained largely deterministic, in the sense that they focused on intrinsic dynamics, direct- and delayed-density dependence (Bjørnstad et al. 1995; Stenseth, Bjørnstad, and Falck. 1996; Stenseth, Bjørnstad, and Saitoh 1996; Framstad et al. 1997), until the development of a model that included a covariate explicitly incorporating the influence of climatic fluctuation on dynamics (Forchhammer, Stenseth, et al. 1998). Although not without limitations, time series analysis has proven broadly useful in analyzing the population dynamics of an array of species, including invertebrates, birds, fish, and small and large mammals. The main appeal of this approach is its utility in identifying and quantifying complex dynamics from analysis of the autocorrelated structure of time series data (Royama 1992; Turchin 2003). Patterns in time series data, such as the dimensionality of time series, or the number of steps backward in time over which there is a significant relationship between current and lagged density, can be used to infer the importance of such processes as species interactions, age structure, fecundity, and mortality to dynamics at a single trophic level. Hence, time series analysis has found considerable favor in attempts to analyze the dynamics of species with complex life histories, or those embedded in multi-trophic-level food webs, but for which data on abundance or density are available only at a single trophic level or as annual measures of abundance.

It was demonstrated in the previous chapter that a simple model of the timing of life history events can be reformulated to quantify the contribution of individual-based phenological responses to climate change to population-scale dynamics. The deterministic skeleton of the model in equation (3.15) is comparable in structure to Lominicki's individual-based, density-dependent model of population dynamics, in which individual heterogeneity was the source of population stability (Lominicki 1988). In the treatments of stability developed throughout the rest of this chapter, however, the focus will remain on models for the analysis of data at the population level. This focus is not to detract from the role of individual variation in population dynamics and stability but rather recognizes the fact that stability is a decidedly population-level phenomenon and the analytical framework for assessing it must by necessity be developed at that level.

In the model developed by Forchhammer, Stenseth, and colleagues (1998), for a simple, two-trophic-level system, climate is assumed to potentially influence consumer dynamics directly or indirectly through resource availability. Additionally, consumers and resources may both experience direct, intrinsic feedback through intraspecific competition (density dependence), as well as through interactions with each other (figure 4.2). An important distinction between the model developed by Forchhammer, Stenseth, and colleagues (1998) and earlier attempts to incorporate climate or weather into models of population dynamics by Moran (1953) and Turchin (1992) is the conceptual visualization of how abiotic variation might influence dynamics, and the quantification of those effects in the former as opposed to an unspecified influence in the latter. The derivation of the model by Forchhammer, Stenseth, and colleagues (1998) from its graphical representation in figure 4.2 is similar to the derivation of the model of life history dynamics in relation to climate, resources, and density in the previous chapter. We begin with unspecified equations for the dynamics of consumer and resource in delay coordinates:

$$N_t = N_{t-1}\exp(a_0 + a_1 X_{t-1} + a_2 Y_{t-1} + c_1 U_{t-1}) \qquad (4.3b)$$

$$P_t = P_{t-1}\exp(b_0 + b_1 Y_{t-1} + b_2 X_{t-1} + c_2 U_t), \qquad (4.3b)$$

in which N_t and P_t represent consumer and resource abundance in year t, X_{t-1} and Y_{t-1} are ln-transformed consumer and resource abundance in year $t - 1$, respectively, and U_t is a term quantifying climatic conditions of interest in year

Figure 4.2. A heuristic model of consumer (in this case, ungulate) population dynamics in relation to the direct influence of climate (φ_{32}), and which incorporates density dependence (α_{22}) and an interaction with resource availability (α_{12}) that also allows for an indirect influence of climate acting through resource availability (φ_{31}). The coefficients describing these interactions constitute the elements of the community matrix and can be estimated from the statistical model in equations (4.3) and (4.4). *Adapted from Forchhammer et al. (1998).*

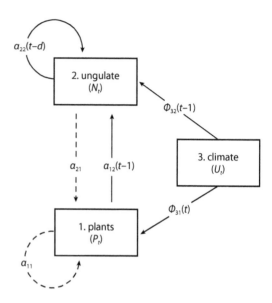

t. As we saw in chapter 3 in the derivation of the statistical model of life history dynamics in relation to climate, we can take the natural logarithm of both sides of equations (4.3a) and (4.3b) and isolate Y_{t-1} in equation (4.3a). The resulting equation is then inserted into equation (4.3b), after which t can be set equal to $t - 1$ and the resulting equation substituted back into equation (4.3a), giving

$$X_t = \beta_0 + (1 + \beta_1)X_{t-1} + \beta_2 X_{t-2} + \omega_1 U_{t-1} + \omega_2 U_{t-2} + \varepsilon_t \quad (4.4)$$

Before simplifying the coefficients to arrive at those used in equation (4.4), however, we should take note of the fact that the coefficients c_1 and c_2 in equations (4.3) derive from the direct influences of climate on consumer and resource, respectively, whereas a_2 and b_2 derive, respectively, from the indirect influences of climate acting through the influences of resources on consumers and consumers on resources (Forchhammer, Stenseth, et al. 1998). Tracking the significance of these coefficients through the application of the model in equation (4.4) to autoregressive time series analysis of data on consumer abundance can help identify whether climate acts directly, indirectly, or in both ways on the dynamics at that trophic level in this system.

Although it may appear altogether different at first glance, equation (4.4) is really just a modification on the $\log_e$ scale of the simple multiplicative model of population growth known as the Gompertz equation (Gompertz 1825). In other words, it assumes a log-linear relationship between population growth rate and density. As well, although developed for analyses utilizing annual estimates of population size or density, the deterministic skeleton of this model is similar to general discrete-time models of population growth based on birth and death rates of individuals (Lominicki 1988). The primary difference between this statistical model of population dynamics and the theoretical models of dynamics in deterministic (Gompertz 1825; May 1974, 1975b) and stochastic (May 1973b) environments, aside from the obvious feature that it incorporates abiotic influences explicitly, is the presence of delayed density dependence. This is represented in equation (4.4) by the term $\beta_2 X_{t-2}$ and is assumed to derive biologically from interactions between consumer and resource (Forchhammer, Stenseth, et al. 1998). In a more complex community, one with, for example, three trophic levels, delayed density dependence could also take the form of a lag-three density term, and in this case would arise in the primary consumer through interactions with its predator and its resource (Stenseth 1995; Stenseth, Falck, et al. 1998).

Stability in this second-order model of population dynamics is dependent on the relationship between statistical density dependence (Royama 1977) in direct and delayed coordinates (Royama 1992; Bjørnstad et al. 1995). This may sound at first like a completely deterministic system akin to that for single-species models in which the relationship between the strength of direct density dependence and the intrinsic rate of population increase determines the form of the dynamics, from highly stable, monotonic damping to chaos (Hassell 1974; Hassell et al. 1976). This may indeed be the case, but only partly so, as closer examination of the relationship between the biological (figure 4.2) and statistical models (equation 4.4) reveals (Forchhammer, Stenseth, et al. 1998). As was the case with the statistical model of the timing of life history events in relation to resources and climate, discussed in the previous chapter, the biological model (figure 4.2) used as a starting point by Forchhammer, Stenseth, and colleagues (1998) in the development of their statistical model indicates that any influence of climate on resource availability captured by coefficient $\varphi_{31}(t)$ will translate into an influence of resource availability on consumer abundance, captured by coefficient $\alpha_{12}(t-1)$, and will in turn be represented statistically as coefficient b_1 in equation (4.3b), which becomes incorporated into the simplified coefficient β_2 in equation (4.4).

The stability properties deriving from the relationship between the coefficients of direct $[(1 + \beta_1)]$ and delayed (β_2) density dependence in equation (4.4) can be visualized in a two-dimensional parameter plane (figure 4.3) (Royama 1977, 1992). As a result of the $\log_e$ transformation of the abundance or density data in this model, the coefficient of direct density dependence includes the value 1; therefore, negative direct density dependence occurs at any value for which the coefficient of direct density dependence is less than one and strengthens with this inequality. It is apparent on examination of the parameter

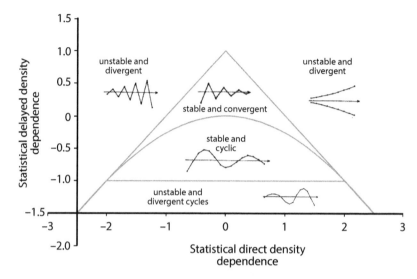

Figure 4.3. The stability properties of the second-order autoregressive (AR(2)) statistical model of population dynamics deriving from the ecological model in figure 4.2 are defined by the relationship between the strength of statistical delayed density dependence on the y-axis in the AR(2) parameter plane and statistical direct density dependence on the x-axis. The types of dynamics that arise from the interaction between varying degrees of delayed and direct density dependence are depicted in each of the five regions of the parameter plane. Note that unstable, divergent cycles arising from very strong delayed density dependence (at values less than -1 on the y-axis) are not usually depicted in representations of the AR(2) parameter plane but are of considerable relevance in the context of the effects of climatic variability on population dynamics. *Adapted from Royama (1977, 1992)*.

plane for the log-linear AR(2) model that stable, cyclic dynamics are possible for only a narrow range of values of negative, delayed density dependence but are possible for a wide range of values of direct density dependence, from negative through null to positive. This is the region bounded to the north by the parabola in figure 4.3 and to the south by the isocline at $\beta_2 = -1$. As delayed density dependence strengthens, below this isocline, the dynamics approach unstable, divergent population cycles (figure 4.3). As delayed density dependence weakens, and over a narrower range of values of direct density dependence, the dynamics display increasing stability, until a convergent population trajectory is achieved; this is the region in the parameter plane bounded to the north by the top of the triangle and to the south by the top of the parabola (figure 4.3). At the endpoints of the spectrum of direct density dependence, if delayed density dependence weakens or becomes positive, the population moves out of the stability region of the parameter space in figure 4.3 toward unstable dynamics, which lead to extinction everywhere outside the triangle (Royama 1992). A strengthening of delayed density dependence or a weakening of direct density dependence will increase the periodicity of cyclic dynamics if present (Bjørnstad et al. 1995). The stability properties of the AR(2) model do not differ substantially in the nonlinear scenario (Royama 1992, 2005).

Considering the potential for climatic fluctuation to contribute to delayed density dependence of consumers through resource availability, there is an obvious potential for a role of indirect climatic influences in long-term population cycles such as the multidecadal dynamics reported for caribou in Greenland (Meldgaard 1986; Forchhammer, Post, Stenseth, et al. 2002; Post and Forchhammer 2004). However, we must not overlook the potential for direct climatic influences on consumer dynamics indicated in the biological model of Forchhammer, Stenseth, and colleagues (1998) by coefficient φ_{32}, which becomes incorporated into the coefficient of direct density dependence in equation (4.4), $(1 + \beta_1)$. Strengthening of the direct climatic influence on dynamics, whether it be a direct negative or direct positive influence, thus has the potential to move dynamics to the verge of instability and thereby chaos and extinction. As alluded to earlier, this illustrates the crucial insight that any strengthening of a climatic influence on dynamics has the potential to destabilize dynamics. Whether that influence is positive or negative would appear to be less relevant to this conclusion than the magnitude of the climatic influence itself.

SIMULTANEOUS THRESHOLDS IN POPULATION-INTRINSIC AND POPULATION-EXTRINSIC FACTORS

Nonlinearities in density dependence and density independence may also alter the stability properties of population dynamics. These nonlinearities can arise through, or be approximated by, threshold effects. The more familiar of these are threshold effects in density dependence, although thresholds in both density dependence and density independence have been documented. Such effects are best understood as a nonlinear response in some dependent variable, such as abundance or density in time $t + 1$, to a linear change in some independent variable, such as density or climatic conditions in time t. Grenfell and colleagues (1992) demonstrated the former utilizing a nonlinear self-excitatory threshold autoregressive (SETAR) model in their analysis of the population dynamics of Soay sheep in the island population off the coast of Scotland. A SETAR model is similar in its basic skeleton structure to the linear autoregressive model described above (equation 4.4), but with an important difference:

the model is split into two distinct regimes, determined by a threshold in either density (Grenfell et al. 1998; Ellis and Post 2004) or population growth rate (Framstad et al. 1997; Post et al. 2002). Hence, a SETAR model is a piecewise linear approximation of a nonlinear autoregressive model that can, in the former case without abiotic covariates, take the form

$$X_t = \begin{cases} \beta_{1,0} + (1 + \beta_{1,1})X_{t-1} + \beta_{1,2}X_{t-2} + \varepsilon_{1,t} & \text{if } X_t \leq \theta \\ \beta_{2,0} + (1 + \beta_{2,1})X_{t-1} + \beta_{2,2}X_{t-2} + \varepsilon_{2,t} & \text{if } X_t > \theta \end{cases} \qquad (4.5)$$

where the threshold density is represented by θ. Similarly, in the latter case, θ may be set in relation to the rate of population increase to analyze dynamics during periods of population increase and decline, which produces a phase-dependent model (Framstad et al. 1997; Post et al. 2002). Note that the suffixes of the coefficients in equation (4.5) indicate that the strength and sign of direct and delayed density dependence may differ above and below the threshold, whether the threshold is set in relation to density or in relation to the rate of increase. In the latter case, this presents the interesting possibility that a population passing through the same density during a phase of increase or a phase of decline may experience different strengths of density dependence despite occurring at the same density while on its way up or down (Post et al. 2002). The differing dynamics observed below and above the threshold are referred to as regimes, with respect to density or rate of increase.

Grenfell and colleagues (1992) identified a threshold density in the population of Soay sheep on the island of St. Kilda below which the strength of density dependence in the growth rate of the population was negligible but above which strong, overcompensating density dependence became operable (Grenfell et al. 1992). Similar disparities in the strength of density dependence between density or growth rate regimes have been identified in rodents (Framstad et al. 1997), wolves (Post et al. 2002; Ellis and Post 2004), and moose (Post et al. 2002). Perhaps more interestingly, and with more obvious relevance to climate change, a subsequent analysis of Soay sheep dynamics in two populations inhabiting separate islands applied a SETAR model to examine the role of weather as a synchronizing agent in the dynamics of these two populations (Grenfell et al. 1998). This analysis revealed that the most parsimonious model of Soay sheep dynamics differed below and above a density threshold. Below this threshold, the population obeyed a first-order density-dependent autoregressive model with random noise, but above this density threshold the dynamics were highly variable and dependent on environmental conditions, specifically spring gales. As a consequence, the likelihood of overcompensatory, density-dependent crashes in this population depended on whether recovery following a previous crash coincided with the occurrence of severe spring gales (Grenfell et al. 1998), another example that appears to accord with the facilitation/promotion hypothesis of the interaction between biotic and climatic factors in population dynamics.

Nonlinearities in climatic influences on population dynamics may be more subtle and difficult to detect analytically. Such subtlety arises from the fact that density or population growth rate may appear invulnerable to slight changes in temperature, precipitation, or some other abiotic factor changing systematically as a result of climate change. This apparent invulnerability may present itself over a wide range of densities and of abiotic factors for which there appears to be little to no relation between the two. For instance, climate change may negatively affect a demographic rate in the population, but this demographic rate may exert little influence on the population growth rate over a wide range of values. Hence, under a scenario of mild, gradual climate change, the growth rate of the population may appear insensitive to climate change. Beyond a certain threshold in the abiotic factor of importance in this scenario, however, the growth rate of the population may undergo a shift, leading to a rapid decline in abundance (Doak et al. 2012).

The analytical framework of SETAR modeling is eminently useful for identifying the stability properties of populations that obey nonlinear dynamics. Application of a SETAR model to data on wolf dynamics using long-term counts of the population on Isle Royale, Michigan, for instance, provided a much better approximation of the observed dynamics than did a linear autoregressive model (figure 4.4a) (Post et al. 2002; Ellis and Post 2004). As well, whereas the linear autoregressive model indicated stable dynamics over the period of observation, the piecewise, nonlinear SETAR model indicated that the wolf population was highly stable during its decline phase but approached instability during its increase phase, as indicated by a weakening of direct density dependence in the model (figure 4.4b) (Post et al. 2002).

The missing piece from the preceding examples, for our purposes in this book, is application of SETAR modeling to investigate the potential for thresholds in climate. We should fully expect the operation of such thresholds, and expect their potential to interact with density dependence. For instance, the dynamics of a population of a given species might show no relation to climatic conditions below a threshold in some abiotic parameter such as temperature but a detectable relationship to climatic conditions above a threshold in that parameter. In a similar vein, we may falsely arrive at the conclusion that climate exerts little or, worse yet, no influence on the dynamics of our population or species of study if, in applying a linear model, we detect no such influence, when in reality a nonlinear model was required to detect the influence of climate beyond some threshold in abiotic conditions. Moreover, if, when applying a linear model under conditions more appropriately modeled using

(a)

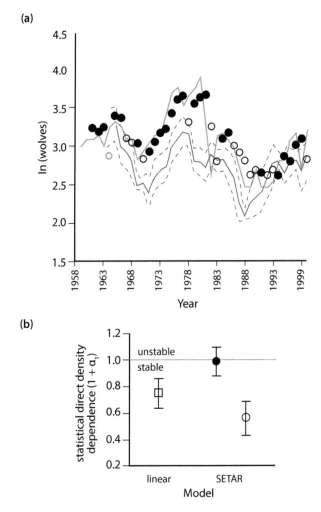

(b)

Figure 4.4. Output from a self-excitatory threshold autoregressive (SETAR) model in panel
(a) of wolf population dynamics on Isle Royale, USA, using one-step-ahead predictions in each
of two regimes of the model defined as the increase phase (solid circles) and decline phase (open
circles) of the population, and letting the model run from a start of six seed values (thin black
line; dashed lines represent 95% confidence bands), compared to actual observed wolf dynamics
(thick black line). The SETAR model indicated stable dynamics during the decline phase and
dynamics bordering on instability during the increase phase, as shown in panel (b). *Adapted from
Post et al. (2002).*

a nonlinear model, we determine that climate exerts a given influence on our population, we may miss the actual point that the influence of climate on the dynamics of our population changes magnitude or even sign around a threshold in abiotic conditions. The picture quickly and easily becomes complicated.

To illustrate this point, let us consider a modification of the threshold autoregressive model in equation (4.5) developed for a recent analysis of the population dynamics of Svalbard reindeer (Tyler et al. 2008). Recall that equation (4.5) is a second-order autoregressive, or AR(2), model that includes the effects of direct (lag-one) and delayed (lag-two) density on density in the current year, with a threshold in density below and above which the strength of density dependence is expected to vary. Tyler and colleagues (2008) developed a first-order autoregressive, or AR(1), model of reindeer dynamics with a threshold in R_t, the rate of population growth. A threshold in R_t was utilized to capture the distinctly different phases of population growth across the entire time series (figure 4.5a). In contrast to most other populations of caribou or wild reindeer throughout the circumpolar north (Vors and Boyce 2009), Svalbard reindeer in this population have been in a strongly increasing phase since the onset of rapid and pronounced Northern Hemisphere warming in the mid-1990s. Prior to 1995, the population experienced a period of irregular oscillation (figure 4.5a). Hence, the a priori assumption was that the intrinsic dynamics of the population likely differed between these two growth phases (Tyler et al. 2008).

The SETAR model employed by Tyler et al. (2008) takes the form

$$X_t = \begin{cases} \beta_{1,0} + (1 + \beta_{1,1})X_{t-1} + \omega_{1,1} C_{t-1} + \varepsilon_{1,t} & \text{if } \tau_t \leq y \\ \beta_{2,0} + (1 + \beta_{2,1})X_{t-1} + \omega_{2,1} C_{t-1} + \varepsilon_{2,t} & \text{if } \tau_t > y \end{cases} \qquad (4.6)$$

in which C_{t-1} represents a lagged climate term quantifying, in this case, an annual index of ablation, and τ represents the threshold in the variable y. The similarity to equation (4.5) is immediately apparent, as are two key differences. The SETAR model in equation (4.6) not only includes a potential nonlinearity in density dependence, represented by the threshold-dependent coefficients of direct density dependence preceding the lag-one density terms, it also includes a potential nonlinearity in climatic influences on density, represented by the threshold-dependent coefficients preceding the lag-one climate terms (Tyler et al. 2008). Notably as well, equation (4.6) relates the threshold to, as mentioned, a variable, y.

The novelty of the analysis by Tyler and colleagues (2008) is the allowance for population phase dependence, in which case y represents the population growth rate, $R_t = X_t - X_{t-1}$, simultaneously with climate phase dependence, in which case y represents the ablation index prior to or after 1995. In other words, the strength of density dependence and the magnitude and sign of the

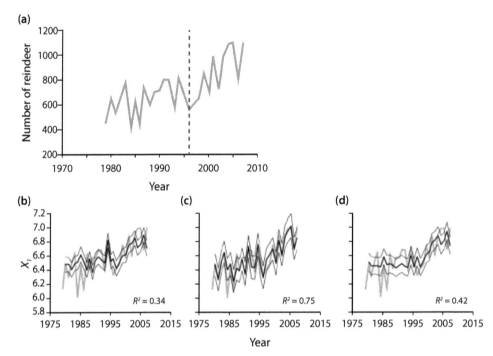

Figure 4.5. Prior to 1996, the Svalbard reindeer population near Longyearbyen on the island of Spitzbergen, Norway, exhibited irregular oscillations (a), but it began to increase after 1996. This pattern suggests the appropriateness of a nonlinear threshold autoregressive model to describe the dynamics of the population. Such a model was applied using simultaneous thresholds in growth rate and climate. The performance of a linear autoregressive model (b) was exceeded by that of nonlinear autoregressive models with the threshold in growth rate (c) and climate (d). In panels (b–d), thick gray lines represent actual observed dynamics of the population, while thick black lines represent model predicted values and thin black lines are bootstrapped standard errors. *Adapted from Tyler et al. (2008).*

climatic influence on dynamics might, according to this model, vary depending on whether the population was increasing or declining and on whether or not climatic conditions were undergoing a trend. As a side note, the ablation index used by Tyler and colleagues (2008) quantifies the magnitude of annual snowmelt, and entered a positive phase beginning in 1996, indicating a trend toward warmer winters with more snowmelt at the study site since then (Tyler and Forchhammer 2009). The nonlinear SETAR model reproduced the observed dynamics of the focal population substantially better than did a linear AR(1) model with the same skeleton structure as the SETAR model in equation (4.6) but without, obviously, the thresholds in growth rate and climate (figures 4.5b–d).

Understanding exactly how thresholds in population growth rate and climatic trends actually operate on populations to influence their dynamics is admittedly difficult. My own understanding of it was aided greatly by a public lecture on the subject by the lead authors of this paper. As Tyler and Forchhammer put it in the lecture, imagine if our society were governed by two sets of laws, one that applied when our population was below a certain size and one that applied when our population had exceeded that threshold size. This might be somewhat tractable in its management, provided we were reminded of where our population currently fell with respect to the threshold. Now imagine that an entirely different set of laws was also in operation, but with a threshold operating in a realm over which we had absolutely no control and which was beyond our capacity to predict. What effect might this have on the function and stability of our society? As you can imagine, it might have a strongly destabilizing effect.

In this analogy, the first set of laws applies to the threshold in population growth rate of the reindeer population, while the latter set of laws applies to the climatic regimes operating on their dynamics. A consequence of the simultaneous operation of dual sets of thresholds in the dynamics of this population is the existence of multiple equilibria (figure 4.6), which, although they are all stable, vary in the strength of their attractors because of the difference among them in their population dynamical return times (Tyler et al. 2008). Such insights would not have been garnered through application of a linear autoregressive model, or even through application of a SETAR model with a single threshold in either population-intrinsic or population-extrinsic factors. The facility with which we as modern ecologists can discuss complex phenomena within the study of population dynamics was almost unimaginable only four to six decades ago. At the outer limit of that timeframe, ecologists had not yet even conceived of a framework for quantifying abiotic influences on population dynamics.

Aside from the analytical innovation of this attempt to model population dynamics in relation to simultaneous thresholds in intrinsic and extrinsic dynamics, this study is of interest because of the insights it generated about the roles of starvation and fecundity in the dynamics of this population. In a linear scenario, that is, a linear AR(1) model equivalent to equation (4.6) but without the thresholds, starvation was density dependent and increased with late winter ablation, whereas fecundity was unrelated to density but increased with ablation (Tyler et al. 2008). The results were quite different, however, in the nonlinear scenarios captured by the SETAR model in equation (4.6). In the growth phase–dependent model, starvation increased with ablation only during population increase, and fecundity was unrelated to density and ablation in

(a) **(b)**

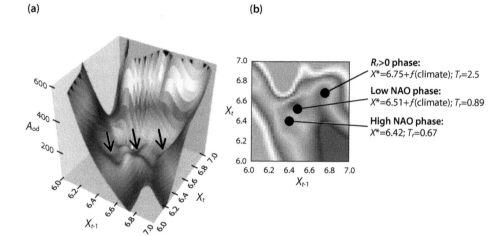

Figure 4.6. Phase-space geometry indicating multiple stable equilibria, revealed by the dual-threshold SETAR model of population dynamics of Svalbard reindeer in Adventdalen. (a) The three-dimensional representation of the state variables (X_t, $X_t − 1$, Aod $t − 1$). X_t and $X_t − 1$ are ln-transformed population size and Aod is the ablation index, for which large values represent increased warming and small values represent cold weather with little ablation early in winter. The three arrows show the location of the local equilibria calculated for the different phases (see b). (b) The two-dimensional plot of (a), in which dots, corresponding to the arrows in (a), indicate the location of equilibria for the $Rt > 0$ phase (black), the low phase of the North Atlantic Oscillation (NAO) index (gray), and the positive NAO index phase (white). The $Rt ≤ 0$ phase equilibrium is not shown because it is never reached, owing to its long return time. *From Tyler et al. (2008), and contributed by M. C. Forchhammer.*

both phases of population growth. In contrast, in the climate phase–dependent model, during periods of declining ablation the only significant relationship observed was an increase in starvation with density. During periods of increasing ablation, however, starvation declined with increased melting, while fecundity was strongly negatively related to density but increased with ablation (Tyler et al. 2008).

In chapter 3 I presented an extensive example focusing on the role of warming in plant phenological dynamics in time and space, and in chapter 6 I will examine the consequences of this advancing growing season for reduced offspring production in caribou in West Greenland. Many, if not most, caribou and wild reindeer populations across the Arctic have been in decline recently, and this has been associated with the recent and rapid warming trend at high latitudes (Vors and Boyce 2009). The role that trophic mismatch might play in such declines is unknown, even as its potential contribution to the dynamics of the population at my own study site remains uncertain (Post and Forchhammer

2008). In light of the fourfold difference in offspring production between years with low and high trophic mismatch (Post and Forchhammer 2008), however, there would most certainly seem to be a potential for trophic mismatch to contribute to dynamics in the Kangerlussuaq population. The story might, however, be quite different for the dynamics of reindeer on Svalbard. In contrast to other populations of caribou and wild reindeer, those on Svalbard do not undertake seasonal or annual migrations between wintering and calving ranges, presumably because resource distribution patterns in the archipelago do not necessitate migration (Tyler and Øritsland 1989). Hence, the potential for trophic mismatch between timing of plant growth and timing of arrival of pre- or newly postparturient females is nonexistent on Svalbard. My guess is that an early onset of spring is advantageous for Svalbard reindeer, which are on site to capitalize on early resource emergence on their calving grounds. Hence, advancing phenology may represent an increasing trophic match for Svalbard reindeer, thereby contributing to higher rates of offspring production and survival, as it might also for other nonmigratory large herbivores such as muskoxen. This would seem to accord with the positive association between ablation, or annual degree of snowmelt, and fecundity in Svalbard reindeer (Tyler et al. 2008).

POPULATION SYNCHRONY AND EXTINCTION RISK

The obvious question that should have arisen in the discussion in chapter 2 on the hypothesized role of climatic warming at the end of the Pleistocene in the extinction of many species of large mammals in the far north relates to the current signal of climate change in population dynamics. If rapid warming during the Pleistocene-Holocene transition precipitated the collapse of multiple species, and if contemporary warming has the potential to do the same, then should we not already be witnessing widespread population responses to ongoing climate change? I believe the answer lies at least partly in the documentation of synchronization of population dynamics by climate.

As touched on briefly above, one of the first attempts to formally incorporate weather into statistical models of population dynamics can be traced to the work by Moran on the dynamics of Canada lynx. Moran realized the importance of drawing a distinction between the process that generates the cycles (Moran 1953a) and the process that generates the remarkable synchrony in the cycle among the widely scattered populations of lynx (Moran 1953b). Whereas the cycle was attributed by Moran to the interaction between the lynx and its primary prey, snowshoe hares, synchrony in the dynamics of lynx across the boreal

forest of Canada was likely driven by abiotic fluctuation—that is, weather. In
Moran's view, cyclicity was a prerequisite for synchrony. The fact that the interaction between lynx and hares produced a fairly regular cycle in the dynamics of
lynx left them predisposed to synchrony if given the appropriate synchronizing
agent. Among the candidates for agents of synchrony in population dynamics
through time and space are roving predators, which hunt among several populations of prey simultaneously and thereby synchronize their dynamics (Ims and
Andreassen 2000); dispersal among populations, which connects them through
exchange of individuals (Lindström et al. 1996; Ranta, Kaitala, and Lundberg
1997); pathogens (Dwyer et al. 2004; Cattadori et al. 2005); and weather, which
influences the dynamics of multiple populations simultaneously over large geographic regions (Moran 1953b; Koenig et al. 1999; Koenig and Knops 2000;
Koenig 2002). In the case of synchronized lynx cycles, Moran favored weather
as a synchronizing agent in his development of the statistical treatment of the
role of environment in correlated population dynamics.

According to Moran (1953a, 1953b), the dynamics of a pair of fluctuating
populations could be represented in AR(2) notation as

$$X_t = \alpha_1 X_{t-1} + \alpha_2 X_{t-2} + \varepsilon_{X,t} \qquad\qquad (4.7a)$$

$$Y_t = \beta_1 Y_{t-1} + \beta_2 Y_{t-2} + \varepsilon_{Y,t}, \qquad\qquad (4.7b)$$

in which X_t and Y_t are ln-transformed abundance in the two populations, respectively, in the current year, and the $\varepsilon_{i,t}$ represent random environmental elements
characteristic of the regions inhabited by populations X and Y. When the $\varepsilon_{i,t}$
are assumed to represent local weather conditions influencing the dynamics of
the respective populations, the correlation between X_t and Y_t will be equivalent
to the degree of correlation between the weather variables $\varepsilon_{i,t}$ (Moran 1953b).
This is known as the Moran effect, or the environmental synchronization of
population dynamics. Of note, this condition holds only if the autocorrelative
structure in the time series, that is, the autoregressive coefficients quantifying
first-order statistical density dependence, are comparable among or between
the populations (Royama 2005). When they differ, stronger environmental correlation may be required to synchronize the populations (Grenfell et al. 1998).

Although the analytical framework for quantifying and analyzing population
synchrony in relation to climate change appears straightforward, in practice,
attributing synchronized dynamics to climate or weather can be problematic
because of the potentially confounding influences of the other synchronizing agents listed above (Bjørnstad, Ims, et al. 1999; Bjørnstad, Stenseth, et
al. 1999; Koenig 1999). In the case of Canada lynx, the best approximation
of the observed level of synchrony among populations across the boreal forest

is achieved by a model incorporating both weather and dispersal (Ranta, Kaitala, and Lindström1997; Ranta, Kaitala, and Lundberg 1997), as is the case with synchronized dynamics of gallinaceous birds in Finland (Lindström et al. 1996). The question of whether and to what extent abiotic factors such as temperature and precipitation can synchronize population dynamics is not solely of academic interest because population synchrony has the potential to contribute to extinction risk (Heino et al. 1997). Documentation of the role of climate and weather in population synchrony is therefore a necessary first step in addressing the potential for climate change to influence extinction risk. The most promising examples of clear roles of climate or weather in synchronized population dynamics unconfounded by other agents have come from analyses of synchrony in masting, growth, and flowering in plants (Koenig and Knops 1998, 2000; Koenig et al. 1999; Koenig 2002; Post 2003b), synchrony in island populations that do not mix (Grenfell et al. 1998), and synchrony between species separated geographically by barriers to movement and without shared predators (Post and Forchhammer 2002).

Answering the question of whether climate can induce population synchrony independently of other synchronizing factors allows us, in and of itself, to say little, however, about the effect climate change may have on population synchrony. For instance, considering the extinction risk posed by synchrony, it is important to determine whether climate change might induce synchrony in populations currently fluctuating independently. Likewise, we would like to know how climate change might alter the strength of synchrony in currently synchronous populations by, for example, weakening or strengthening it. Both questions require application of a nonstationary analytical framework. The time-windowing approach applied in an analysis of all possible pairwise correlations of the Canada lynx data illustrated that populations move in and out of synchrony with each other through time (Ranta, Kaitala, and Lindström 1997; Ranta, Kaitala, and Lundberg 1997). Hence, stating that populations are in synchrony or, by contrast, out of phase with each other provides a rather static view of the phenomenon of synchrony, one that depends on the specific temporal window through which the dynamics of the focal populations are examined. The time-windowing approach allows a temporally dynamic assessment of synchrony in which a window of some predetermined and constant width is moved across a set of time series, with the correlation between or among the populations calculated for each segment of the data that fall within the window at each step (Ranta, Kaitala, and Lindström 1997).

Such a nonstationary approach has been used in other applications to demonstrate a link between the El Niño–Southern Oscillation phenomenon and cholera outbreaks in Bangladesh (Pascual et al. 2000; Rodó et al. 2002).

Similarly, synchrony among populations of caribou inhabiting the west coast of Greenland has increased as Northern Hemisphere temperatures have increased (figure 4.7a) (Post and Forchhammer 2004). In this example we were unable to rule out a contributing influence of dispersal among these populations because they are not isolated and, as we know, caribou are highly mobile and engage in long-distance seasonal migrations. Notably, however, the degree of spatial correlation, analogous to synchrony, in temperature data recorded at weather stations located within the same regions inhabited by these caribou populations has also increased along with synchrony among the caribou populations (figures 4.7a and b). The increase in spatial autocorrelation of local temperatures along the west coast of Greenland correlated with an increase in Northern Hemisphere temperatures (Post and Forchhammer 2004). Although circumstantial, this body of evidence suggests the possibility that warming has the potential to increase synchrony among populations. It furthermore suggests an additional manner in which the tension versus facilitation/promotion hypotheses are relevant to the study of ecological response to climate change. In the case of caribou populations along the west coast of Greenland, we can regard regional synchrony as an emerging property of the interaction between the climatic influence on dynamics of multiple populations simultaneously, and the biotic influence of migration of individuals among those populations. Hence, in this case, regional synchrony would appear to represent an example consistent with the facilitation/promotion hypothesis.

The role of climate change as a potential agent of extinction through synchronization of population dynamics can be visualized through the following analogy. Imagine a pot full of bouncing balls. The balls represent populations of a given species, and the collection of balls bouncing, straight up and down, in and out of the pot, represents the entire suite of populations that constitute the species as a whole. The up-and-down movement of the balls can be seen to represent fluctuations in abundance in the populations. A ball bouncing upward represents an increasing population and a ball bouncing downward represents a declining population. Let us imagine a situation in which all of the balls bounce in and out of the pot independently (figure 4.8a). This scenario represents a state of asynchronous population dynamics. Now let us imagine that the pot has a sliding lid and that we are tasked with capturing all of the balls in the pot by sliding the lid into place. This would be difficult, if not impossible, because there is a good chance that at least one of the balls will be outside the pot at any given moment. Let us now imagine a different scenario, one in which all of the balls bounce in and out of the pot at the same time (figure 4.8b). This scenario represents a state of perfectly synchronous population dynamics. The balls need not bounce to the same height in this scenario, provided that they

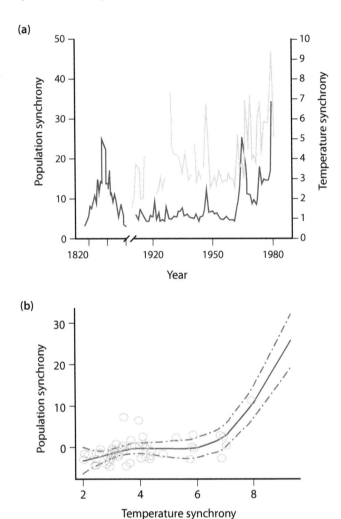

Figure 4.7. (a) Increase in spatial autocorrelation among populations of caribou (black) distributed along the west coast of Greenland in association with an increase in spatial autocorrelation of local temperatures (gray) at the sites inhabited by caribou in those populations. (b) The association between population synchrony and temperature synchrony is nonlinear and related to Northern Hemisphere warming, suggesting that further warming has the potential to precipitate disproportionate increases in spatial synchrony among the caribou populations. *Adapted from Post and Forchhammer (2004).*

Figure 4.8. Conceptualization of the relationship between population synchrony and extinction risk. A set of balls bouncing independently
and asynchronously into and out of a lidded pot
(a) is more difficult to capture inside the pot than
is a set of balls bouncing in synchrony into and
out of the pot (b). The behavior of the balls in
this example is analogous to the fluctuations of
all the populations comprising a given species.
Balls bouncing up represent population increases
and balls falling represent population declines.
If the pot represents an ecological state such as
extinction, it is more likely that all populations
representing a species will go extinct simultaneously if they are fluctuating in synchrony.

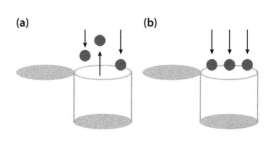

move at sufficiently different rates to remain in complete synchrony as they
bounce in and out of the pot. In this case, the populations may fluctuate at different levels of maximum abundance, much as the Canada lynx fur harvest data
indicate, but do so in synchrony with one another. Now imagine again attempting to slide closed the lid on our pot to capture all the balls inside of it. This
would be easily accomplished, provided one's sense of timing was accurate
enough. Now let us see the pot for what it really represents, an ecological state
such as extinction, and the closing of the lid for what it really represents, an
extrinsic event or process such as climate change. The analogy is imperfect, but
the image it produces is vivid: climate change is much more likely to result in
global extinction of all populations of a given species if those populations are
fluctuating in synchrony than if they fluctuated independently.

EROSION OF POPULATION CYCLES

Population cycles are a ubiquitous feature of animal species inhabiting northern or other highly seasonal environments (Keith 1963). Analyses of spatial
gradients in cyclicity, or the extent to which and regularity with which populations cycle, have been undertaken for both small (Bjørnstad et al. 1995; Stenseth, Bjørnstad, and Saitoh 1996) and large mammals (Post 2005). These have
demonstrated, in both cases, gradients of increasing cyclicity as one moves
from southern to more northern populations. Explanations for this pattern
include increased specialist predation in northern compared to southern environments and increased seasonality in more northern compared to southern environments. According to the former, the predominance of specialist

predators in northern environments promotes the phase coupling of predator and prey dynamics, resulting in pronounced cycles in the dynamics of both (Hanski and Korpimäki 1995). According to the latter, in the case of small mammals, snow cover provides refuge from predation and renders population dynamics more readily driven by fecundity and reproductive success, thereby introducing delayed density dependence (Hansson and Henttonen 1985; Steen 1995). As was illustrated in our examination of the stability properties of the linear, second-order autoregressive model, population cycles can arise when delayed density dependence is strong relative to direct density dependence, but not overly strong (Royama 1992). Moreover, as was made clear in the derivation of the second-order autoregressive model of consumer dynamics in relation to density dependence and resource availability, delayed density dependence can arise through an influence of climate on resource availability (Forchhammer, Stenseth, et al. 1998). In other words, outside the deterministic scenario in which it has most commonly been employed, the AR(2) model is capable of generating cyclic dynamics as a consequence of the interplay between and among climate, resources, and density dependence.

This is the understanding we need to apply in examining the recent assertion that population cycles of some northern species have begun to erode. When we consider the close association between climate and trophic interaction cycles in northern environments, it should not be surprising that climate change might have the potential to alter the cyclicity of populations at high latitudes (Ims and Fuglei 2005). This section requires us as well to return to the notions of stationary and nonstationary dynamics that were introduced in the discussion of synchrony in population dynamics. In that section, we saw how the categorization of fluctuations in pairs of populations as synchronous employed a necessarily stationary conceptualization of dynamics. Such a conceptualization implies that dynamics are fixed: the nature and structure of the dynamics of a population are invariant through time and independent of the window bracketing the population for study. By contrast, the nonstationary conceptualization of population dynamics argues that the structure and nature of population dynamics either change through time or, similarly, depend on the length and location of the window of observation along the time series (Henden et al. 2009). Several recent analyses have challenged the notion of stationarity in population dynamics and even the universality of cyclicity in northern environments.

Vole and lemming cycles have been so thoroughly studied (Elton 1924, 1927; Solomon 1949; Chitty 1952; Chitty and Phipps 1966; Krebs et al. 1969, 1973; Hansson and Henttonen 1985; Brown and Heske 1990; Hanski et al. 1991, 1993; Akcakaya 1992, Bjørnstad et al. 1995; Boutin et al. 1995; Hanski and Korpimäki 1995; Korpimaki and Krebs 1996; Bjørnstad, Stenseth, et al. 1999; Stenseth

1999; Turchin and Batzli 2001) that any attempt to summarize the highlights of this body of research here would inevitably fall short and border on abject trivialization. Nonetheless, perhaps the most noteworthy feature of the population cycles of small mammals is the seemingly insatiable attraction they have held for population ecologists for decades. This attraction derives mainly, it can be fairly stated, from the prevalence of population cycles in this taxonomic group. As was noted earlier in this chapter, the proclivity for cyclic behavior increases along a latitudinal gradient, from south to north, along with the roles of specialist predators and snow cover in dynamics (Bjørnstad et al. 1995).

One of the longest-studied populations of small rodents is the population of grey-sided voles (*Myodes rufocanus*) at Kilpisjärvi, Finland. Since approximately 1990, this population, which cycled with a regular period of three to four years for most of the fifty-year time series, has displayed dampening dynamics (figure 4.9a). What is notable about this apparent erosion of cyclic behavior during the last decade of the time series is that the population has apparently settled into damped oscillations at a very low density. Concurrent with this decline has been a decline in numbers of a specialist predator on voles at the same site, the arctic fox (*Alopex lagopus*). One hypothesis explaining the loss of cyclicity in the vole population at Kilpisjärvi is that the northward expansion of red foxes (*Vulpes vulpes*) has produced two effects: a reduction in the arctic fox population due to competition, and a reduction in the vole population through generalist predation in which predator numbers do not cycle with prey numbers (Ims and Fuglei 2005). In this case, it appears that a climate-change-related change in cyclic dynamics in voles reflects the indirect effect of increasing temperatures through the northward expansion of red foxes.

An even more striking example of loss of cyclicity derives from another population of grey-sided voles in northern Sweden. As with the population in Finland, voles in this population displayed a regular period of three to four years in their dynamics between approximately 1970 and 1985 but have undergone a shift toward chronically low densities with highly damped variation since then as the cycle has collapsed (figure 4.9b). In this case as well, a role of recent climate change can only be inferred from the large scale of the collapse of cyclicity in an area of approximately 10,000 km² over which these data were collected (Ims et al. 2008). A more convincing case for a role of climate change can be made for the collapse of population cycles in field voles (*Microus agrestis*) in northern England. The dynamics of this population have undergone an abrupt shift from regular, multi-annual cycles from 1985 to 2000 to aperiodic, divergent dynamics since then (figure 4.9c). Throughout this period, the persistence of snow cover has declined steadily, suggesting an adverse effect on overwinter survival, fecundity, or both of a shortening of the length of the

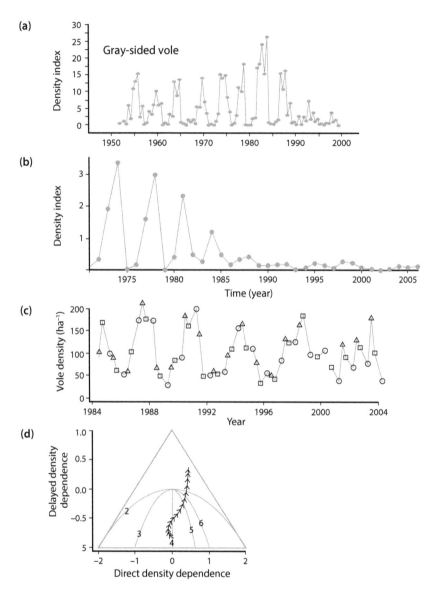

Figure 4.9. Erosion of population cycles in northern environments in species well known for such cycles, including populations of grey-sided voles in Kilpisjärvi, Finland (a) and in Sweden (b), and the field vole population in Kielder Forest, northern England (c). In this last example, the dynamics have moved within the AR(2) parameter plane (see figure 4.3) from stable and cyclic toward unstable and divergent (d). *Adapted from Ims and Fuglei (2005) and Ims et al. (2008).*

annual period with snow (Ims et al. 2008). This suggestion is bolstered by the evident transition of dynamics in this population from the highly stable region of the AR(2) parameter plane toward the region of divergent instability (figure 4.9d). This transition has occurred almost exclusively as a consequence of the weakening of delayed density dependence, although with a slight weakening of direct density dependence as well (Ims et al. 2008).

Multiple additional examples of altered population cycles in response, either directly or indirectly, to climate change in voles, lemmings, gallinaceous birds in Finland, and the larch budmoth (*Zeiraphera diniana*) in Central Europe have appeared since either the mid-1980s or early 1990s (Ims et al. 2008). Such taxonomically and geographically diverse examples would seem to argue strongly for a role of climate change in the loss of cyclicity as a general feature of population dynamics. If climate change has indeed contributed to a dampening of population cycles in northern populations of small mammals, then its interaction with density dependence would appear to support the tension hypothesis because the loss of cyclicity has been associated with a weakening of both direct and delayed density dependence. Similarly, an analysis of the dynamics of multiple populations of caribou distributed along a latitudinal gradient indicated a clear negative association between the strength of density dependence and density independence (i.e., climate) among populations (Post 2005). Moreover, this analysis revealed opposing associations between latitude and the strengths of density dependence and climate in models describing their dynamics: southern populations displayed strong density-dependent limitation and weak climatic limitation, while northern populations displayed weak density-dependent limitation and strong climatic limitation (Post 2005), further indicating a tension between biotic and climatic factors in population dynamics and cycles.

GLOBAL POPULATION DYNAMICS, POPULATION DIVERSITY, AND THE PORTFOLIO EFFECT

We have already seen that climatic fluctuation may induce synchrony in fluctuating populations, and that climate change may increase the strength of regional synchrony in population dynamics. Increasing synchronization of populations of Chinook salmon in the Snake River basin of the U.S. states of Oregon and Washington in recent decades is likely to reduce the portfolio productivity, or overall production of the combined populations, of this stock considerably (Moore et al. 2010). Examples of increasing population synchrony in salmon and caribou, both of which are culturally and economically important species with a long history of human exploitation, should be of concern because of the enhanced risk of global extinction deriving from population synchrony. Spatial and temporal heterogeneity in population dynamics and response to climate change may act as a buffer against global extinction risk because of the role of source-sink dynamics. The important qualifier here is the emphasis on *may*. As I have argued earlier, stability theory suggests that it is not only the direction of the climatic influence on a population but also its magnitude that determines its potential for eroding population stability, and thereby contributing to extinction risk. Hence, while heterogeneity in population response to climate may reduce the risk of global extinction due to climate change this is no guarantee because even if populations are not driven down in synchrony by a climatic trend or event, they may wink out one by one over time if the climatic influence on their individual dynamics strengthens to the point that the variance in density it induces exceeds the stabilizing influence of density dependence (*sensu* equation 4.3b).

Assessing global extinction risk deriving from climate change requires, at a minimum, comprehensive analyses of the contribution of climate to the dynamics of multiple populations throughout the distribution of the species of interest. Layers of complexity are readily added to such an assessment, including embracing the nonstationary perspective described above, which can be applied to examine the extent to which, if any, the contribution of climate to population dynamics has changed as climate has changed. Our analysis of the strength of the contribution of the North Atlantic Oscillation to the population dynamics of caribou along the west coast of Greenland, for instance, revealed no difference in the strength of its influence in the nineteenth century compared to the twentieth century (Post and Forchhammer 2004). A more spatially comprehensive, though temporally more restricted, analysis of the dynamics of thirty-three populations of caribou and wild reindeer throughout the distribution of this species revealed widespread response heterogeneity to Northern Hemisphere temperature anomalies (NHTAs) (Post et al. 2009a). Applying the same skeleton autoregressive model shown in equation (4.4), we detected negative climate coefficients—quantifying the influence of NHTAs in local dynamics—in twenty-seven of these populations, but considerable variation in the magnitudes of these coefficients (Post, Brodie, et al. 2009). This analysis was similarly applied to thirty-two time series on the dynamics of populations in the *Cervus* species complex, which includes elk, wapiti, and red deer. The analysis revealed even greater variability among populations in the direction and magnitude of response to warming, in some cases over very small spatial scales, such as within the Rocky Mountains region of the United States (Post, Brodie, et al. 2009).

As will be explored in greater depth in ensuing chapters, Elton (1958) supposed that species-rich communities and the food webs composing them should be more stable or resistant to disturbance than their species-poor counterparts;

ensuing developments in theoretical ecology predicted that species diversity would contribute to ecosystem stability (May 1973a). The species-wide analyses described just above lend empirical support to the extension of this reasoning, which suggests that heterogeneity in the response of populations to climate change may lend species-level stability to dynamics induced by climate change. More recently, a series of biocomplexity studies has also suggested that, just as biodiversity is predicted to confer stability on ecosystem function, population diversity may confer stability on the persistence of species in fluctuating environments (Schindler 2001; Schindler et al. 2005, 2010).

At the ecosystem level, species richness may confer stability because of functional redundancy among species or because of the buffering effect of species interactions on maintenance of community stability (Suttle et al. 2007). In the case of functional redundancy, the loss of one species may be "absorbed" by the ecosystem if another species performs a similar role in ecosystem function as the species that was lost. In the case of the buffering effect of species interactions, increasing complexity renders a greater level of connectivity among interacting species, so that loss of a single species has a smaller chance of leaving any other species isolated within the food web.

Similarly, population diversity may render species less vulnerable to extinction in the face of climate change. This may occur in one of two ways. First, genetic and life history heterogeneity among populations representing the species may leave them differentially vulnerable to constraints on productivity imposed by climate change, pathogen exposure, land-use changes, or direct exploitation by humans. Alternatively, population diversity in the form of phenotypic variation in body size, age structure, or age at first reproduction may reflect habitat differences among populations, which may in turn generate response heterogeneity to climate change, pathogen exposure, land-use changes, or direct exploitation by humans. The diversity of population structure and life history strategies displayed by the hundreds of populations of sockeye salmon (*Oncorhynchus nerka*) that constitute the Bristol Bay, Alaska, salmon fishery enhance, for example, the long-term stability of this heavily exploited system (Schindler et al. 2010). The populations of this stock exhibit a wide range of spawn dates, from mid-July through early October, which introduces considerable spatial variability in the timing of salmon runs among the districts using the Bristol Bay fishery; as well, the size of the harvest varies considerably on an interannual basis among the districts owing to spatial heterogeneity in the dynamics of the subpopulations (Schindler et al. 2010). This heterogeneity likely buffers the entire Bristol Bay harvest against environmental variation, and reduces by half the overall variability among years in total harvest in comparison to the variability that would be observed in this system if it consisted of a single large population (Schindler et al. 2010). Hence, population heterogeneity is essential to species persistence, especially in the context of environmental variability such as climatic fluctuation and change. As we will see in chapter 6, however, the consequences for community stability of population response to environmental variability are far less straightforward.

The Niche Concept

One of the earliest applications of niche theory in quantitative ecology addressed the seemingly simple question of the extent to which the niches of two species can overlap and allow co-occurrence or coexistence of the species (May and MacArthur 1972). This question grew out of the then recent development of the notions of limiting similarity and niche packing, according to which coexistence among species with similar resource requirements was assumed to be promoted through minimization of niche overlap through divergence in habitat utilization patterns or character displacement (MacArthur and Levins 1964, 1967). The answer is highly relevant in the context of climate change, or of any environmental change in general. As we will see in the next chapter, community stability may be eroded by environmental fluctuation if such fluctuation is of sufficient magnitude to destabilize dynamics at one or more trophic levels within a vertically structured community. In the context of niche theory, the niches of two co-occurring species may overlap almost entirely in deterministic, unvarying environments but may overlap to lesser and lesser degrees as the magnitude of environmental variation increases (May and MacArthur 1972). Hence, fluctuation in abiotic conditions such as mean annual temperature may once again be seen as just as important, if not more so, to the persistence or maintenance of the degree of niche overlap that is tolerable for co-occurring species as the trend in abiotic conditions itself.

The concept of the niche is one of the most intellectually powerful and intuitively appealing notions in ecology, and its development and refinement reflect the true complexity of this central concept, one that unifies ecology and evolution. The development of the concept of the niche can be traced to three seminal works. The earliest was that by Grinnell (1917) and consisted of an observational study of the factors limiting the distribution of the California thrasher (*Toxostoma redivivum*). Grinnell undertook a painstaking study of the habitat associations of the California thrasher throughout its range in California. He concluded that the distribution of the species was determined by its close association with a narrow range of environmental conditions, mainly temperature and precipitation regimes, that defined its suitable habitat. Grinnell (1917) also noted that "no two species regularly established in a single fauna have precisely the same niche relationships." The association of the habitat requirements of a species with its niche has subsequently become known as the "Grinnellian niche." The Grinnellian niche concept can be applied to some extent in understanding the role of habitat in explaining or determining the distribution of species.

Elton (1927) subsequently modified Grinnell's niche concept to define it from the perspective of the species. In this perspective, the niche is defined as the organism's role in its environment, or, more commonly, as the manner in which the species makes its living (Elton 1927). This is a trophic-centric view of the niche, and one that is not as easily resolved into definitions based on the local presence or absence of a species throughout its distribution. The concept of the Eltonian niche has since found utility in defining the position of a species within the food web in which it is embedded, and thus depends on the roles of other species for context. In this view, we may consider the niche of a species as being defined—at least in part, if not nearly entirely—by interactions with other species. Needless to say, this complicates the application of the Eltonian niche concept in studies of the effects of climate change on the persistence, redistribution, or extinction of species. Doing so would require exhaustive amounts of information on variation throughout the distribution of a species on the strengths and nature of interactions with competitors, pathogens, predators, and resources.

Hutchinson (1957) provided us with the most useful concept of the niche by rendering it quantifiable, at least in theory, where most research on niche dynamics has been conducted. According to Hutchinson, the niche comprises the n-dimensional hypervolume of resources that a species can potentially exploit (Hutchinson 1957). The subset of this hypervolume that a species occupies in the presence of competitors, or other species with partially overlapping resource demands, is known as its realized niche, whereas that which it could occupy in isolation from such competitors is its fundamental niche. As we will see below and in the following chapter, the Hutchinsonian view of the niche underpins the discipline of modeling the presence or absence of species based on environmental correlates of this n-dimensional hypervolume in which positive contributions to future generations are possible. This approach, variously known as environmental niche modeling or bioclimatic envelope modeling, assumes that a few measurable environmental variables, such as temperature and precipitation, capture sufficiently the relevant niche components to accurately predict the presence of species currently, and to project their occurrence as these conditions shift in time and space with climate change.

GRINNELLIAN NICHES AND CLIMATE CHANGE

As we will explore in chapter 6, whether communities are understood to exist as predictably interacting species assemblages or whether they are understood to exist merely as coincidental aggregations of species whose habitat requirements result in distributional overlaps (Ricklefs 2004) can be viewed as a consequence of whether we consider Hutchinsonian or Eltonian niches versus Grinnellian niches to be the drivers of community assemblage. Recent evidence indicates that avian communities in California have shifted their distributions to follow their Grinnellian niches with recent climate change.

Resampling of surveys undertaken by Grinnell in the Sierra Nevada Mountains of California provides a basis for evaluating shifts, or lack thereof, in the distributions of species surveyed by Grinnell almost a century later, under different climatic and, perhaps, habitat-associated conditions (Tingley and Beissinger 2009; Tingley et al. 2009). Data on site occupancy collected by Grinnell in the period between 1911 and 1929 provided a baseline assessment of the Grinnellian niche space of fifty-three bird species in two dimensions: temperature and precipitation regimes where they occurred. These parameters grossly oversimplify the niche volume of each species and ignore the potentially important influences of species interactions on distribution and, thereby, occupancy status. Nonetheless, it could be argued that the presence or absence of competitors for each of the species surveyed might also be influenced to some extent by abiotic parameters such as climatic conditions.

Tingley and colleagues (2009) resampled Grinnell's routes to determine whether the distributions of the fifty-three species of birds observed in Grinnell's original surveys had changed, with species altering their local patterns of site occupancy in accordance with geographic shifts in abiotic conditions that overlapped their distributions in the original surveys. The rationale behind this approach is simple: if species' distributions are determined by habitat requirements, and if the distributions of the habitats utilized by each species are in turn determined by geographic variation in precipitation and temperature regimes, then any change in precipitation and temperature that elicits changes in habitat distributions should also result in comparable shifts in the distributions of the species dependent on those habitats. As Tingley and colleagues (2009) explain, tracking of Grinnellian niches under climate change may take the form of any of three possible scenarios: conditions within a locally occupied site may remain within a species's niche envelope, in which case occupancy will remain constant; conditions within a site may move outside a species's niche requirements, in which case local extinction may result from increased mortality, reduced productivity, or both; or conditions characterizing the niche envelope may expand geographically, promoting, in the case of vagile species, spread and colonization of new sites that have become amenable. At the scale of species' ranges, these three scenarios would result, respectively, in range stasis, range contraction, and range expansion.

The interaction between dynamics at the site and distributional scales in determining the range shifts of species in response to climate change can be visualized with a simple example (Tingley et al. 2009). Suppose environmental conditions at a suite of sites within a region are plotted in their climate space according to average historical annual precipitation and temperature at those sites (figure 5.1, solid gray circles). Then those same sites can be plotted in climate space according to their current precipitation and temperature means (figure 5.1, solid black circles). In this case, suppose the realized Grinnellian niche of a species, estimated from climatic conditions throughout its historical range, encompasses a wide range of temperatures but a narrow range of precipitation (figure 5.1, gray oval). Sites within this climatic niche envelope would be occupied by that species, both historically (figure 5.1, solid gray circles) and currently (figure 5.1, solid black circles). Note, however, that there may be movement of some sites out of this climatic niche envelope and movement of other sites into this climatic niche envelope with climate change. In the case of this hypothetical species, site 2, previously occupied, has moved outside the niche envelope because of increases in precipitation at that site. However, although temperature has also increased at the site, average conditions still remain within the climatic niche envelope of the species. In contrast,

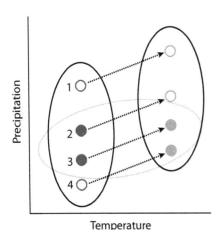

Figure 5.1. Hypothetical response of an organism's niche in two climatic dimensions to climate change that affects both dimensions. Numbered circles indicate sites within a region, bounded by the black ellipse, which occurs within the climate space defined by the relationship between precipitation and temperature. The gray horizontal ellipse defines the boundary of the niche in this climate space of the species of interest. Under current climatic conditions (the ellipse on the left), the species' niche does not exist at site 1. Following an increase in temperature and precipitation (the ellipse on the left), conditions characterizing the niche of the species still do not occur at site 1. In contrast, the species's niche requirements occur at sites 2 and 3 under current climatic conditions, but do so at site 3 only after climate change. At site 4, the species's niche requirements exist only after climate change. Hence, under current conditions, the species may be found at sites 2 and 3, but, following climate change, it will go extinct at site 2 and may colonize site 4. *Adapted from Tingley et al. (2009).*

site 4, previously unoccupied because it fell outside the precipitation limits for this species, has now become occupied (or favorable for occupation) owing to increased precipitation at that site. In this example, it appears that the distribution of the species within the region delimited by the black oval in figure 5.1 has shifted southward within this hypothetical region (figure 5.1) owing to range contraction away from the center of the region and colonization of its southern boundary.

An appealing feature of this approach is that it can be used to assess not only whether species track their environmental niches but also which components of the niche they track (Graham et al. 1996). For instance, the vector defined by the distance and direction of separation between the niche centroid (figure 5.2, dark gray cross) and the historical distribution centroid in climatic niche space (figure 5.2, dark gray star) can be compared to the vector defined by the distance and separation between the historical distribution centroid and the current distribution centroid (figure 5.2, light gray star) in climatic niche space to determine whether a species has moved toward its niche centroid along one or both axes defining the niche (Tingley et al. 2009). In this hypothetical example, both the precipitation and temperature components of the vector from the historical distribution to the niche centroid are positive, as are both components of the vector from the historical distribution to the current distribution centroids (figure 5.2). In this case, we could conclude that the species has tracked both the precipitation and temperature components of its Grinnellian niche with climate

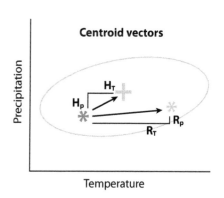

Figure 5.2. Before and after the change in climate depicted in figure 5.1, the distributional or geographic centroids in climate space of the species of interest are depicted by the dark gray and light gray stars, respectively. Comparing the locations of these in climate niche space to the centroid (gray cross) of the niche space (ellipse) of the species in climate space allows inferences to be drawn regarding the movement of the species in niche space in response to climate change. This is done by comparing the elements defining the vectors (black arrows) from the original distributional centroid (dark gray star) in climatic niche space to the niche centroid (gray cross), to those defining the vector from the original distributional centroid (dark gray star) to the new distributional centroid (light gray star) resulting from climate change. The labels H_p and H_T denote the precipitation and temperature vector components from the historical range to the climatic niche centroid, respectively, while the labels R_p and R_T denote the precipitation and temperature vector components from the historical range to the current range centroid. If these agree in sign, then the species has tracked its niche in both dimensions; if they agree in sign in only one dimension, then the species can be said to have tracked its niche in only that dimension. If neither agrees, then the species has not tracked its niche in those dimensions.
Adapted from Tingley et al. (2009).

change (Tingley et al. 2009). More difficult to infer, however, is the extent to which these abiotic changes have driven changes in habitat availability, quality, or both. Moreover, determining the scales at which these processes operate within the distribution of the species of interest is difficult to resolve from this example. Presumably, the process of habitat change in response to climate change that results in the spatial redistribution of a species's Grinnellian niche and the species itself occurs both locally (at the scale of the individual organism's breeding and foraging home range) and at larger scales (across the entire distribution of the species).

Tingley and colleagues (2009) applied their analysis to fifty-three species of birds surveyed originally by Grinnell and subsequently resurveyed one century later and found that nearly 91 percent of the species tracked their Grinnellian niches as climate changed. Three examples illustrate variation in the extent to which and the manner in which species may track environmental niches under climate change. First, the lazuli bunting tracked changes in both the precipitation and temperature components of its climatic niche (figure 5.3a). This is evident in the fact that positive precipitation and negative temperature components of the vector from its historical distribution to niche centroid were matched by the same sign in both components of the vector from its historical distribution to its contemporary distribution. Townsend's solitaire, on the other hand, tracked the temperature component, but not the precipitation component, of the vector from its historical distribution to its niche centroid (figure 5.3b).

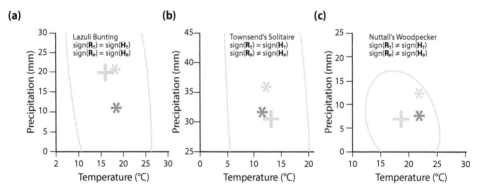

Figure 5.3. Three examples of niche responses to climate change by birds breeding in the U.S. state of California, according to the approach to identifying niche tracking developed in figures 5.1 and 5.2. In panel (a), the Luzuli Bunting tracked its niche in both dimensions of precipitation and temperature. In panel (b), the Townsend's Solitaire tracked its niche along the temperature axis but not along the precipitation axis. In panel (c), Nuttall's Woodpecker did not exhibit evidence of niche tracking along either axis. See the caption for figure 5.2 for descriptions of H_P, H_T, R_P, and R_T. *Adapted from Tingley et al. (2009).*

Nuttall's woodpecker, in contrast, tracked neither changes in temperature nor changes in precipitation, and appears to have moved to the outer limits of both components of its Grinnellian niche (figure 5.3c). This last example suggests the utility of this approach for identifying extinction risk or highlighting factors of conservation concern that contribute to erosion of niche space, in this case mostly a failure to track precipitation changes. One conceptual limitation of employing the Grinnellian niche concept in this manner, however, relates to the fact that defining a species's niche according to environmental or habitat requirements implies the possibility of vacant niches, which might predispose scientists using this approach toward detecting or inferring range expansion and invasion under climate change.

NICHE VACANCY

If a species is able to colonize previously unoccupied habitat as a result of climate change, and thereby become established in a community previously lacking that species, does this process imply the existence of vacant niches? From a Grinnellian perspective, the answer seems to be an obvious yes. Several scenarios can be invoked to support this possibility. For instance, climate change may—and indeed already has and will continue to—promote shifts in plant species' distributions and competitive outcomes that shape plant community composition. This process in turn influences habitat availability for animal species, creating new pockets of suitable habitat in previously marginal or unsuitable areas, eliminating it in others, or resulting in connectivity between occupied and unoccupied pockets of suitable habitat. It is tempting to regard this last scenario as being indicative of a niche vacancy that may become exploited or occupied as a result of climate change. But this perspective unnecessarily simplifies the niche concept because it fails to take into consideration the role of interactions with other species in determining the niche of a species. As well, it disregards the possibility that local communities can accommodate invasion from the regional species pool through a narrowing of niche breadths (Ricklefs 1987, 2004), a concept that will be expanded upon in the next two chapters. Nonetheless, the exploitation of previously unexploited host species or expansion into new habitat by invasive species, as in the case of the emerald ash borer (*Agrilus planipennis*) in North America (Sobek-Swant et al. 2012), the redlegged earth mite (*Halotydeus destructor*) in Australia (Hill et al. 2012), or the invasive fish rotan (*Perccottus glenii*) in the Palearctic, appears consistent with the notion that occupation of vacant niche space may be promoted in some instances by climate change (Dukes and Mooney 1999; Wiens and

Graham 2005). Indeed, invasive species may be at an advantage over endemic species in keeping pace with the direction and rate of change of climatic conditions because their rates of spread are often much greater than the expected velocity of climate change (Burrows et al. 2011; Hulme 2012).

The notion of niche vacancy derives originally from a seminal paper on the evolutionary ecology of two sympatric species of blackbirds (Hardy 1967). In a lowland marsh region of central Mexico, Hardy (1967) speculated over the explanation for the existence of two closely related species of red-winged blackbirds that segregated ecologically but bred sympatrically. Noting that one of the species was a recent colonist, and that this colonization apparently occurred following the extinction of a species of grackle with similar habitat requirements, Hardy (1967) suggested that a niche vacancy for the colonizing blackbird had been created by the extinction of the grackle. This example relates more closely to the concept of the Hutchinsonian niche, especially the distinction Hutchinson drew between the fundamental and realized niches of a species. Although I happen to be of the opinion that adhering to a Grinnellian niche perspective in explaining the distribution and abundance of organisms predisposes one to predicting range shifts, colonization, or extinction risk based on habitat alteration resulting from climate change, this is not to say that adhering to a species-centric view of the niche—such as that embodied by the Hutchinsonian niche model—predisposes one against predicting changes in niche occupancy as a result of climate change. Indeed, this may result from the processes of niche packing and niche evolution.

NICHE EVOLUTION

In the simplest sense, the question of whether contemporary climate change will precipitate the next great extinction event is really one of whether net loss of species will exceed the rate at which diversification occurs. The difference between rates of extinction and speciation will vary among ecoregions and bioclimatic zones, but we may still ask some general questions. For instance, has the rate of speciation in any taxonomic group varied with climate in the past? How about the rate of extinction? The field of paleontology has a long history of promoting the concept of niche conservatism, in which prolonged periods of stasis in diversity and community composition appear to be the norm (Simpson 1944). These periods are disrupted by relatively brief periods of punctuated diversification (Eldredge and Gould 1972, 1997) despite highly variable environmental conditions. This apparent conundrum, the juxtaposition of long-term evolutionary stasis against a background of unstable environments, has been explained as relating to the stabilizing influences of natural selection (Eldredge 1995). Tracking of suitable habitat under changing environmental conditions may also contribute to evolutionary stasis despite episodic, gradual, or rapid climate change (Wake et al. 2009).

The Cambrian explosion is perhaps the best-studied instance of a major speciation event in the history of life on Earth. Radiation of the flora of the Cape region of South Africa during the Cenozoic appears to have been climatically driven, although in this instance geography likely played an important role in the relationship between climate and diversification because of the high number of endemic species in this region and its geographic isolation (Linder 2003). Major events in hominid diversification in Africa appear as well to have been related to shifts in climatic regimes on the continent (Vrba et al. 1996). These examples suggest that rapid, punctuated changes in climatic conditions can be important in spurring major speciation events or increases in taxonomic diversity at higher levels of organization. However, any potential relationship between climatic change or fluctuation and speciation events or rates of diversification must invariably relate ultimately to niche evolution. Grinnell's pioneering work on species distributions and the niche concept formalized the axiom of niche exclusivity by concluding that multiple species cannot share niche requirements (Grinnell 1917). Several decades later MacArthur (1957), echoing Hutchinson's musings mentioned at the beginning of this chapter, used niche theory to evaluate potential explanations for species diversity in animal communities, focusing on birds. Although MacArthur's starting point was an interest in finding explanations for the observed diversity of bird species coexisting at or near equilibrium abundances, his arguments are still applicable to a discussion of the importance of niche evolution to any potential increase in diversification rates resulting from climate change.

MacArthur (1957) advanced three hypotheses to explain patterns of abundance and diversity of coexisting species. According to the first, diversity can be explained by the existence of continuous, nonoverlapping niches. In this case, the environment, and some unspecified set of resources it comprises, is divided up along a continuum. The widths, or area in two dimensions, of the units composing the complete environment are proportional to the abundance of each of the species in the community (MacArthur 1957). This predicts an inverse relationship between abundance and rarity of species; that is, common species are more abundant than rare species, a pattern that MacArthur noted is observed for many temperate and tropical regions (MacArthur 1957). The second hypothesis states that diversity can be explained as owing to the existence of continuous, overlapping niches. In this instance, the presence or absence of a species occurs independently of the presence or absence of all others, so that niche exclusivity is not operable in this environment. The problem with this hypothesis is that it tends to underestimate the abundance of the rarest species. The third and final hypothesis explains diversity as resulting from the existence of "particulate," nonoverlapping niches. In this example, MacArthur described niches as urns into which particles, representing units of abundance, are placed at random, without regard to what has been placed in any of the other urns (MacArthur 1957). As with the second hypothesis, the problem with this one is that it fails to predict the abundance of the rarest species, though more severely so than the second hypothesis (MacArthur 1957). MacArthur concluded that the best explanation for the observed diversity of coexisting bird species was the first hypothesis, the existence of continuous, nonoverlapping niches.

The relevance of this example to climate change and niche evolution is that it suggests niche vacancy may be rare or even nonexistent. In the case in which diversity is determined by the availability of continuous niches, or the extent to which resources are partitioned along some axis, there may be some degree of flux in niche breadth even near equilibrium (Ricklefs 1987), but this is not the same as niche vacancy. For instance, we may imagine a niche or resource continuum along which three species are distributed according to the preference displayed by individuals within each species for that resource (figure 5.4) (MacArthur and Levins 1967; May and MacArthur 1972). Even though MacArthur's assessment favored the hypothesis of continuous, nonoverlapping niches in explaining diversity, let us assume that there is some, perhaps negligible, degree of overlap among individuals at the tails of the distributions of each species along this resource axis. In this case, the distance d between the peaks of the distribution curves for adjacent species quantifies the extent of interspecific competition for that resource, while the distribution about the mean within the curve for each species, w, quantifies the extent of intraspecific competition for that resource (density dependence) (figure 5.4) (MacArthur and Levins 1964, 1967; May and MacArthur 1972).

Individuals whose strategy is to minimize w, intraspecific competition for this resource, will tend to forage near the tails of the distributions and, if interspecific competition is minimal there, may enjoy greater reproductive success (MacArthur and Levins 1967). Over time, this will lead to a gradual change in the distribution of individuals of this species along the niche axis, extending the tails of the distribution. Obviously, this process will eventually be held in check by intensifying interspecific competition, d, at which point selection will again favor individuals who forage away from the tails of the distributions, resulting in an intensification of intraspecific competition (MacArthur and Levins 1967). Even though there is flux in this scenario, there are no unoccupied niches along this resource continuum. As a consequence, it would

(a)   (b)

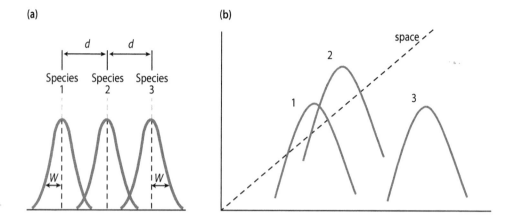

Food particle size

Figure 5.4. (a) Distribution of three hypothetical species along a resource utilization curve, in which the *x*-axis defines the resource state, such as food particle size, with each of the three species displaying a nearly unique food preference that defines the respective niches along this dimension. The distances *d* between the peaks of the adjacent resource utilization curves define the degree of resource overlap, and competition, between adjacent species, while the spread about the curve, *w*, for any one species defines the degree of intraspecific competition for that resource. In panel (b) a second dimension, space, has been added to illustrate the effect it may have on the degree of niche overlap among the same three species. In this case, species 2 now overlaps more with species 1 along the food size axis but has segregated from both species 1 and 3 along the space axis, where species 1 and 3 overlap completely. *Panel (a) modified from May and MacArthur (1972).*

seem unlikely that selection for preference for this resource somewhere along the axis that is already occupied by one of the three existing species would promote invasion of this niche space by a new species.

However, we can expect that changes in climate might influence the abundance and availability of this resource, with consequences for niche breadth, as follows (May and MacArthur 1972). Let us first imagine that climate change increases the availability of this resource, whether we define this resource as some nutrient, habitat component, or forage component, at either or both ends of the continuum. In this case, selection might favor those individuals in species 1 and 3 that occupy the lower and upper ends, respectively, of this continuum. As a result, the distributions of both species 1 and 3 might expand to the left and right, respectively, along the niche axis. This might in turn promote a broadening of the niche space occupied by species 2. Conversely, climate change may narrow the abundance and availability of this resource at either or both ends of the continuum. In this case, both species 1 and 3 would be forced away from their respective ends of the axis, placing greater pressure on species 2, which might eventually result in extinction through competition or a narrowing of the distributions of all three species along this axis. In the first case, expansion of the resource, invasion of this continuum by a new species at one of the extremes may be promoted because, in this instance, a vacant niche was exploited.

We can add another dimension, space, to this simplified niche to illustrate the potential effects climate change might have on the spatial dynamics of niche evolution or loss. According to the concept of niche complementarity, the coexistence of species that overlap along one resource axis is facilitated by segregation along another axis (Hutchinson 1957). Adding space to the uni-dimensional niche axis in figure 5.4 potentially allows for a greater degree of overlap among the three species along the original resource axis, which, for the sake of illustration, we may consider as quantifying the size of food particles preferred by individuals of each species.

Segregation in space may be facilitated or constrained by climate change and its interaction with landscape heterogeneity. This may occur in at least three ways. First, climate change may relieve spatial constraints on the dis-tribution of species along the primary resource axis in both directions along the space axis (figure 5.4b). Such a change would favor individuals in species 1 and 3 that forage along the most remote edges of the space axis, expanding the distributions of both species away from the middle of this axis. Individuals of species 1 and 3 employing this strategy would be favored only if forag-ing along these edges incurred lower intraspecific competition and resulted in greater offspring production and survival. Expansion of species 1 and 3 in opposing directions along the space axis may in turn facilitate an expansion of species 2 in one or both directions along the space axis, depending on the relative outcomes of intraspecific versus interspecific competition in all three species at the zones of contact. Alternatively, expansion of the spatial gradient of resource distribution might also open up niche space for evolution of or col-onization by a fourth species; obviously, the latter process would occur much more quickly than the former, and therefore it seems more likely to occur.

In the second scenario, climate change may precipitate an expansion in one direction along the spatial gradient but no change in the other direction. For simplicity, let us assume in this example that such a change has resulted in an expansion in one direction at the outer edge of the space utilized by species 3. In this case, species 3 may expand into this unoccupied space, and species 2 and 1 may or may not similarly expand in this direction along the space axis, depending, once again, on the relative strengths of intraspecific versus inter-specific competition in all three species.

Finally, climate change may further constrain all three species along the space axis by reducing spatial heterogeneity in the distribution of the primary resource. This could occur in two ways: by restriction of resource distribution in space at one or at both ends of the space axis. In the former case, one of the "edge" species may be forced to shift its utilization of the primary resource into more overlap with species 2. In the latter case, both species 1 and 3 would be forced to shift inward along the space axis, increasing the intensity of both intraspecific competition within each species and interspecific competition between species 2 and each of species 1 and 3. In this instance, extinction of one the three species seems likely, with species 2 the most plausible candidate. The relevance of these scenarios to a discussion of the implications of changes in habitat availability and species' distributional shifts resulting from climate change will be pursued in chapter 7.

PHENOTYPIC PLASTICITY AND EVOLUTIONARY RESPONSE TO CLIMATE CHANGE

Before moving this chapter toward the concept of niche conservatism, let us give further attention to the subject of evolution. The preceding discussion suggests extinction is a likely outcome in the absence of some form of adaptive niche response to climate change. An alternative to this, however, is that species may display plasticity in phenotypic response to climate change or microevolutionary response to climate change. The former has already been discussed in the context of phenological response to climate change in chapter 3 and will be discussed further in chapter 7 in the context of altitudinal and elevational redistributions. The microevolutionary response to climate change, while potentially also reflected in directional shifts in the timing of key life history events, will take the form of changes in other traits, such as thermal tolerance, that promote the persistence of species in the absence of distributional responses to climate change (Skelly et al. 2007). While phenotypic-plasticity responses to climate change may be analytically difficult to disentangle from microevolutionary responses (Gienapp et al. 2008), several examples of the latter have been documented recently, suggesting that a relatively rapid evolutionary response to climate change may be more ubiquitous than was thought a decade ago. For instance, in approximately three decades, populations of the wood frog (*Rana sylvatica*) in the northeastern United States displayed evolution in thermal tolerance and temperature-dependent rates of larval development (Skelly and Freidenburg 2000; Freidenburg and Skelly 2004). Similarly, despite low heritability, dates of parturition in a population of red squirrels

(*Tamiasciurus hudsonicus*) in North America advanced at a rate of six days per generation in response to rising spring temperatures, and it was estimated that 87 percent of this change was due to phenotypic plasticity and 13 percent to evolutionary response (Reale, Berteaux, et al. 2003; Reale, McAdam, et al. 2003). Although attributing such changes to an evolutionary response to climate change as opposed to phenotypic plasticity remains difficult, analytical advances in this field are developing rapidly. For example, application of a multivariate state-space model to time series data on dates of migration by sockeye salmon on the Columbia River in the northwestern United States revealed that approximately two-thirds of the advance in migration timing over sixty years was attributable to an evolutionary response to an increase in July mean water temperature on the river (Crozier et al. 2011).

Examples such as these emphasize the potential for adaptive behavioral or physiological responses to climate change in the context of thermal limits (minima and maxima) that reduce the risk of local or global extinction. However, constraints on an adaptive, microevolutionary response to climate change may limit the rate of response relative to that of climate change. Such constraints are typically considered to include limits to phenotypic plasticity, the time lag in the genetic response to climate change, limits imposed by the degree of genetic variation within and among populations, and loss of genetic variation (Skelly et al. 2007; Gienapp et al. 2008). I would argue that this list overlooks additional, potentially important constraints on evolutionary response to climate change. Foremost among these are climatic variability and species interactions. In the case of the former, the variability in climatic conditions from year to year may be viewed as a penalty or disadvantage to a microevolutionary response to the trend in climatic conditions. If the potential for an evolutionary response to climate change exists within a population, it will be expressed most rapidly under directional selection. Extreme fluctuation in climatic conditions over the short term may select for opposing genotypes from one extreme event to the next. Even in the absence of extreme fluctuation, any high-frequency variation in climatic conditions, such as on an interannual basis, will inhibit the rate of directional selection under the climatic trend on which such fluctuation is superimposed. In the case of the latter, our understanding of microevolutionary response to climate change would be improved by giving more consideration to the constraints imposed by species interactions on niche evolution and niche space invasion. Niche breadth or niche centroids can only expand or shift, respectively, if they are not limited by species interactions either in the first place or under novel conditions. One of the primary differences between the fundamental niche and the realized niche is, after all, the set of constraints imposed by species interactions. Experimental investigations of, for example, evolution in thermal tolerance, particularly those conducted under laboratory conditions, may overlook such constraints. The example in the preceding section on niche breadth dynamics illustrates the manner in which an adaptive, directional response to climate change may be countered by limits imposed by the presence of, or interactions with, other species at the boundaries of niche space.

NICHE CONSERVATISM

Whereas it seems intuitive that niche evolution under climate change has the potential to drive diversification or promote the adaptive dispersal of species, we may not readily consider niche conservatism as a driver of diversification under climate change. Niche conservatism—the propensity for species to retain the distribution or other traits such as resource utilization of their ancestral forms—is not intuitively considered an agent of diversification under climate change. In contrast, species groups or clades that display rapid evolution in physiological traits related to adaptation to changing climatic conditions should be less vulnerable to extinction under climate change (Holt 1990; Webb et al. 2002). This tendency would contribute to diversification under climate change if the evolution of traits related to climatic adaptation promoted the utilization of disparate environments and this in turn promoted speciation by reducing competition (Holt 1990, 2009; McPeek 1996; Case et al. 2005). Alternatively, diversification may be facilitated by niche conservatism under the scenario in which populations of a species become isolated as a result of climate change, and thereafter undergo allopatric speciation (Kozak and Wiens 2010). These two views are to some extent in opposition because the former suggests diversification may be driven by interspecific interactions and adaptations oriented toward limiting competition, whereas the latter argues that isolation will result in diversification.

An analysis of phylogenetic data for 250 species of plethodontid salamanders occurring in the Western Hemisphere reveals clear segregation among species belonging to eastern North American, western North American, and neotropical clades along two principal components axes representing abiotic conditions describing the environments inhabited by these species (figure 5.5) (Kozak and Wiens 2010). Notably, not only do the North American clades segregate in climate niche space from the neotropical clades, within the neotropical clades there is considerably less overlap among clades within climate niche space (figure 5.5) (Kozak and Wiens 2010). In other words, neotropical species of salamanders display greater segregation from one another in climate niche space than do those in the other two regions, which appear to display

(a)

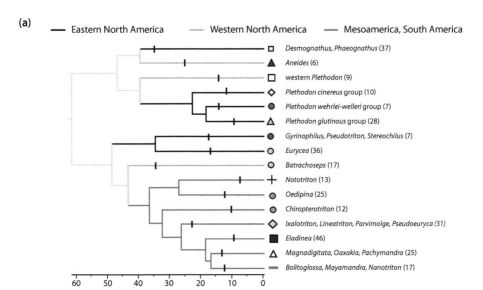

(b)

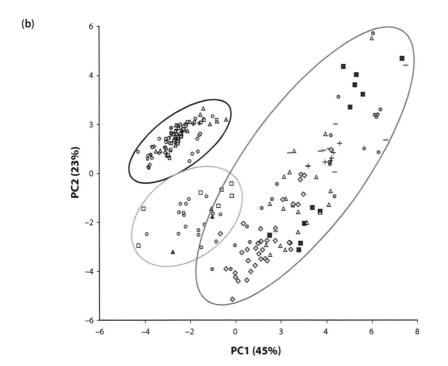

Figure 5.5. Distribution of species belonging to 16 clades of plethodontid salamanders within a climatic niche space defined by a principal components analysis. The species segregate along both axes, and are grouped by ellipses according to eastern North America, western North America, and the neotropics. *Adapted from Kozak and Wiens (2010).*

a greater degree of niche packing. Perhaps not surprisingly, the authors also reported a positive association between species richness and rates of diversification within clades (Kozak and Wiens 2010). Rates of diversification, in turn, were positively correlated with rates of climatic niche evolution (figure 5.6) (Kozak and Wiens 2010). Of note, both rates of climatic niche evolution and diversification displayed negative associations with the extent of clade overlap. A path analysis of these relationships indicated that only the extent of clade overlap, the rate of climatic niche evolution, and species diversification rate within clades were significantly interrelated (figure 5.7).

The authors' interpretation of these relationships is intriguing. They suggest that it is the process of changes in climatic niches of closely related species, not interactions among closely related species with presumably similar resource demands, that drives species diversification in this group (Kozak and Wiens 2010). In fact, in this example, species interactions are seen as inhibiting rates of climatic niche evolution and diversification (figure 5.7) because clade overlap hinders occupation of novel climatic niches. The strength and appeal of this interpretation derive from the context established by the authors in applying their conclusions to Janzen's classic treatment of the role of topographic barriers and temperature gradients in patterns of species diversity in the tropics.

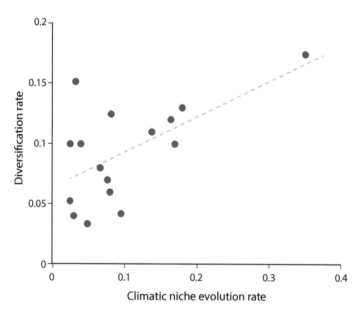

Figure 5.6. Increase in the species diversification rate among plethodontid salamanders belonging to the 16 clades in figure 5.5 in association with their rate of climatic niche evolution. *Adapted from Kozak and Wiens (2010).*

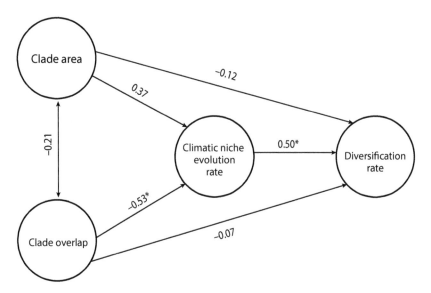

Figure 5.7. Among the possible relationships driving the diversification rate in plethodontid salamanders in the Western Hemisphere, only degree of clade overlap, the rate of climatic niche evolution, and the diversification rate were significant. *Adapted from Kozak and Wiens (2010).*

Here, tropical diversity is ascribed to specialization into narrow zones of temperature tolerance in tropical species as a result of long-term climatic stability (Janzen 1967). Such specialization promotes species packing along elevational gradients in the tropics because temperature regimes change rapidly as one moves upward in elevation on tropical slopes (Janzen 1967). Kozak and Wiens (2010) speculate that the high degree of temperature zonation typical of the tropics provides the basis for high rates of climatic niche evolution and, thereby, diversification and species richness. As will be discussed in chapter 7, Janzen's temperature segregation and specialization may, however, also leave tropical species especially prone to extinction on mountaintops.

MODES OF NICHE RESPONSE TO CLIMATE CHANGE

If the niche of a species is viewed as measurable along a pair of environmental axes, it is immediately apparent that the fundamental niche of that species may or may not take form within the current environmental conditions that prevail at any given site throughout the distribution of that species (Jackson and Overpeck 2000). Here we may consider, for example, a species whose fundamental niche overlaps completely with environmental conditions along one axis but is represented only along a narrow range of conditions on the other environmental axis (figure 5.8). Currently existing environmental conditions are, under this scenario, termed the realized environmental space (Jackson and Overpeck 2000) and encompass the range of abiotic conditions characterizing a site at present. In the example presented in figure 5.8, the realized environmental conditions overlap completely with the fundamental niche of this species along the *y*-axis, which might, for purposes of illustration, represent total annual precipitation at that site. In contrast, along the *x*-axis, the fundamental niche is represented by only a narrow range of values for environmental variable 1 in figure 5.8, which might, for example, represent mean annual temperature. According to Jackson and Overpeck (2000), the zone of overlap between the realized environmental space and the fundamental niche space along both environmental axes is the potential niche space of the species at that location and at that time. Within this zone of overlap, the realized niche of the species may be a subset of the potential niche space, depending on limitations established by biotic interactions, that is, interactions with other species (Jackson and Overpeck 2000).

Climate change may alter the realized environmental niche space of a species in at least four ways, depending in part on whether the environmental variables defining the environmental niche space are correlated with each other (Jackson and Overpeck 2000). All of these have consequences for the persistence

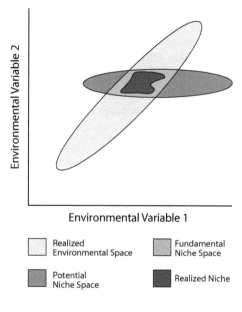

Figure 5.8. The fundamental niche (dark ellipse) of a species may be defined by the relationship between two environmental variables. Realized environmental conditions (lightly shaded ellipse) at a given site may, however, include only a narrow swath of the fundamental niche space, constraining the possibility for existence of the species at that site by the availability of potential niche space. This potential niche space may be further restricted at that site by the presence of other species with which the focal species interacts, producing the realized niche. *From Jackson and Overpeck (2000).*

of populations at local sites and, as will be explored in the next chapter, for the persistence of species associations at local sites. In the case of independent environmental conditions, one variable may shift in response to climate change while the other remains constant; the result would be a shift of the entire realized environmental space along the axis of the variable undergoing change (figure 5.9a). When the environmental parameters are unrelated, multiple outcomes of climate change are possible. For instance, realized environmental space may increase as a consequence of coherent changes in both environmental variables, which would be the case, for example, with increases in both temperature and precipitation (figure 5.9b). Alternatively, changes in one environmental variable may be slighter than changes in the other, so that the realized environmental space remains relatively unchanged along one axis but shifts along the other (figure 5.9c). As well, values of both environmental variables may decline together, resulting in a decrease in environmental space; this may be the case, for example, with declines in sea ice and snow cover in the Northern Hemisphere described in chapter 1. Finally, one variable may decline while the other increases (figure 5.9d), as in, for instance, increased aridity with increasing mean annual temperatures in some systems. As explained by Jackson and Overpeck (2000), the local persistence of a species at a given site encountering such changes depends on several factors, including the persistence of a potential niche within the newly formed environmental space at that site or the capacity for dispersal to new sites if conditions defining the potential niche space fail to persist at that site. I would add that the species must also be able to contend with the added presence of new species at the site whose potential niche space moves into that of the species occurring there originally.

Shifts in realized environmental space, and their effects on potential niche space, thus have consequences for the persistence of both local populations of a given species at a site and species associations at that site. However, changes in realized environmental space may have more important consequences for species assemblages than for the population persistence of any given species. This is because the zone of co-occurrence of species within the niche space, defined by the overlap between existing environmental conditions and the fundamental niches of those species at that site, will always be a subset, for any individual species, of the zone of overlap between realized environmental space and the functional niche of that species (Jackson and Overpeck 2000). As well, climate change has the added potential to alter species' occurrences along spatial or environmental gradients. For instance, under current climatic conditions, the position and distribution of the fundamental niches of an array of species in environmental space defined by two variables of interest might result in the occurrence of those species in a predictable fashion along a gradient

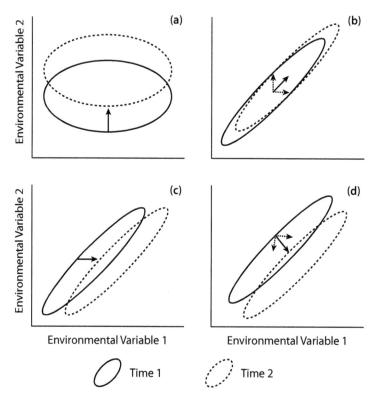

Figure 5.9. Examples of potential changes in the realized environmental space depicted in figure 5.8 in response to climate change. Panel (a) illustrates how the realized environmental space might change as a result of independent changes in uncorrelated environmental variables, with, in this case, only environmental variable 2 changing. Panels (b–d) illustrate the changes in the realized environmental space that might ensue when the environmental variables are correlated with each other. *Modified from Jackson and Overpeck (2000)*.

that captures simultaneous variation in both environmental variables through space (figure 5.10) (Jackson and Overpeck 2000). In the example shown in figure 5.10, the species arrayed along this environmental gradient under current climatic conditions would be encountered in the following order: species a, species a and b together, species b, species b and c together, species c, species c and d together, and finally species d alone. Following movement of the realized environmental space as a result of climate change, species would subsequently be encountered along this same environmental gradient in the following sequence: species c, species c and d together, species d, species a and d together, species a, species a and b together, and finally species b alone (figure 5.10) (Jackson and Overpeck 2000).

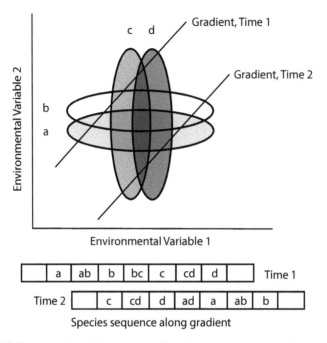

Figure 5.10. The manner in which a sequence of species (denoted a, b, c, and d) observed along an environmental gradient may change in response to changes in the realized environmental space through time in response to, for example, climate change. The environmental gradient indicated by the black line might represent, for example, environmental conditions defined by the relationship between two environmental variables before (Time 1) and after (Time 2) climate change. The sequences of the species encountered during both Time 1 and Time 2 are given in the text. *Adapted from Jackson and Overpeck (2000).*

We must be cautious, however, in interpreting a shift in the association of a species with environmental conditions or other niche parameters as constituting evidence of niche evolution or the adaptability of niche breadth along one or more niche axes. Consider, for instance, the following example. Let us assume that a species of interest occupies a realized niche that is a small subset of its fundamental niche in two dimensions (figure 5.11). The realized niche of this species is contained entirely within the realized environmental space occurring at the site where the species is observed (figure 5.11). In this example, the realized niche is only a subset of the potential niche this species could occupy, but this fact need not reflect the confining influence of biotic factors such as ongoing species interactions (Jackson and Overpeck 2000) but may relate instead to the existence of a relict population. Following climate change, however, the environmental space of the focal species has shifted along one

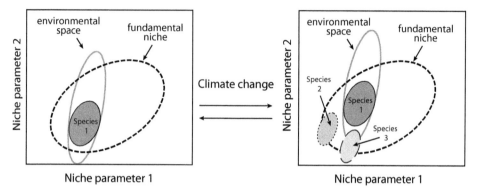

Figure 5.11. The environmental envelope or realized niche of species 1 appears to have shifted along both axes of its niche parameters as a result of climate change. Notice that species 2 and 3 now occupy at least part of the fundamental niche of species 1. However, the fundamental niche of species 1 has in fact remained conserved, neither contracting nor expanding along either niche axis. Only the realized niche of species 1 has shifted within its fundamental niche space in response to movement of its realized environmental space, and perhaps as well in response to invasion of that space by species 2 and 3. Without knowing the boundaries of species 1's fundamental niche, we might mistakenly conclude that species 1 has undergone a niche contraction or expansion.

axis, that of niche parameter 2, and, as a consequence, the realized niche of this species has shifted along this axis as well. However, the realized niche of this species has not changed with respect to its position within its realized environmental space, nor has it colonized new niche space outside its former functional niche, which has remained conserved. Perhaps as a consequence of this shift along the axis of niche parameter 2, two new species have partially occupied the fundamental niche of the first species, but we may never know whether this has occurred as a result of shifts in their respective realized environmental spaces, as a result of movement of the first species within its functional niche space, or both. As well, we cannot determine from the end result whether species 1 might also have adjusted itself within its fundamental niche as a response to the invasion of this space by the two other species.

The entire situation can, of course, be viewed in reverse, in which case diversity has declined at the focal site as a consequence of climate change. No matter the direction in which this is viewed, we cannot conclude, based on shifts in association of a species with certain environmental or niche parameters, that it has undergone niche evolution. Nor can we easily determine in this case whether changes in species associations have driven niche changes or reflect them, and we should use caution when doing so. Such caution would seem to be particularly relevant in conclusions derived from comparisons of patterns evident in paleo versus contemporary data, as well as those derived from environmental niche models. In the former case, the example of the distribution of tapirs during the Late Pleistocene in North America is instructive. Today, the genus comprising American tapirs is limited in distribution to tropical and subtropical zones extending southward from Mexico to equatorial South America, but during the Late Pleistocene the genus occurred in North America as far north as northern Missouri, Pennsylvania, and southern New York state (Graham 2005). Are we to infer, based on the association of this genus with tropical lowland forests today, that the occurrence of tapirs at higher latitudes during the Late Pleistocene indicates that those environments were also tropical during the Late Pleistocene? Such a conclusion would rest on the inference that the current distribution of tapirs is reflective of their fundamental niche, the distribution of which has undergone severe contraction since the Late Pleistocene. In contrast, it appears that tapirs were in fact associated with spruce forests during the Late Pleistocene in North America (Saunders 1988), indicating that the current distribution of the genus reflects a constriction of the realized niche of the genus within the conserved fundamental niche owing to changing biotic interactions (Graham 2005).

BIOCLIMATIC ENVELOPE MODELING
AND ENVIRONMENTAL NICHE MODELS

As we will see in chapter 7, the Grinnellian niche concept rather than the Eltonian niche concept underpins almost exclusively efforts to predict range shifts in response to future climate change (Soberon 2007; Soberon and Nakamura 2009). According to the Grinnellian niche concept, the distribution of a species is assumed to be determined by its habitat requirements (Grinnell 1917), whereas the Eltonian niche concept relates to the functional role of a species (Elton 1927). In predicting species' range shifts, the habitat requirements of a species are, in turn, assumed to map onto abiotic conditions such as temperature and precipitation regimes that determine resource availability through, for example, edaphic conditions and nutrient turnover. The discipline concerned with predicting species' range shifts in response to climate change has grown from predictive habitat distribution models (Guisan and Zimmermann 2000) and is variously known as bioclimatic envelope modeling, species distribution modeling, and environmental niche modeling (Peterson et al. 2011). Extensive treatments of this subject (e.g., Peterson et al. 2011) provide a more detailed assessment of its underlying assumptions and utility than will be given here, where the focus will be on a handful of case studies that illustrate the application of niche theory to the study of ecological responses to climate change.

Commonly, environmental niche modeling begins with the construction of a descriptive model of the current distribution of a species of interest, based on abiotic factors such as temperature and precipitation that explain spatial variation in habitat types with which the species is currently associated (Guisan and Zimmermann 2000). Climate model projections for the region of interest are then employed to project changes in habitat distribution or composition, and these are used to project changes in the distribution and abundance of the species of interest, often under additional assumptions of dispersal capacity (Guisan and Zimmermann 2000, Guisan and Thuiller 2005). Obviously, quality and breadth of the species-level data are essential to the success and robustness of the model projections. The species-level data employed in this approach commonly take the form of presence-absence data, but simple occurrence data, locations derived from museum specimens, and, in rare cases, estimates of abundance based on transect sampling may be used (Guisan and Thuiller 2005). Abundance estimates would be the most desirable data because they afford the ability to address questions of changes in population size, but they are also difficult to obtain and come with their own caveats and limitations resulting from sampling procedures. Presence-absence data appear at first sight to be more amenable to collection over large spatial scales such as at the scale of species' distributions, but they too are accompanied by their own set of limitations. Absence data in particular are troublesome and prone to errors of omission for several reasons. For instance, lack of a record of a species at a given locale does not necessarily mean that the species does not occur there at least part of the year. As well, even under the best of circumstances, not all localities are sampled with equal frequency or effort, and in some cases false absences may actually result from a lack of sampling (Peterson et al. 2011). The environmental data employed fall into three categories based on their expected influences over the presence or absence of species at a given site: limiting or regulating factors that control the ecophysiology of the species, including temperature, hydro-regime, and soil edaphic characteristics; disturbance factors; and resources, including available energy and water (Guisan and Thuiller 2005).

Among many assumptions required by environmental niche modeling, the two of primary importance in my estimation are the assumption of pseudo- or quasi-equilibrium and the assumption of niche conservatism (Guisan and Thuiller 2005). In the former, a species is assumed to be at near equilibrium with its environment. This is necessary because data used in this approach are inherently time-static, representing a snapshot of the species in relation to its environment. However, the assumption that suitable habitat for a given species

is at equilibrium, or is saturated in the sense that all potentially suitable habitat is occupied by that species, presents obvious limitations to projecting species distributions under future climate scenarios. For example, pockets of habitat may be currently unoccupied because of dispersal limitations or local extirpation of species, while some currently occupied locations may be on the verge of local extirpation as a result of recent habitat alteration but are currently occupied because of lag effects in the local extinction process (Wiens et al. 2009). Hence, in such cases, errors of omission and commission may arise (Wiens et al. 2009). In the case of limitations deriving from the assumption of niche conservatism, the functional role of a species in its environment and in relation to other species is assumed to be fixed and predictable, based on the measurable environmental parameters listed above. This may be the greatest pitfall of environmental niche modeling because of the difficulty of estimating niche breadth on the basis of presence-absence data for species whose realized niche may embody a fraction of their potential or fundamental niche, which may be exceedingly difficult to envision, much less quantify. The main problem in this case may be that niche dimensions that are overlooked or not quantified may be the actual parameters determining the distribution and abundance of the focal species (Wiens et al. 2009).

An early application of environmental niche modeling focused on projecting the distributional shifts of eighty species of trees in the eastern United States in response to expected climate change (Iverson and Prasad 1998). In this analysis, Iverson and Prasad (1998) applied regression tree analysis to Forest Inventory Analysis data consisting of abundance and range observations for trees on more than 100,000 plots to determine the environmental factors associated with the then-current distributions of their focal species. Based on the patterns derived from these analyses, Iverson and Prasad (1998) projected that two potential scenarios of climate change resulting from a doubling of atmospheric CO_2 concentration could elicit northward range shifts of 100 km in almost half of the species considered, and northward shifts of up to 250 km in seven of the species considered.

Peterson and colleagues (2002) undertook an even more taxonomically sweeping assessment of faunal responses to projected climate change in Mexico. This assessment made use of data on 1,870 species of animals, including all 175 species of butterflies in the families Pieriedae and Papilionidae, all 416 species of mammals, and all 1,179 species of birds that occur in Mexico, and projected species turnover throughout the country on the basis of two future climate scenarios and three dispersal scenarios (Peterson et al. 2002). Based on their analysis of all 1,870 species, Peterson and colleagues (2002) reported that species' distributions appeared equally likely to expand or contract in response to climate change if dispersal capacity was assumed to be universal among all species. When, however, dispersal capacity was reduced or negligible, distributions tended to contract under climate change (Peterson et al. 2002). Under the most drastic scenario of no dispersal ability, approximately 2 percent of species are expected to lose at least 90 percent of their habitat, the threshold habitat loss for extinction, as a result of climate change (Peterson et al. 2002). This suggests that, in Mexico alone, up to thirty-seven animal species could go extinct during the next century of climate change if unable to disperse to suitable habitat. These results varied somewhat between the two climate scenarios considered, though they remained qualitatively similar. The main difference related to the proportion of species expected to suffer at least 90 percent habitat loss; this rose to 11 percent, or 205 species, under the more dramatic climate change scenario (Peterson et al. 2002). These results appear to have serious implications for the role of human disturbance of habitat through deforestation, agriculture, urban sprawl, and other modifications in the persistence and geographic extent of species' ranges over the next century of climate change. Although suitable habitat may remain for most species even under the most dramatic climate change scenario, the more pertinent issue is whether those species remain able to gain access to suitable habitat even if they are fully capable of dispersing.

Estimates of current levels of species richness varied widely across Mexico, with the highest concentrations of species occurring toward the Mexican subtropics in the southeast of the country (figure 5.12) (Peterson et al. 2002). Whereas colonizations, under the assumption of dispersal into areas of contiguous habitat, are expected to be most pronounced in south-central Mexico, local extinctions are expected to be mainly concentrated in interior Mexico (figure 5.12). Although species turnover in general is expected to be most pronounced in northern and central Mexico (figure 5.12), this appears to relate mainly to the assumption that species' ranges will move primarily northward, and to the lack of inclusion of northward immigration by species from Central America (Peterson et al. 2002). Local colonizations and extinctions may also, of course, create novel local niche space for other species that can utilize colonizing species in exploitative interactions, and for those that may capitalize on expansion of their functional niche space following the extinction of competitors.

In addition to criticism arising from the limitations imposed by the assumptions highlighted above, one of the most persistent critiques of environmental niche modeling has been its failure to take into account the role of species interactions in determining the distribution and abundance of organisms (Davis et al. 1998; Pearson and Dawson 2003). A recent attempt to model the distributional response to climate change of the clouded Apollo butterfly in Europe

(a)   (b)

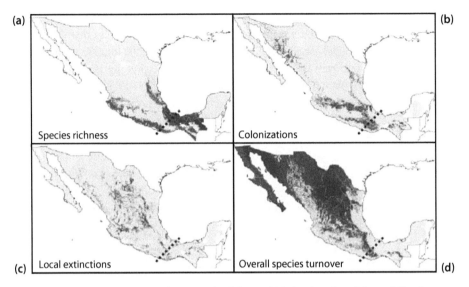

(c)   (d)

Figure 5.12. Estimated current faunal species richness of Mexico, based on niche modeling, in panel (a), followed by expected intensity of colonization into local communities resulting from climate change in panel (b), the resultant expected intensity of local extinctions in panel (c), and overall species turnover in panel (d). *Adapted from Peterson et al. (2002).*

addressed this shortcoming by testing competing hypotheses about the contribution of biotic interactions to species distributions and their response to expected changes in climate (Araújo and Luoto 2007). In this study, data on the current distribution of the Apollo butterfly in Europe were obtained from an atlas of European butterflies and recorded on model grid cells as present or absent. Because this butterfly species associates with four species of host plants used during the larval stage of the butterfly life cycle, the distributions of these four host plants were assumed, under the hypothesis of biotic interactions, to be of importance in determining the distribution of the clouded Apollo butterfly and its response to expected climate change. Araújo and Luoto (2007) then used data on six climate parameters assumed to be of importance in determining the distribution of both the butterfly and its host plant species in a series of generalized additive models (Hastie and Tibshirani 1999) to compare the fit of several candidate models explaining the current distribution of the clouded Apollo butterfly in Europe. The models included various combinations of the butterfly, alternative climate parameters, and host plants. Models with the lowest explanatory power were those that included only host plant distributions, and explanatory power increased with models that included climate only, and finally climate and the host plants (Araújo and Luoto 2007). Hence, this portion of the analysis demonstrated that accounting solely for biotic interactions produced the weakest descriptive model of butterfly distribution compared to a model based on climatic conditions, but the climate model of butterfly distribution was significantly improved by including the additional influence of host plant distribution (Araújo and Luoto 2007). Furthermore, spatial patterns of the predicted distribution of the clouded Apollo butterfly under expected future climatic conditions differed considerably among models that included only influences of climate, only influences of host plants, and the combined influences of climate and host plant distributions (Araújo and Luoto 2007). Hence, the landmark nature of this study within the field of environmental niche modeling is its fairly conclusive demonstration of the importance of considering biotic relationships—that is, species interactions—alongside the effects of abiotic conditions in efforts to predict the distributional responses of species to expected climate change.

In a similar undertaking, Heikkinen and colleagues (2007) incorporated data on the distribution of, and association with, woodpeckers into species distribution models of tree nesting owls in Finland. Woodpeckers were assumed in this exercise to facilitate the presence of boreal owls because their own foraging behavior results in excavation of holes in trees that are subsequently used by owls as nesting sites. Heikkinen and colleagues (2007) attempted to address the hypothesis that species distribution–environment relationships vary with the scale of observation. According to this hypothesis, abiotic factors such as landscape characteristics and regional climatic variation may be more important in determining the presence, abundance, and distribution of species at large spatial scales, whereas interactions with other species may be more important in determining species distributions at local scales (Wiens 1989). By incorporating climatic variables, landscape cover, and the distribution of woodpeckers into their analysis, Heikkinen and colleagues (2007) compared model predictions of the distribution of the four focal owl species to observational records at two spatial scales of 10 km resolution and 40 km resolution. Comparative models included those with climatic variables only; climate and land cover; or climate, land cover, and the occurrence of facilitative woodpecker species. The inclusion of land-cover data improved the explanatory power of the models for all four species of owls over that of the models that considered only climate as an explanatory variable, and the inclusion of occurrence data on woodpecker species improved the explanatory power of these models even further (Heikkinen et al. 2007). Notably, improvements to the power of the models to explain and predict the distribution of boreal owls were more evident at the 10 km scale of resolution than at the 40 km scale of resolution, suggesting that, although the inclusion of biotic interactions in environmental niche models improves their accuracy at predicting patterns of species distributions at macroscales, such interactions do indeed appear to have better explanatory power at finer spatial scales (Heikkinen et al. 2007).

Both of the examples described above considered only facilitative inter-actions, those in which the presence of one species promotes or maintains the presence of another. Interference interactions, such as competition, are, of course, of potentially great importance in determining the presence or absence of species at local sites, and may in some cases work at odds with facilitative interactions at the same sites. In tundra plant communities, dominant species such as crowberry (*Empetrum nigrum*) may exert differential effects on other, less common plant species through competition and facilitation. Taking si-multaneous account of such interactions in species distribution models should substantially improve the predictive capacity of such models over those that consider only abiotic factors such as climatic conditions. Pellissier and col-leagues (2010) employed such an approach in an attempt to identify factors explaining the distributions of twenty-nine tundra plant species in northern Norway, but went a step further by incorporating plant functional traits that might also reveal mechanisms through which species interactions determine patterns of co-occurrence. The functional traits taken into consideration in-cluded leaf dry matter content, stem specific density, specific leaf area, and canopy height. Both specific leaf area and canopy height were inferred to re-flect species-specific growth rates, which were presumed to be of importance in the establishment of individuals under a high-nutrient regime (Pellissier et al. 2010). In contrast, leaf dry matter content and stem specific density were included as indices of the inherent capacity for resource sequestration, which was assumed to be of importance in the establishment of individuals under nutrient-poor conditions, while leaf dry matter content was additionally con-sidered indicative of the capacity for stress resistance related to defoliation (Pellissier et al. 2010).

Including the presence of crowberry as a predictor variable improved the predictive power of species distribution models above those accounting for cli-matic influences only to varying degrees for the twenty-nine focal species, and in fact reduced the predictive power of the models for twelve species (Pellis-sier et al. 2010). Moreover, the only functional trait under consideration that contributed to the power of the species distribution models to explain patterns of co-occurrence with crowberry, whether those patterns were positive, nega-tive, or neutral, was leaf dry matter content (Pellissier et al. 2010). Species that covaried negatively with crowberry were typified by low leaf dry matter con-tent, whereas those that positively co-occurred with crowberry were character-ized by high leaf dry matter content. Low leaf dry matter content presumably renders such species less able to compete for resource uptake with crowberry, and vice versa (Pellissier et al. 2010). The integration of traits representative of competitive ability into environmental niche models that also incorporate potentially competitive or facilitative species should substantially improve the ability of such models to predict changes in species distributions over large spatial scales under various scenarios of climate change.

CHAPTER 6

Community Dynamics and Stability

Returning briefly to the starting point in chapter 4, we are reminded that, in May's simple two-species community, stability was achieved through interaction between the two species in a deterministic environment, but was potentially eroded by environmental variability. Although not presented as such, this is representative of the tension hypothesis, in which biotic interactions counteract abiotic influences on the stability of the biological system.

In this chapter we will explore in greater detail the implications of climate change for community composition, dynamics, and stability and examine further examples in which climatic variability mediates interactions among species, in some cases degrading community stability and in other cases promoting it. As will become readily apparent, the conclusions we are able to reach regarding the consequences of climate change for community persistence are fraught with qualifications. Indeed, ecological theory offers contrasting predictions regarding the consequences for species coexistence and community stability of environmental variability. For instance, short-term instabilities in community composition owing to, for example, high-frequency environmental disturbance may promote the long-term coexistence of species by preventing competitive exclusion of one species by another (Chesson and Huntly 1989). Other work suggests, however, that the stability of biological communities in stochastic environments is only possible if there is sufficiently strong self-regulation at one or more trophic levels, or if self-regulation is strong while species interactions are weak, because environmental erosion of population stability at one trophic level may contribute to instability of the community as a whole (May 1973a; Post and Travis 1979). Contrasting views on the role of environmental variability, such as that represented by climate change, in community dynamics and stability thus relate at least to some extent to the conflation of the notions of lateral and vertical species interactions. This point is pursed later in this chapter, along with a plea for greater specificity in the use of the term "species interactions" as it applies to exploitative versus interference interactions.

The prediction relating environmental variability to the erosion of population stability, and the consequences of this relationship for community stability, are not, however, equivalent to stating that communities will of necessity be stable if there is strong self-regulation at one or more trophic levels within the community. Hence, we may wonder how much alteration of self-regulation, or density dependence, by climate change would be necessary to alter the stability properties of one of many interacting populations to such an extent that the dynamics of the entire community would be altered (Holt 1977). In other words, the potential may exist for changes in the strength of species interactions within the community to compensate for the altered stability of any individual population within the community. This notion was developed by Hutchinson in a seminal paper on coexistence in phytoplankton communities, in which he argued that factors preventing competitively dominant species from reaching equilibrium would simultaneously promote the coexistence of multiple species with different competitive abilities (Hutchinson 1961). Subsequently, Tilman (1996) concluded from empirical work that population-level instability may be necessary for community-level stability. The argument behind this conclusion, based on a long-term field experiment relating the interannual dynamics of productivity of individual plant species to variation in the long-term productivity of the community as a whole, is that population-level instability prevents dominance of the community by a single or few species (Hutchinson 1961; Tilman 1992).

COMMUNITIES DEFINED THROUGH
LATERAL AND VERTICAL STRUCTURING

Before addressing the implications of climate change for community dynamics and stability, we must first arrive at an understanding of how to define a community. Traditionally, a community is defined, in an ecological context, as a predictably occurring assemblage of interacting species. Nonetheless, despite this ostensibly broad definition, it is common to place numerous restrictions on how the term community is used, or to employ distinctions such as terrestrial versus aquatic communities, plant versus animal communities, pelagic versus near-shore communities, and so on. Such distinctions aid in simplifying the interactions that must be taken into consideration when attempting to quantify species interactions or identify the factors limiting recruitment into communities, determining maintenance of community composition, or influencing community dynamics. It is important to acknowledge, however, that such distinctions are arbitrary if not exclusive to the minds of scientists employing them (Paine 1988).

I am of the conviction that any discussion of community composition, community dynamics, and community stability should be informed by distinguishing between laterally structured communities (*sensu* Hubbell 2001), that is, communities comprising species at the same trophic level or within the same guild, and vertically structured communities, or the collection of species interacting across trophic levels. Although this distinction appears to go largely overlooked in community ecology, it is fundamentally important because different factors may be responsible for structuring and governing the dynamics of communities along these different organizational gradients. It is particularly important to note that these terms do not reflect similar usage meant to refer to regional species diversity and local species assemblages, respectively (Ricklefs 2008), a distinction that is explored below.

A near distinction between lateral and vertical communities was approached in the classic paper by Hairston, Smith, and Slobodkin, but their contribution focused on the difference among trophic levels in a vertically structured community in the factors limiting abundance in each trophic level (Hairston et al. 1960) rather than on the differences between vertically and laterally structured communities. Lateral community composition and dynamics relate more to interference interactions, such as competition, while vertical composition and dynamics relate more to exploitative interactions, such as predation, herbivory, and parasitism. Even a generality such as this, however, requires qualification because the composition and stability of communities of exploited species may be altered by vertical interactions such as herbivory or predation differentially influencing the productivity, abundance, and competitive dominance of interacting species within the lateral community (Paine 1969; Holt 1977). Hence, in the context of this book, our conclusions about the implications of climate change for community composition, dynamics, and stability are likely to depend on whether we are considering laterally or vertically structured communities. This, however, brings us to an even more fundamental question about the nature of biological communities in general: do they represent true ecological entities?

REGIONAL VERSUS LOCAL DIVERSITY AND THE COMMUNITY CONCEPT

In the preceding section, a community was defined as a predictably occurring assemblage of interacting species with some geographic reference. Typically, ecologists employ this definition when discussing and investigating the structure and dynamics of what is more properly called the local community (May 1986; Ricklefs 1987, 2004). The local community consists of a subset of the regional species pool at a given site. The regional species pool, also referred to as the regional community, constitutes the continental or subcontinental species assemblage, the diversity and species composition of which are products of biogeography, history, chance events, and allopatric speciation (Ricklefs 1987, 2009b; Webb et al. 2002). Although competitive interactions have long been assumed to be of primary importance in determining the subset of species making up the local community (MacArthur 1955, 1957; MacArthur and Levins 1964; Tilman 1982, 1988), equability in the strength of selective agents operating among local populations, and consequent sympatric speciation, also contribute to structuring local communities (Post and Pettus 1966, 1967; McPeek 1996; Sætre et al. 1997).

Recently, the local community perspective, arguably the main if not the exclusive focus of the discipline of community ecology (Diamond and Case 1985; Agrawal et al. 2007a) has been the subject of increasing debate (Ricklefs 2008, 2009a; Brooker et al. 2009; Paine 2010). The core of the criticism of the community concept as traditionally embodied in ecology is its emphasis on local processes and the lack of any reliable or predictable spatial component (Ricklefs 2008). According to Ricklefs (2008), to understand local coexistence and species diversity, ecologists must focus on regional processes shaping species distributions. Two competing models of the processes generating local diversity from regional diversity embody this distinction, the saturation model and the regional enrichment model.

The saturation model of local diversity emerged from the predictions of niche theory concerned with the concept of limiting similarity (Hutchinson 1957). According to the notion of limiting similarity, niche packing along one axis of resource utilization eventually results in competitive exclusion, thereby limiting membership in local communities by species derived from the regional species pool (Hutchinson 1957, 1961; MacArthur and Levins 1964, 1967; Tilman 1994; Chesson 2000), which is shaped by longer-term, larger-scale processes (Ricklefs 2004, 2009b). According to the saturation model, the addition of any new species to the local community from the regional species pool is thus balanced by the local extinction of previously established species (figure 6.1) (Ricklefs 1987). In contrast, the regional enrichment model challenges the notion of limiting similarity on the basis that community equilibrium is a fluid state, and that communities are likely in constant flux between states of equilibrium in species numbers and composition (Ricklefs 1987). According to this model, local species richness is a product of niche overlap and niche specialization (figure 6.1), both of which can accommodate variation in regional diversity because niche dimensions vary according to the number of species occurring locally (Ricklefs 1987).

The regional enrichment model of community ecology thus encourages consideration of larger-scale—in both time and space—processes in the dynamics and composition of local communities. Typically, these processes are considered

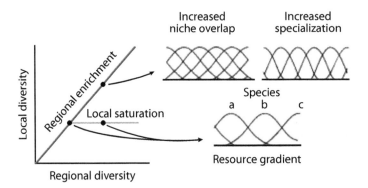

Figure 6.1. The saturation model and regional enrichment models describing two possible means by which local community composition, consisting of species "a", "b", and "c," and species richness relates to diversity at the regional scale. According to the local saturation model, further additions to intact communities from the regional species pool are met with local extinction or facilitated by it. In contrast, the regional enrichment model suggests that species additions to local communities should be met with increasing niche specialization or niche overlap. *Adapted from Ricklefs (1987).*

to include, in addition to those listed above, climatic variation (Ricklefs 2008). For instance, distribution-scale analyses suggest that species-level patterns of distribution and abundance reflect the influences of climate and exploitation interactions driven by predation and pathogens but do not show evidence of a role of interference interactions such as competition, whereas population-level patterns of distribution and abundance do (Anders and Post 2006; Post, Brodie, et al. 2009; Ricklefs 2011). Hence, climate change and its regional variation, as described in chapter 1, are natural candidates for inclusion in investigations of community dynamics in time and space. As we will see below, however, our conclusions and predictions concerning the influence of climate change on community dynamics and stability are likely to depend on whether our focus is placed on laterally or vertically structured communities.

EXPLOITATION AND INTERFERENCE INTERACTIONS

Disagreement over the nature of biological communities and the scales of operation of factors contributing to the generation of, limits to, and maintenance of species diversity may relate to a large extent to our use of the general term *species interactions*. This term embraces a wide array of interspecies dynamics and is used consistently to embody any form of interaction among or between species found together. I would urge ecologists to be more judicious in the use of the term and to refer specifically to interactions as being characterized by exploitation or interference. It is probably fair to characterize most interactions between or among species as antagonistic (Ricklefs 1987), although positive species interactions play an essential role in the dynamics of communities in stressful environments (Bertness and Leonard 1997). Hence, in one of two ways, individuals of one species usually, through their interactions with other species, act to reduce the numbers of individuals of those species: either through exploitation or through interference. Exploitative interactions are those in which one species consumes or utilizes as a resource another species. Obvious examples include predation, parasitism, or other pathogenic relationships; herbivory; and even mutualism, in which two species exploit one another to the selfish benefit of each. Interference interactions are those that involve or result in competition over some resource or group of resources, whether or not the species interfering with one another actually encounter each other in time and space. These two types of interactions may also be distinguished as vertical or horizontal interactions, respectively (Loreau 2010).

The nature of such interactions, and the broad differences between them, aid us in understanding their relative contributions to the structuring of species assemblages that are more easily referred to simply as communities. Perhaps the most important distinction we can recognize between exploitative and interference interactions is that the former are often species-specific, whereas the latter are not. This distinction in turn has important implications for the roles of the two types of interactions in determining, limiting, and maintaining regional versus local diversity and community composition (Paine 1966; Schoener 1968, 1974). Exploitation interactions, such as consumption of prey by predators, consumption of plants and plant tissues by herbivores, and infection of hosts by pathogens, to name a few, tend to be specific to species or functional groups but can nonetheless exert community-level influences (Spiller and Schoener 1994; Schmitz 1998; Schmitz et al. 2000). Interference interactions, by contrast, may often occur without regard to the species being interfered with through competition and tend to be highly localized (Hubbell 2001). Hence, whereas exploitative interactions such as pathogen-host dynamics and predator-prey interactions are important in influencing the regional species pool from which local diversity and species assemblages derive (Ricklefs 1987; Werner and McPeek 1994; McPeek 1996, 1998; Wellborn et al. 1996, Skelly 1997; Skelly et al. 1999), interference interactions are viewed more properly as an outcome of the establishment of species with overlapping or nearly overlapping resource demands within local sites (Hubbell 1979; Hansen and Hubbell 1980; Condit et al. 1996, 2000; Wills et al. 1997; McKane et al. 2002; Volkov et al. 2003). The extent to which local sites, and local populations, are integrated

spatially over larger scales, and hence the scale over which interference interactions become important, will, of course, depend on dispersal and connectivity,
which are themselves influenced by climate change in many systems (Clark
1998; Higgins and Richardson 1999; Cain et al. 2000; Hannah 2011; Zhang et
al. 2011; Dobrovolski et al. 2012).

GLEASONIAN AND CLEMENTSIAN COMMUNITIES

Classically, communities have been regarded as the level of biological organization at which species interact. In this view, a community comprises individuals of
multiple species interacting in predictable aggregations through time and space.
It is immediately obvious that this conceptualization of the community is unre-
alistically static: communities change in character and composition through both
time and space. But if we are willing to make even this obvious acknowledgment
of the ephemeral nature of communities, how much further do we need to go in
our thinking of the limitation of the community concept before the concept itself
begins to erode? An alternative to the classic perspective is that communities
do not exist as entities of biological organization but rather represent points of
overlap among species, and that these points of overlap may involve ecological
coincidence and will be ephemeral over some scale of time (Ricklefs 2004).
According to this view, any spatial predictability to species' co-occurrence is an
artifact of the predictability of species' individual habitat or niche requirements
as they are manifested across landscapes (Ricklefs 2004, 2009b).

These two philosophically opposing perspectives on the nature of biological
communities pose fundamental differences for the predictability of community-
level responses to rapid environmental change such as climate change. In the
first view, facilitation and competition will undoubtedly play an important role
in the success or failure of species assemblages to track environmental change.
According to the latter view, there is a likelihood of so-called "non-analogue"
communities arising as a consequence of individual species tracking—or fail-
ing to track—environmental changes (Graham et al. 1996).

The concept of the community in ecology likely had its origins in early attempts to explain the apparent stability of grassland species assemblages in a
climax state (Clements 1916). Clements viewed competition as the main force
driving the succession of plant communities to a stable state through interactions
whose outcomes were determined by a dominance hierarchy in which some
species were superior competitors and thus would eventually become more
abundant (Clements 1936). Absent any changes in climate, such climax communities could remain intact, embodying species interactions with predictable

outcomes, indefinitely. Clements clearly considered the community as an entity, one whose constituent species functioned together in much the same way that organs work together to maintain the integrity of the organism. With any change in climate, therefore, the Clementsian community would be expected to move, intact, in the direction of favorable environmental conditions.

The archetypal contrast to the Clementsian community concept is one in which the community is regarded as a collection of species with highly individualistic niche requirements that happen to overlap in time and space. This view has its origins in work by Gleason on the nature of structural uniformity in plant species associations (Gleason 1926). Although Gleason was prepared to acknowledge the apparent stability of species associations in some plant communities, he did not regard these associations as permanent or predictable across space or through time. For instance, Gleason noted that the idea of a characteristic species, that is, one that is diagnostic of a community, seemed to work well enough in far northern regions, where species diversity is low and the numbers of individuals of a few species in a community could be quite high and predictable. At the same time, he argued, there was very low predictability across the landscape in species occurrences in the Tropics, where a given species was less likely to be found in association with another given species from one site to the next because of the incredibly high diversity of species packed into the landscape. In the Gleasonian view, communities were in constant flux as individual species responded to spatial and temporal variation in conditions favorable to their existence. In the Gleasonian framework, communities should not be expected to respond to climate change as intact entities. Rather, we should expect individualistic species' responses to changing climate to determine whether communities persist or reemerge newly configured under climate change.

How important is it to our understanding of the nature of biological communities, and to our ability to understand how they will respond to climate change, that the origins of the concept of the ecological community lie in studies of plant species interactions? I would say it is very important, but not necessarily because of the focus in these early studies on plant community ecology. Some of the same limitations to our understanding of community dynamics and stability could be obtained from studies of animal communities. The primary challenge imposed by this focus relates to our understanding of the role of resource competition in structuring communities, and this has developed mainly from the study of plant community ecology (Tilman 1982, 1988). This focus, I would argue, has inherently shaped our view of the role of species interactions in community dynamics as horizontal and of the importance of organism-environment interactions as having to do with influencing the outcome and balance of horizontal species interactions.

NON-ANALOGUES: THE COMMUNITY
IS DEAD—LONG LIVE THE COMMUNITY

If communities behave as Clementsian superorganisms, we should expect them to shift, intact, with changes in the boundaries to and distributions of the environmental conditions that determine their limits. If, on the other hand, communities are better understood according to the Gleasonian paradigm, then we should not expect them, a priori, to respond as communities to climate change. Rather, communities as we know them will not track the environmental conditions that predicted their existence in the first place, but the populations of species they comprise will, potentially in disparate fashion. In other words, climate change, and the ensuing environmental changes precipitated by it, should have the potential to result in non-analogue communities.

The non-analogue community concept describes the phenomenon of the existence of communities in the past—as well as the potential for the existence of communities in the future—that have no contemporary analogue (Graham et al. 1996). An example is found in an analysis of changes in the distributions of several species representing the small mammal community over the past 20,000 years in North America following late Quaternary climate change (figure 6.2). In this example, distributions of the focal species responded to warming and the associated retreat of the Laurentide ice sheet with different rates and directions of change. For instance, the heather vole, collared lemming, meadow jumping mouse, Franklin's ground squirrel, water shrew, and masked shrew all dispersed northward rapidly (figure 6.2b). By comparison, the southern red-backed vole, American pika, prairie vole, meadow vole, and southern bog lemming dispersed northward and westward, but at a slower pace (figure 6.2c). Other species, including the least shrew, eastern chipmunk, short-tailed shrew, and marsh rice rat, responded with slow eastward movements (figure 6.2d); while the northern pocket gopher, montane vole, long-tailed vole, black-tailed prairie dog, and plains pocket gopher showed slow westward contractions (figure 6.2e). In contrast, the eastern woodrat and the northern pygmy mouse underwent no distributional shifts (figure 6.2f).

Graham and colleagues (1996) concluded in their analysis that the faunal assemblage of an entire subcontinent responded in Gleasonian fashion to climate change; that is, species tracked changes in environmental conditions individualistically. This conclusion appears most convincingly supported when one focuses on the community represented by the heather vole, southern red-backed vole, least shrew, and eastern woodrat (figures 6.2b–d). The distributions of these species overlapped considerably during the late Quaternary, but today only two of them, the heather vole and southern red-backed vole, are found

Figure 6.2. Individualistic responses of small mammal species to climate change in North America during the Late Pleistocene. Panel (a) shows the spatial coverage and distribution of sites from which the FAUNMAP samples used in this analysis were derived. In all other panels, the current distribution of each species is indicated by the shaded region of the map, and their distributions based on fossil evidence are indicated by symbols during the full glacial (circles), late glacial (crosses), and Late Holocene (triangles). In panel (b), rapid northward dispersal following warming is evident in the heather vole. In panel (c), more gradual northward and westward redistribution is evident in the red-backed vole. In panel (d), an eastward shift is evident in the least shrew. In panel (e), a westward, gradual contraction of the distribution of the northern pocket gopher is evident. In panel (f), the distribution of the eastern woodrat displays relative stasis. Hence, species formerly making up an intact community during the Pleistocene exhibit nonoverlapping distributions today as a result of their diachronic and spatially disaggregated responses to past climate change, resulting in non-analogue communities. *Adapted from Graham et al. (1996).*

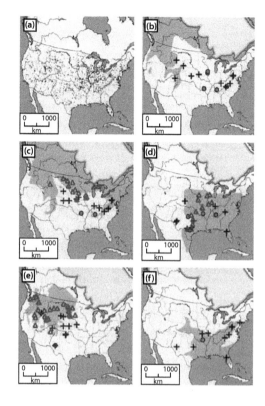

together. Clearly, had there been a Clementsian shift in the small mammal community of the late Quaternary, most or all of the species constituting it would have shifted their distributions in similar fashion. Similar non-analogue communities are evident in analyses of late Quaternary vegetation dynamics reconstructed from macrofossil samples retrieved from ancient packrat middens (Betancourt et al. 1990). These reconstructions have shown a shift in what is now the southwestern United States from plant communities characterized by evergreen forests 20,000 YBP to mixed evergreen and desert scrub communities in the early Holocene, to communities similar to contemporary desert scrub approximately 4,000 YBP (McAuliffe and Van Devender 1998). These changes in plant community composition were also mirrored by changes in associated arthropod species assemblages at the same sites over the same period (Hall et al. 1988).

These examples raise the possibility that future climate change may promote the development of communities that have no contemporary counterpart. Such communities would be unrecognizable based on our expectations for the co-occurrence and interactions of species as we know them today. This is the concept of the non-analogue community (Graham et al. 1996), more recently referred to as the "no-analogue" community (Williams and Jackson 2007).

If species respond individualistically to climate change and the associated changes in environmental conditions that contribute to their niche space, non-analogue communities are likely to arise. In addition to the examples just mentioned, other analyses of fossil data from late Quaternary samples have disclosed non-analogue assemblages of species of plants, mollusks, mammals, and beetles (Williams and Jackson 2007). These communities include species that are found today, but in different assemblages, and thus have no contemporary analogue. Obviously, non-analogue communities may also arise as a consequence of diachronic shifts in species' distributions into communities not formerly containing them, as in the examples of the northwestward shifts of the heather and southern red-backed voles (figure 6.2). Hence, non-analogue communities may arise as a consequence of climate change in two ways: through the dissolution of existing communities as species respond to environmental change according to their individual resource and habitat requirements, and through the formation of novel communities as species previously occupying distinct ranges come into contact with one another.

Inferring the existence of past non-analogue communities is perhaps more straightforward than projecting the potential for the development of future non-analogue communities but is not without its own difficulties. Fossil assemblages indicate overlap in species' distributions and the co-occurrence of species but indicate nothing about species interactions, a key component of the community concept, or about the role of species interactions in a community's response to climate change. Environmental changes that accompanied the end of the Last Glacial Maximum are presumed to have precipitated species redistributions (Williams and Jackson 2007), but environmental conditions are only part of the realized niche of a species, which also includes the presence or absence of other species that may be competitors for limiting resources.

Projecting the potential for the development of non-analogue communities under future climate change relies, at present, on projections of the development of non-analogue climates. This pursuit relies in turn on a Grinnellian niche–based assumption of the relationship between climatic conditions and habitat requirements of the species composing a given community. Here, a community is formed, or exists, only under conditions in which the climatically determined fundamental niches of each species overlap in the climate envelope space (figure 6.3) (Williams et al. 2007). In the example in figure 6.3, under current (late twentieth-century) climatic conditions, species 1 and 3 form a community, but the depicted shift in climatic conditions projected

Figure 6.3. Development of non-analogue communities in response to climate change as a result
of changes in the climatic space that influences
species distributions under the Grinnellian niche
paradigm, and thereby influences patterns of species
co-occurrence in local communities. In this example,
species' fundamental niches in climate space are
indicated by shaded ovals. Co-occurrence of species
is indicated where these niche ovals overlap. Hence,
under twentieth-century climatic conditions delineated
by the first black ellipse, communities might consist
of species 1 and 3 together, but not species 2 and 3
together. Under twenty-first-century climatic conditions, however, communities might consist of species
2 and 3 together, but not species 1 and 3 together.
Furthermore, species 4 is likely to go extinct in the
twenty-first century because its climatic niche space
no longer exists. *Adapted from Williams et al. (2007).*

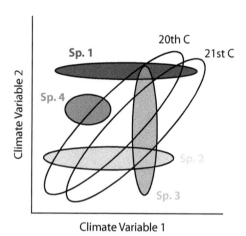

for the twenty-first century will result in the formation of a novel community
comprising species 2 and 3 and the dissolution of the original community of
species 1 and 3 (figure 6.3) (Williams et al. 2007).

Projections of the evolution of novel climates and the disappearance of existing climates over the next century vary according to the climate change scenario employed. Allowing carbon emissions to proceed at their current rate
results in the widespread disappearance of current climates across the Tropics
and the replacement of these with novel climates (figures 6.4a and c) (Williams
et al. 2007). Although curbing carbon emissions to maintain an atmospheric
CO_2 concentration of 550 ppm reduces the likelihood of the development of
novel climates and the disappearance of existing climates across the globe,
even under this more optimistic scenario the Tropics will be hardest hit (figures
6.4b and d) (Williams et al. 2007). Because the Tropics are characterized by
the highest rates of species endemism on Earth (Brooks et al. 2006; Kier et al.
2009), the potential for the development of non-analogue communities in these
regions appears particularly high over the next century.

It is important to note that this exercise does not assume niches will evolve
with climate change, nor does it assume that species interactions determine, to
at least some extent, the boundaries of the realized niche of a species in climate
envelope space. Nor does it allow for the possibility that climatic conditions
mediate species interactions at the community level, or that species interactions
mediate community response to climate change (Suttle et al. 2007). Examples
of the former accrue from studies of arctic char (*Salvelinus alpinus*) and brown
trout (*Salmo trutta*) in Scandinavia and from an analysis of long-term data on

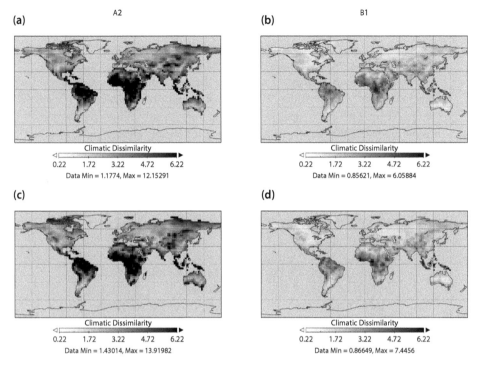

Figure 6.4. Development of novel climatic conditions and the disappearance of existing climatic conditions under two climate change scenarios. The first scenario, A2, derives from an expected increase in atmospheric CO_2 concentration to 856 ppm by the year 2100, while the second, more conservative scenario, B1, derives from an expected increase in atmospheric CO_2 concentration to 549 ppm by 2100. The risk of developing novel climatic conditions under these two scenarios increases with increasingly dark shading in panels (a) and (b). The risk of disappearance of existing climatic conditions increases with increasingly dark shading in panels (c) and (d). *Adapted from Williams et al. (2007).*

the dynamics of two species of sympatric *Ficedula* flycatchers in the Czech Republic, while examples of the latter emerge from a multi-annual experiment in the California grasslands and from an analysis of long-term data on the dynamics of wolves and moose on Isle Royale. These examples are discussed in the following section in the context of promoting the perspective of interference and exploitation interactions in laterally versus vertically structured communities. To date, the subdiscipline within climate change ecology that is concerned with the development of non-analogue communities has focused overwhelmingly on laterally structured communities, as the examples given above indicate. Insofar as the regional community perspective (Ricklefs 2004, 2008) argues that interference interactions such as competition cannot explain regional-scale patterns of diversity and species assemblage, which relate more clearly to exploitation interactions (Ricklefs 2011), our discussion of the role of climate change in the development of non-analogue communities would be better informed by giving greater consideration to vertically structured communities. Moreover, this sub-discipline has mainly concerned itself with the development of non-analogue communities as a result of the dissolution of species assemblages by climate change. Yet in some cases species interactions may actually buffer communities against the potentially destabilizing influences of climate change and variability. Hence, in the next section I draw comparisons and contrasts between the role of climate change in the dynamics of laterally versus vertically structured communities, and between the role of climate in species interactions that influence community dynamics versus the role of species interactions in mediating the influence of climate change on community dynamics.

THE ROLE OF CLIMATE IN MEDIATING SPECIES INTERACTIONS VERSUS THE ROLE OF SPECIES INTERACTIONS IN MEDIATING COMMUNITY RESPONSE TO CLIMATE CHANGE

I have argued above for greater specificity in the use of the term species interactions to indicate whether interference interactions or exploitation interactions are of interest. Here I present four examples of the manner in which a focus on interference and exploitation interactions in laterally structured and vertically structured communities, respectively, can generate insights into community response to climate change.

Arctic char and brown trout exhibit widely overlapping distributions in Scandinavia as a result of human introductions, and display nearly identical thermal performances (Helland et al. 2011). An analysis of the competitive impact of the presence of arctic char on food consumption by brown trout in eighteen lakes spanning a climatic gradient in central Scandinavia revealed that the annual timing of surface ice melt was a major determinant of the negative effect of arctic char on brown trout (Ulvan et al. 2011). Although brown trout are considered more aggressive than arctic char, the latter species is better able to maintain positive growth in darkness and the ice-covered lake conditions characteristic of winter in Scandinavia (Helland et al. 2011). As a result, the length of the annual ice-covered season on Scandinavian lakes is an important driver of the impact on brown trout of competition with arctic char (Ulvan et al. 2011), suggesting that continued warming may promote competitive exclusion of arctic char by brown trout (Urban et al. 2011). This effect extends beyond individual performance to population dynamics: analysis of a twenty-five-year time series of the abundance of both species inferred from catch-per-unit effort data from a lake where they co-occur in south-central Norway revealed a negative association between brown trout population growth rate and char abundance only in years with a prolonged ice cover (Helland et al. 2011). A more comprehensive analysis of presence-absence data for arctic char and brown trout from over 1,500 lakes in Norway concluded that arctic char have outcompeted brown trout in cold lakes with low productivity, whereas brown trout appear to have outcompeted arctic char in warm, productive lakes (Finstad et al. 2011). Hence, interference interactions between these two species are mediated by climatic conditions.

Collared and pied flycatchers coexist over a narrow band of overlap in the distributions of both species in the Czech Republic. Throughout most of their respective distributions, collared and pied flycatchers do not co-occur, but there is a narrow zone across Central and Eastern Europe where they are sympatric (Sætre et al. 1997). In this area of coexistence, collared flycatchers favor warm, low-lying areas, while pied flycatchers favor subalpine areas that are colder (Sætre et al. 1999). Where they come into contact, the pied flycatcher is inferior to the collared flycatcher in interspecific territoriality displays and in competition for nesting sites (Sætre et al. 1993, 1997), and at one study site in the Czech Republic with multi-annual data on the population dynamics of both species, the pied flycatcher occurs at chronically lower densities than the collared flycatcher (Sætre et al. 1999).

An analysis of the role of climatic fluctuation in the dynamics of both species in populations where they coexist revealed that the dynamics of the superior competitor, collared flycatchers, were driven mainly by the intrinsic processes of density dependence and nestling mortality, with a modest contribution of climate and no influence of interspecific competition (Sætre et al. 1999). By contrast, the dynamics of the inferior competitor, pied flycatchers, were driven mainly by interspecific competition and a weaker effect of the intrinsic processes of density dependence and nestling mortality, with no influence of climate (Sætre et al. 1999). In this case, it appears that climatic fluctuation, by exerting just enough of a limiting influence on the dynamics of the superior competitor, prevents collared flycatchers from outcompeting pied flycatchers. Hence, climate may not only contribute to patterns of the distribution and abundance of species in the Grinnellian niche framework, it may also play an important, if in this instance subtle, role in species interactions such as competition that are important in determining community composition. As in the example of the arctic char and brown trout, climatic variation in this instance is an important driver of interference interactions in laterally structured communities.

In another example, a multi-annual watering experiment in California grass-lands tested the hypothesis that the timing of the water treatment relative to the phenological schedules of the species composing the plant community would be of importance to the potentially differential responses of the species to this form of climate manipulation (Suttle et al. 2007). Of the three watering treat-ments applied, only the late winter/early spring treatment, which effectively prolonged the winter growing season, produced significant effects. Initially, only nitrogen-fixing forbs increased their productivity in response to spring watering, but this in turn facilitated an increase in winter-growing grasses, the species group with the most vernal phenology in the plant community (Suttle et al. 2007). Eventually, enhanced productivity of the grasses in response to this treatment led to suppression of the growth of leafy forbs because of the interference effect of grass litter on those species. As a consequence, the initial increase in productivity across plant functional groups and species richness was eventually reversed as grasses suppressed any potential productivity re-sponse of other plant types to the watering treatment (Suttle et al. 2007). In contrast to the two examples above, in this case, interference interactions medi-ated community-level response to simulated climate change.

In a further example, let us consider the dynamics of wolves and moose interacting in a clear case of exploitation interactions in a vertically struc-tured community. Long-term data on the population dynamics of wolves and moose on Isle Royale, Michigan, reveal that an outbreak of canine parvovirus (CPV) in 1980–81 led to an immediate and severe reduction in the number of wolves on the island (Peterson et al. 1998), after which the moose popu-lation increased rapidly (figure 6.5a). Autoregressive analysis of the moose population time series revealed that prior to the CPV outbreak in the wolf population, moose dynamics were strongly negatively related to wolf abun-dance, whereas after the CPV outbreak and the decline in the wolf population, moose dynamics were unrelated to wolf abundance (figure 6.5b) (Wilmers et al. 2006). In contrast, before the CPV outbreak, moose dynamics were only weakly and negatively related to wintertime climatic conditions determined by the North Atlantic Oscillation (NAO), but after the CPV outbreak moose dynamics were more strongly, and positively, related to the NAO (figure 6.5b) (Wilmers et al. 2006). A phase-dependent autoregressive model of moose dy-namics was previously applied to determine whether the stability of moose dy-namics in this population was related to the phase in the wolf dynamics (Post et al. 2002). In this approach, the time series on moose abundance was analyzed according to phases of the wolf population. This analysis revealed that during the wolf increase phase, moose dynamics were characterized by proper multi-annual cycles (Post et al. 2002). In contrast, during the wolf decline phase, moose dynamics approached the boundary of instability in parameter space for the AR(2) model described in chapter 4 (figure 6.5c) (Post et al. 2002). This movement of herbivore dynamics toward the boundary of instability with a strengthening of the climatic influence on them further reinforces the theoretical prediction by May, given at the beginning of chapter 4, that environmental variability has the potential to destabilize population dynamics regardless of the direction of its influence. This example further illustrates that exploitation interactions in a vertically structured community determined the response of the community to climate change and suggests that a stabilizing influence of predation on moose dynamics was in tension with the destabilizing influence of climatic variation.

As a final example, let us consider both exploitation and interference interactions in a community of grazers and plants that is simultaneously both vertically and laterally structured (Post 2013). This example originates in an ongoing, now ten-year warming experiment combined with an herbivore exclusion experiment at my study site in Greenland. The composition of the plant

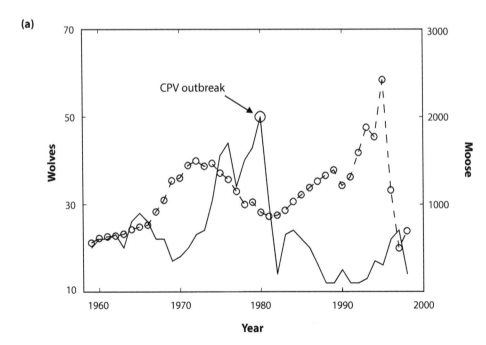

Figure 6.5. (a) Population dynamics of wolves (solid line) and moose (broken line with symbols) in Isle Royale National Park in the U.S. state of Michigan. The outbreak of canine parvovirus (CPV) in the wolf population precipitated an immediate decline in wolf numbers and a gradual increase in moose numbers. (*continued*)

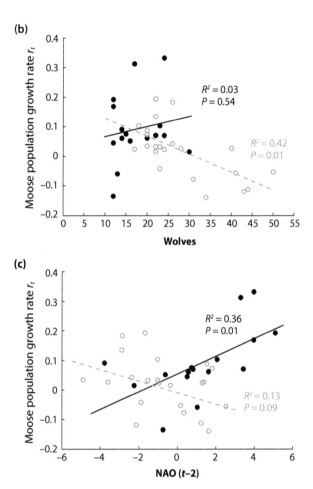

Figure 6.5. (*continued*) (b) Relationship between moose population growth rate and wolf abundance in Isle Royale National Park before (open circles and dashed line) and after (sold circles and line) the outbreak of canine parvovirus in the wolf population. (c) Shift in the relationship between moose population growth rate and winter climatic conditions from before the outbreak of canine parvovirus in the wolf population (open circles and dashed line) to after the outbreak (solid circles and line). *Adapted from Wilmers et al. (2006).*   (*continued*)

community at the beginning of this experiment was dominated by graminoids, mainly *Poa pratensis*, with dwarf birch as the second-ranking species in the community (figure 6.6). After the second year of warming and a third year of herbivore exclusion, the community experienced a severe outbreak of the larvae of a noctuid moth, which defoliated most of the aboveground vegetation. Recovery after this outbreak resulted in divergent communities. Where

(d)

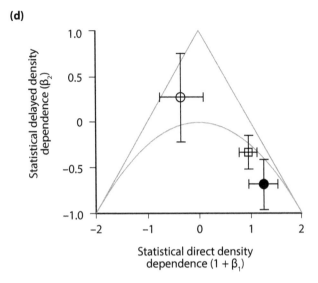

Figure 6.5. (*continued*) (d) The AR(2) parameter plane for the second-order autoregressive model of population dynamics described in chapter 4 showing the relationship between delayed and direct density dependence in the dynamics of the moose population on Isle Royale during the wolf increase phase (solid circle) and during the wolf decline phase (open circle). The inverted triangle defines the boundary between stable (inside the triangle) and unstable dynamics (outside the triangle). *Adapted from Post et al. (2002).*

herbivores had been exclosed, the community quickly became dominated by dwarf birch, but where herbivores were allowed to maintain their pressure on the community, it returned to its graminoid-dominated state. Warming exerted only a slight effect on these divergent patterns of recovery from the moth outbreak (figure 6.6). Hence, in this example, exploitation interactions in the form of grazing by large herbivores were important to the maintenance of community composition among plant functional groups in a laterally structured community under the influence of warming.

PHENOLOGY AND THE EPHEMERAL NATURE OF COMMUNITIES

At the beginning of this chapter, a community was defined as a predictably occurring assemblage of interacting species. This predictability can be assigned a value in time and space. But over what scales of time is a community recognizable and predictable in its composition? The difficulty in recognizing the

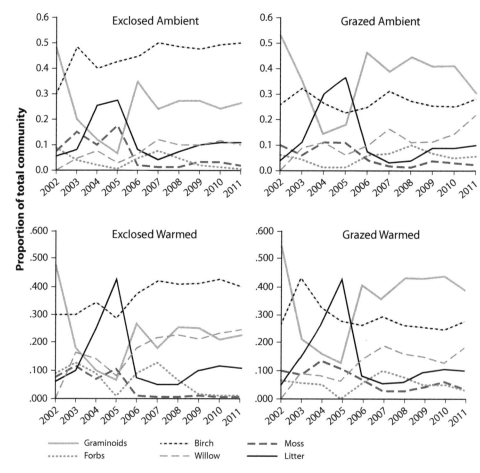

Figure 6.6. Changes in relative community composition according to plant functional groups at the study site near Kangerlussuaq, Greenland, during an ongoing warming and herbivore exclosure experiment. An outbreak of moth larvae in 2004–5 was followed by recovery of the plant community along different trajectories under exclosed conditions (left panels) compared to grazed conditions (right panels). Whereas community composition on all plots was dominated by graminoids initially, recovery following the outbreak resulted in domination of the community by dwarf birch on exclosed plots but not on grazed plots.

existence of communities as ecological entities lies in the ephemeral nature of species assemblages. At the same time, however, on hearing the term *community,* a concrete image likely comes to mind of an easily recognizable collection of species. One such compelling example of a predictably occurring and well-defined assemblage of interacting species is the invertebrate and inquiline fauna inhabiting some species of North American pitcher plants.

Among the eleven species of pitcher plants in the genus *Sarracenia*, one species, *S. purpurea*, attracts a diverse array of macroinvertebrates, and the species composition of the macroinvertebrate community is remarkably consistent across the distribution of this species of carnivorous plant (Ellison et al. 2003). As macroinvertebrates are attracted to the pitcher plant and fall in, they become entrapped in the water held by the plant, and eventually succumb. They are then broken down and consumed by the inquiline community of the pitcher plant. This inquiline community includes several bacteria, nematodes, rotifers, protozoans, fungi, and algae, and its composition is also consistent throughout the range of *S. purpurea* (Ellison et al. 2003). Nonetheless, each of these functional groups has its own emergence phenology, as does each of the species within each functional group. At what point in the annual life cycle of *Sarracenia* pitcher plants, then, does the inquiline community become a community? Presumably, for instance, the nematode and algal components of this community display different emergence phenologies. Similarly, the macroinvertebrates providing a resource base for the plants themselves also undergo differential emergence phenologies that determine the schedule of their appearance in the pitcher plants and thereby their availability for the inquiline invertebrates responsible for their decomposition. If that is the case, then these phenological dynamics surely also vary across the latitudinal gradient occupied by *Sarracenia* pitcher plants, even though the composition of the inquiline community is constant across this gradient.

Similar considerations apply in plant communities. As an example, let us return to a consideration of the composition of the plant community at my study site in Low Arctic Greenland that was touched on briefly above. Ignoring for the time being the experimental treatments of herbivore exclosure and warming, we can turn our focus to the composition of the plant community under ambient conditions exposed to herbivory, the natural state of this community. Data from a single year of this study, collected at the time of peak biomass in July, reveal that the community comprises primarily grasses and sedges, followed by dwarf birch and gray willow, with less than 10 percent of the community consisting of forbs (figure 6.7a). A closer look at the forb component of the community reveals an additional seven species contributing to community composition, with the most abundant being horsetail, chickweed, polygonum, and cerastium (figure 6.7b). Hence, the community appears to be fairly typical of Low Arctic sites: a graminoid-dwarf shrub–dominated community interspersed with an array of forb species.

In practice, however, the extent to which these species interact as a community is influenced to a great extent by the differing phenological dynamics of the individual species. The representation of plant community composition in figure 6.7 is based on point-frame data collected on a single day in late

(a)

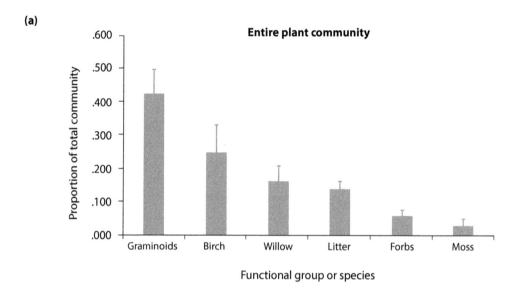

(b)

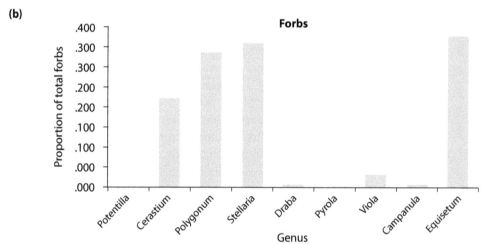

Figure 6.7. Composition of the entire plant community (a) at the study site near Kangerlussuaq, Greenland, according to functional groups, and of the forb community (b) according to genera. These data represent time-static conceptualizations of the plant community that are typical of generalized statements of community composition but may have little relevance in relation to dynamics on seasonal or interannual scales.

summer during one year of the study. Data collected during the same year, but on a daily basis from the beginning of the growing season through its peak, reveal considerable differences among these species in their emergence schedules (figure 6.8). Early in the growing season, for instance, the community consists solely of two species of graminoids, a grass and sedge (figure 6.8).

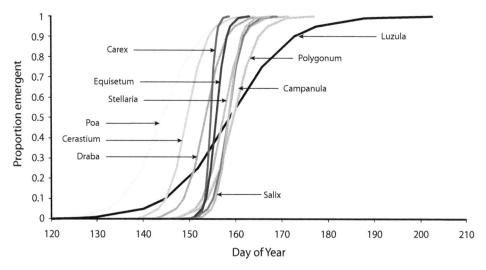

Figure 6.8. Phenology and the community concept may not be compatible. Shown here are the species-specific emergence curves for all plant species represented in the community at the study site near Kangerlussuaq, Greenland, based on daily monitoring of phenology plots in 2009. The different dates of onset of emergence and different rates of progression of emergence of the species present at the site may lead one to wonder at what date the community becomes recognizable. As well, we may wonder to what extent interference interactions among these species, if in operation, are altered as the growing season progresses and the local species collection changes as a result. It is also worth noting that this figure would likely vary considerably among years.

This is the case for approximately the first two weeks of the growing season. Subsequently, forbs such as *Cerastium* and *Draba* begin to emerge. Hence, for the first three and one-half weeks of this short growing season, the entire plant community is represented by four species. Beginning in early June, there is a flush of emergence by additional forb species, followed finally by a leaf bud burst in the two shrubs, dwarf birch and gray willow, that contribute most to the aboveground biomass of this community (Post and Pedersen 2008; Post et al. 2008a). It is only late in the growing season, on approximately the first day of July, that the full suite of species composing this community is emergent and potentially interacting with one another.

These observations suggest that phenology plays a critical role in the ephemeral nature of communities on short time scales, that is, within a growing season. Phenology may exert similar control over community composition from year to year. To visualize this, we may consider what the Clementsian view of communities implies with respect to the role of phenological responses to warming in the composition of communities. For instance, if we assume that species' phenologies respond uniformly to changes in temperature, we might expect the phenology of the entire community to advance and retreat from one year to the next with temperature fluctuations. As well, under this paradigm, we might expect the phenology of the entire community to advance over the long term in response to a warming trend. As was emphasized in chapter 3, however, species adjust their life histories individualistically to warming, with some advancing their phenologies more so than others, and some even delaying theirs. Hence, just as with compositional changes in communities in response to warming, phenological changes within the community appear to follow the Gleasonian paradigm of individualistic shifts. The implications for exploitation interactions of individualistic phenological responses to climate change among species at different trophic levels are the focus of the next several sections in this chapter. First, however, a brief reevaluation of the notion of resource availability is warranted.

THE GREEN WORLD HYPOTHESIS, AND PHENOLOGY AS AN INDEX OF RESOURCE AVAILABILITY

The potential contribution of advances in the plant growing season or spring emergence of invertebrates to life history variation in resident and migratory bird populations was recognized in the early reports of advances in egg-laying dates and amphibian spawning mentioned above (Beebee 1995; Crick et al. 1997; Forchhammer, Post, et al. 1998; Crick and Sparks 1999). It is to our advantage as ecologists to view phenology as an index of resource availability for the species consuming those whose phenology is under observation. I adopted this view early in my own career when, as a doctoral student, I initiated a three-year project monitoring the onset and progression of the plant growing season on two adjacent calving ranges used by caribou in Alaska and a third one in Greenland (Post 1995). Concurrently, on each calving range, I collected data on the timing and progression of the caribou calving season and offspring production in each population (Post et al. 2003).

My initial interest in collecting these data was to investigate the role of the timing of resource availability in offspring production in a migratory herbivore. According to the green world hypothesis that grew out of the seminal work by Hairston, Smith, and Slobodkin (1960), herbivores should not be limited by forage availability because vegetation is superabundant. In my opinion, the ensuing decades-long debate ignited by the landmark HSS paper (Hairston et al. 1960) over top-down versus bottom-up limitation of herbivore populations has been generally misguided. The question lies not in whether herbivores are limited by forage availability or by predation but rather, in the case of forage limitation, whether such limitation is due to the quantity (biomass) of vegetation or its window of availability as determined by plant phenology. The world may be green, but the period during which vegetation is of peak nutritive value must also be taken into consideration in discussing the role of forage availability in limiting herbivore abundance, especially in highly seasonal environments. In such environments, digestibility and the nutrient content of plant tissues decline rapidly with progression of phenology and the growing season (Klein 1990).

The relevance of phenology as an index of resource availability lies mainly in the application of this perspective to understanding the consequences of climate change for consumers. In the case of the timing of onset and progression of the plant growing season, there may be a strong climatic signal in interannual variation and longer-term trends in plant phenology, but not necessarily in the phenology of primary consumers. Similarly, there may be considerable climatically driven interannual variation and trends in the emergence dates of invertebrate species that serve as the prey base for their predators but a weaker to nonexistent climate signal in the phenology of those predators.

As an example of the former, we may consider multi-annual data on the onset and progression of the plant growing season collected at the Kangerlussuaq study site in Greenland since 1993 (figure 6.9). These data were obtained from permanent monitoring plots observed daily or near daily beginning in mid- to late May each year. I have expressed the data in figure 6.9 as the mean proportion of species emergent in relation to the day of the year. To obtain these estimates, we work backward from the final number of species recorded as emergent on all plots at the end of the sampling period in each year to calculate the daily proportion of that final number emergent on each day of observation prior to the final day (Post and Klein 1999). At first glance, it is immediately apparent that there is considerable variation among years in the timing of onset of the plant growing season and the rate at which it progresses. In fact, the date of 5 percent emergence varied by approximately nine days during the first seven years of this study (Post and Forchhammer 2008), but now varies by approximately two weeks (figure 6.9). In the Arctic, where the growing season is short and progresses rapidly, such variation represents a considerable challenge for herbivores interested in capturing the period of peak nutritive value of the plants.

ASYNCHRONY AND TROPHIC MISMATCH

In contrast to the reports summarized above of numerous species of migratory and resident birds migrating earlier and laying eggs earlier in association with

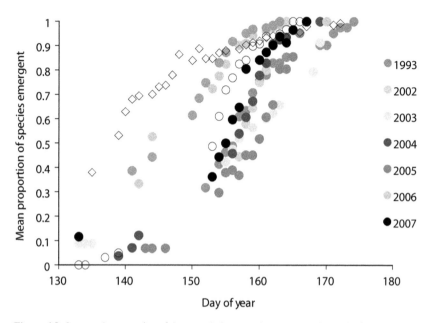

Figure 6.9. Onset and progression of the annual plant growing season at the study site near Kangerlussuaq, Greenland. Data were derived from monitoring permanent plant phenology plots on which the number of species emergent is recorded on a daily or near daily basis. The onset of the growing season, quantified as the date on which 5 percent of species are emergent, best correlates, to date, with mean April temperature (Post and Forchhammer 2008; Post et al. 2008b), though data on the timing of snowmelt, an important driver of the onset of the growing season elsewhere in the Arctic (Høye et al. 2007), are not recorded at the Kangerlussuaq site.

rising spring temperatures, a report surfaced in 1998 asserting that a population of great tits (*Parus major*) breeding in the Hoge Veluwe in the Netherlands had not advanced its mean date of egg laying over the same period, from 1973 to 1995 (Visser et al. 1998). In the same location, however, the date of peak biomass of the main prey of great tits in this population, caterpillars, had advanced over the same period by approximately ten to twelve days, and displayed a close association with mean spring temperature: warmer springs were characterized by earlier dates of peak caterpillar biomass and, presumably, earlier dates of peak resource availability for great tits (Visser et al. 1998). This seminal paper demonstrated that phenological advancement with warming may not be equivalent among all phases of the breeding season, and warned of the potential for the development of a mismatch between the timing of reproduction and peak resource abundance, as well as the potential consequences of this trophic mismatch for population viability.

A thorough exposition of the history of development of the match-mismatch hypothesis and its evolution into the trophic mismatch hypothesis has been given elsewhere (Kerby et al. 2012), but certain points related to the development of this idea are worth emphasizing. Even though their analysis included documentation of numerous examples of birds advancing their laying dates between the early 1970s and the mid-1990s in the United Kingdom, Crick and colleagues (1997) also cautioned that "birds may be adversely affected if they become unsynchronized with the phenology of their food supplies." Such foresight seems almost prescient in hindsight, especially in light of the data reported by Visser and colleagues (1998) the following year. The results presented by Visser and colleagues (1998) may seem incongruent with earlier and subsequent reports of advancing laying dates in other European populations and species, but it is important to keep in mind that all of the studies described above also mentioned populations in which observers detected either no trend in laying dates or no relation between laying dates and temperatures. Hence, the potential for trophic mismatch in other populations certainly existed at the time of publication of the example from Hoge Veluwe. McCleery and Perrins (1998), for instance, reported an advance in egg-laying dates of great tits in a population in the United Kingdom from 1971 to 1997, but this trend represents a relationship in only a subset of the entire time series of phenological data at their disposal. Over the entire period during which data on the timing of egg laying by great tits at Marley Wood near Oxford, United Kingdom, had been collected, from 1947 to 1997, there was no apparent trend in the data (McCleery and Perrins 1998). Similarly, for the subset of data acquired prior to 1971, there was no trend in egg-laying date in this population. Correlations between mean laying date and a warmth sum, calculated as the sum of the maximum daily temperatures from March 1 to April 25 annually, did not differ between the two periods. This indicates both that the apparent influence of spring temperatures on laying dates has not changed with warming and that the lack of a trend in egg-laying dates in the earlier period and overall can likely be attributed to the lack of a trend in warmth sum both overall and during the window period 1947–70 (McCleery and Perrins 1998).

As well, the value of the contribution of the development of the trophic mismatch hypothesis to the field of climate change ecology cannot be overstated. It stands as a prime example of the subtle, indirect consequences of climate change that may easily go undetected but that ultimately may have an important role in the response of populations to rapid climate change. The hypothesis has embodied multiple definitions throughout its evolution (Kerby et al. 2012) but can most simply be described as an asynchrony between the timing of offspring production and the timing of peak resource availability that may have adverse consequences for offspring production. Trophic mismatch, then, moves the consideration of phenology from something that is regarded as a response to climate change, or an index thereof, to something that can be regarded as a driver and integrator of ecological response to climate change or lack thereof (Post and Inouye 2008). It also urges us to regard phenology as an index of resource availability in time and space, rather than simply as the timing of life history events that bear only on the ecology of the species whose phenology is under investigation.

A powerful visualization of the role of phenology in resource dynamics serves to illustrate this latter point. Let us assume that resource availability for a consumer is closely associated with the phenology of its forage species. Then seasonal resource availability can be viewed as a function of time, with, in seasonal environments, low availability of the resource during the nonproductive part of the year and a clearly discernible peak in availability during the productive part of the year (figure 6.10a, dashed line). Resource availability varies through time, with the beginning of the period of annual productivity characterized by negligible to low availability of the resource, but as the season progresses, invertebrate prey or plant forage species begin to emerge slowly and then more rapidly until peak emergence is achieved before it tails off again. Optimally, the timing of offspring production by the consumer of this resource should overlap the resource availability curve (figure 6.10a, solid line) so that the period of peak resource demand associated with rearing offspring coincides with the peak of the period of resource availability (Both et al. 2009).

As we have seen, warming may advance the emergence phenology of invertebrates and plants, and we can visualize this in figure 6.10b as movement of the resource phenology schedule and its peak toward the left along the x-axis. If the reproductive phenology of the consumer species is similarly responsive

Figure 6.10. Development of a mismatch between the timing of resource availability (dashed line) and consumer demand during the breeding season (solid line). (a) The onset and peak timing of resource availability are matched closely by the onset and timing of the peak demand for resources needed to provision offspring. The shaded area indicates the frequency of individuals reproducing successfully. (b) Warming has elicited an advance in the timing of the onset and peak availability of resources to a greater degree than the timing of resource demand, which is less sensitive—or potentially insensitive—to temperature change. The resulting trophic mismatch has reduced the frequency of successful reproduction. *Adapted from Both et al. (2009).*

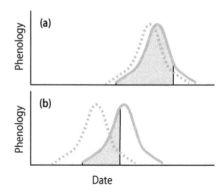

to temperature cues, as in the case of amphibians, the timing of onset and peak reproductive activity may similarly track this advance in the timing of onset and peak availability of resources. If, however, the timing of reproduction in the consumer species is not temperature dependent, or if it is only weakly temperature dependent, the warming-driven advance in the timing of resource availability may not be matched by a comparable advance in the timing of offspring production by the consumer. This is represented in figure 6.10b as a developing asynchrony between the peak of resource availability, which has advanced with warming, and the timing of offspring production, which has not advanced to the same extent, resulting in a trophic mismatch such as that documented for great tits in the Netherlands (Visser et al. 1998).

Such a mismatch may be compensated for in highly mobile consumers by capitalization on spatial heterogeneity in the timing of onset and peak resource availability. As we saw earlier in this chapter, there is evidence from studies conducted at large spatial scales that variability across space in phenological events such as initiation of the plant growing season increases with warming. We can think of this variability across the landscape in the timing of resource availability as an insurance policy against restrictions in the timing of resource availability at any one site. For instance, a consumer whose success in provisioning offspring is closely linked to synchronizing offspring production with the peak timing of resource availability may be able to compensate, to some extent, for the mismatch depicted in figure 6.10b by expanding its foraging horizon upslope to include sites where resource phenology is delayed in comparison to lower-elevation sites. Large herbivores, for instance, may migrate elevationally in spring to maintain intake of newly emergent, highly nutritious plant material as the growing season advances (Thing 1984; Skogland 1989; Albon and Langvatn 1992; Hebblewhite et al. 2008). Benefits to offspring provisioning and survival would accrue from such a response because elevational migration, or any use of landscape heterogeneity that maintains exposure of the consumer to resources at or near their peak of availability, would allow the consumer to compensate spatially for temporal constraints on resource availability.

This can be visualized by considering a series of resource dynamics curves separated along axes of time and space (figure 6.11). In this scenario, the timing of peak resource availability is designated on each of the three resource dynamics curves as "a," "b," and "c," respectively. Given the separation of these resource peaks along the temporal axis, the consumer will maximize resource intake by moving from one peak to the next as each curve begins its decline, or, if the consumer misses one peak, it can utilize variation along the space axis to arrive at the next site in time to realize its resource peak.

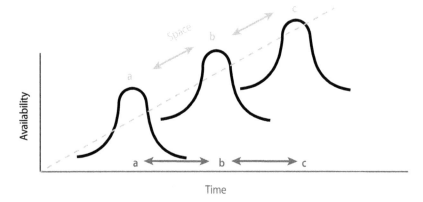

Figure 6.11. The alteration of resource availability in time and space by climate change as visualized for three peaks of resource availability (labeled a, b, and c). Along the x-axis, the three resource curves may represent resource availability through time as three species of forage plants emerge, reach peak nutritive content, and then mature. The differing phenological schedules of these three plants may be altered by climate change as they either advance or delay their emergence in response to warming. If all three peaks advance, the consumer species will experience a temporal constriction of the period of peak resource availability. Along the z-axis, the three resource curves are segregated spatially, as is the case in which a single or multiple species occur along an elevational gradient and display different emergence phenologies as a consequence. In this situation, warming may elicit an advance in the resource phenology curves at all three sites, constricting the spatial heterogeneity in the timing of peak resource availability among them. This latter scenario represents a spatial component of trophic mismatch.

In theory, the strategy of using space to compensate for constraints in time should be operable at any spatial scale over which the consumer is capable of foraging, but is, on the basis of the studies described above, likely to be more effective at larger spatial scales. This is because warming may not only advance resource phenology, it may also reduce spatial variability in resource phenology. This is depicted in figure 6.11 as the movement of the three resource dynamics curves along the space axis. Ultimately, for the consumer, the most detrimental situation may be one in which resource phenology not only advances along the temporal axis until all three resource peaks occur on or near the same date but also compresses them along the space axis so that resource phenology becomes more synchronized across the landscape.

There is some, though limited, observational and experimental support for the scenario of spatial compression of resource phenology. I conducted a two-year experiment at my field site near Kangerlussuaq that was designed to test the hypothesis that warming increases spatial variability in plant phenology. The warming experiment was conducted using open-top chambers (OTCs)

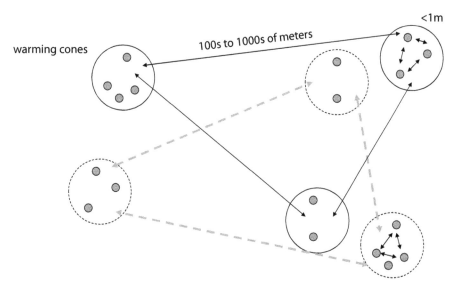

Figure 6.12. Design of the experiment used to investigate the consequences for spatial hetero-geneity in plant phenological dynamics of warming. Large solid circles represent open topped warming chambers, or cones, and large dashed circles the associated control plots. These were distributed throughout the study site at Kangerlussuaq, Greenland, at distances of hundreds of me-ters to just over 1 km from each other. The timing of plant phenological events was observed and compared within plots and among them to determine whether the increase in spatial variability in plant phenological events observed in other studies conducted at the scale of thousands of square kilometers was mirrored at smaller scales relevant to the foraging horizon of mammalian herbi-vores. In fact, at both of these smaller scales, the opposite pattern was detected (Post et al. 2008b).

distributed across the landscape at a scale of approximately 3 km² (figure 6.12). All plant phenological events were recorded within the OTCs and adja-cent control plots, and the variability in their timing was compared within and among all OTCs, that is, at scales of less than 1 meter up to kilometers. Warm-ing significantly reduced the spatial variability in phenological events both within and among the warming chambers compared to the control plots (Post et al. 2008b). Moreover, several years of observations of plant phenological dy-namics on a set of permanent plots distributed across the landscape at the same spatial scale revealed a compression of spatial variability in plant phenology in association with observed spring warming, corroborating the results of this experiment (Post et al. 2008b). An association between spatial variability in plant phenology and offspring production by caribou at the study site suggests that compression of resource availability in space by warming may contribute to reduced offspring production in this species, although this correlation was significant only after taking into account the temporal mismatch between timing of offspring production and plant phenology (Post et al. 2008b). Nonetheless, these observations suggest that resource compression in time associated with advancing resource phenology may place constraints on offspring production and survival if advances in consumer reproductive phenology do not keep pace with those of their resources.

As noted earlier, quantifying the contribution of phenology to demographic variation and population dynamics has been a substantial challenge to date (Miller-Rushing et al. 2010), but should be an obvious objective of trophic mismatch studies. Populations of pied flycatchers (*Ficedula hypoleuca*) in the Netherlands suffered population declines of up to 90 percent between 1987 and 2003 (Both, Bouwhuis, et al. 2006). During the same period, flycatchers had not undergone an advance in their timing of arrival at spring breeding grounds in the Netherlands but had advanced their egg-laying dates, though not to the same extent that timing of peak availability of their caterpillar prey had advanced in response to warming on site (Both, Sanz, et al. 2006). Both and colleagues (2006) demonstrated that in populations of flycatchers breeding in areas where the date of peak caterpillar abundance had advanced the most, flycatcher populations had declined by approximately 90 percent. In contrast, those flycatcher populations breeding where the date of caterpillar abundance had advanced the least had declined by approximately 10 percent (Both, Sanz, et al. 2006). That this pattern does not relate to a general deterioration in food supply is suggested by the fact that caterpillar biomass was greatest in association with earliest dates of peak abundance (Both, Sanz, et al. 2006). Hence, Both, Sanz, and colleagues (2006) demonstrated an association between population declines and trophic mismatch related to warming, although the mechanism underlying the declines in flycatcher populations was not identified.

Plant phenology at my study site near Kangerlussuaq, Greenland, displays considerable interannual variation in the timing of onset and progression of the growing season. This variation is evident in a plot of the proportion of species emergent on daily or near-daily basis on permanent phenology monitoring plots over all of the years of study to date (figure 6.13a). The estimated date of 5 percent emergence, or the date on which 5 percent of species have emerged, which is used in this case as a measure of the timing of the onset of the plant growing season at the site, averages day 138.1 over the period covered by the data (figure 6.13a), while the date of 5 percent births in the caribou population averages day 152.7 over the period covered by those data (figure 6.13b), 1977–2010. Data on the timing of calving by caribou in this population have been collected concurrently with phenology data beginning in 1993, and annually

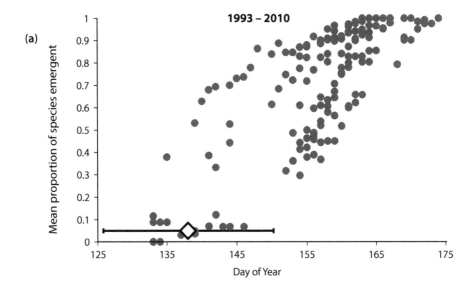

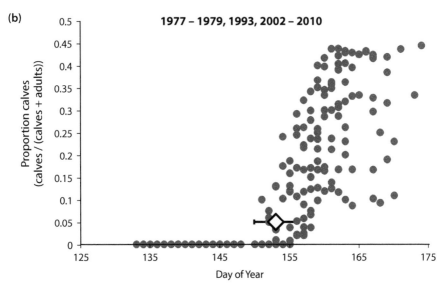

Figure 6.13. Data on the timing of onset and progression of the plant growing season at the study site near Kangerlussuaq, Greenland (a), and the onset and progression of the caribou calving season at the same site (b) indicate considerably greater interannual variation in the timing of the annual growing season than in the timing of the caribou calving season. The diamonds in each panel indicate the mean (± 1 SE) date of 5 percent emergence in panel (a) and 5 percent of births in panel (b). *Data on caribou calving from 1977–1979 from Thing and Clausen (1979).*

since 2002. Additional data on the timing of calving in the same population at the same study site were collected from 1977 to 1979 by Thing (1984). What is striking about these two sets of data is the amount of variation among years in the timing of onset of the plant growing season (12.2 days) compared to the variation among years in timing of onset of the caribou calving season (3.1 days), a fourfold difference. Whereas plant phenology is highly variable among years, caribou calving phenology apparently varies only slightly. The difference reflects the fact that while plant phenology is acutely sensitive to variation in local conditions, the reproductive cycles of seasonally breeding mammals are regulated by changes in photoperiod, and therefore are little sensitive to changes in their immediate environment (Goldman 2001; Lincoln et al. 2003). Thus, caribou display the typical strategy for northern ungulates in which mating is timed to ensure that calving coincides with what on average are the most favorable conditions for supporting the costs of lactation and offspring provisioning (Skogland 1989). Notably, this species has evolved unusually strong photoperiodic regulation of annual cycles of physiology and behavior (van Oort et al. 2005, 2007; Lu et al. 2010) precisely to inure individuals against the vagaries of northern climate, but by the same token, individuals cannot easily track directional environmental change.

During the years over which the data are continuous, 2002–10, the timing of the plant growing season has displayed a significant ($r = -0.61$) trend toward earlier onset (figure 6.14a). Over the same period, caribou calving has displayed a much weaker, nonsignificant ($r = -0.36$) trend toward earlier onset (figure 6.14b), while caribou offspring production has declined significantly ($r = -0.61$) (figure 6.14c) and offspring mortality has increased significantly ($r = 0.68$) (figure 6.14d). The index of trophic mismatch we developed earlier relates to the resource state at the time of calving and is calculated as the proportion of forage species observed to be emergent midway through the caribou calving season (Post and Forchhammer 2008). This index is closely related to both the onset ($r = -0.55$) and progression ($r = -0.76$) of the plant growing season; as plant growth begins earlier and the growing season reaches its midpoint earlier, the trophic mismatch for caribou is exacerbated. There is a negative association between caribou calf production and this index of trophic mismatch (figure 6.15a), and a weaker, positive association between caribou calf mortality and the index of trophic mismatch (figure 6.15b) (Post and Forchhammer 2008; Post et al. 2008b). Although correlative, there appears in this instance to be a series of associations between rising spring temperatures, earlier plant phenology, increasing trophic mismatch, and declining offspring production and increasing offspring mortality by caribou at the study site.

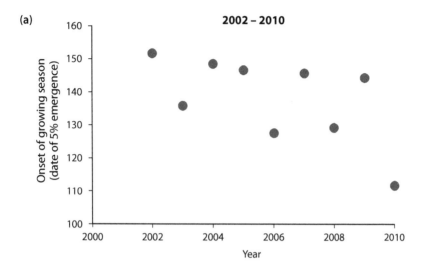

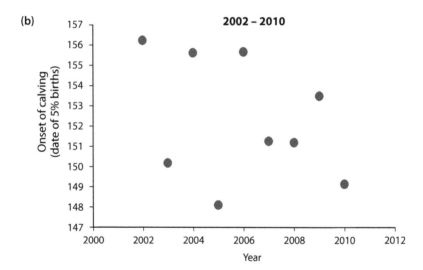

Figure 6.14. Advancement of the onset of the plant growing season (a) at the study site near Kangerlussuaq, Greenland, and the lack of such advance in the onset of the caribou calving season (b) at the same site over the same period. Although the timing of the annual season of plant growth at the study site correlates closely with mean April temperature, the timing of the caribou calving season does not. This disparity promotes trophic mismatch between caribou and their forage. Over the same period, offspring production by caribou at the study site has declined steadily (c), while offspring mortality has increased (d), possibly as consequences of trophic mismatch related to recent warming at the site described in chapter 1. *Adapted from Kerby and Post (in review).*   (*continued*)

(c)

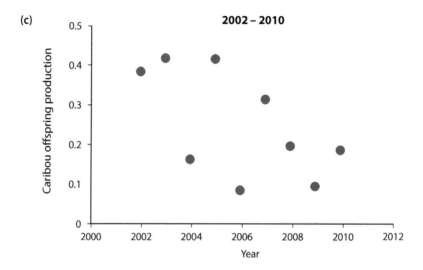

(d)

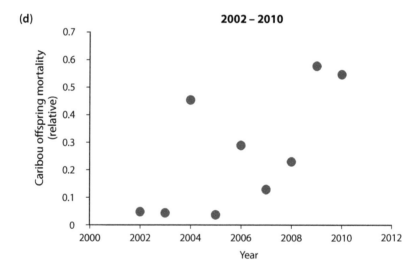

Figure 6.4. (*continued*)

THE CAFETERIA ANALOGY OF TROPHIC
MISMATCH IN TIME AND SPACE

Although imperfect, the cafeteria analogy of trophic mismatch in time and space has proven effective in my classroom in conveying the concept of trophic mismatch as it applies to mistiming, as originally developed, and to the

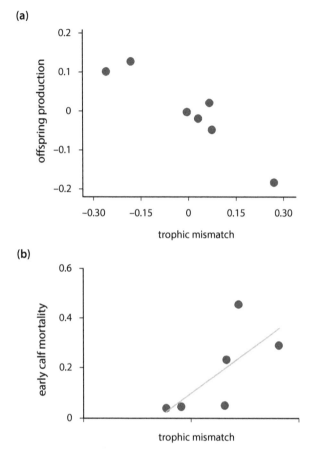

Figure 6.15. Offspring production by caribou at the study site near Kangerlussuaq, Greenland, declines in association with increasing trophic mismatch (a) between the timing of the onset of plant growth and the timing of caribou calving. As trophic mismatch increases, so does early, neonatal offspring mortality (b). Quantification of the index of trophic mismatch and offspring mortality is described in detail elsewhere (Post and Forchhammer 2008; Post et al. 2008b).

mismatch between consumer and resources deriving from spatial dynamics in resource phenology. Imagine there is a cafeteria at your workplace with fixed opening hours centered on the lunch period. You, like many other employees at this workplace, are allowed by your employer to leave for lunch between 11:30 and noon each day. Because the cafeteria is open between 11:00 and 13:00, there is plenty of time for you to arrive there and obtain a full meal. Now imagine that, for some reason, the cafeteria changes its opening hours so that it serves food only from 10:00 until 11:30 each day, and then begins to close slightly earlier each day. Unfortunately for you, your employer has not changed the timing of your lunch break because it is unaware of this change or because it has optimized the lunch break of its employees according to other criteria.

At first, you may be able to catch the very tail end of the cafeteria's new opening hours, but even then you are unlikely to be able to obtain a full lunch under these conditions, which represent a developing trophic mismatch in time, but eventually you find yourself completely out of synch with the new, earlier and shorter opening hours. Fortunately, you know of several other cafeterias or restaurants on the block where you work, all of which have different opening hours, and so, by capitalizing on this variability in space, you are able to maintain resource consumption during the period of your resource demand. But next you discover that, for reasons unknown to you, all of the cafeterias in the neighborhood have begun to operate under the same advanced opening hours as the one in your workplace, as well as having begun to shorten their opening hours simultaneously by closing earlier. Under these conditions, the spatial heterogeneity in resource availability you had been able to utilize to compensate for temporal constraints on resource availability has been reduced. This, of course, represents the developing spatial mismatch.

GLEASONIAN DYNAMICS AND STABILITY IN LATERALLY STRUCTURED COMMUNITIES

Good evidence for a Gleasonian response of laterally structured communities to climate change derives from two studies of changes in the composition of communities of small mammals and birds in California. In the first example, surveys of the small mammal fauna of Yosemite National Park in California were conducted by Grinnell between 1914 and 1920 as part of a more comprehensive assessment of the distribution and species diversity of vertebrates in California (Moritz et al. 2008). As a baseline for comparison with subsequent surveys, the Grinnell data are a potential gold mine for investigations of species' responses to climate change, changes in human land use, and other stressors. In Grinnell's Yosemite mammal survey, data were collected along elevational gradients on mountain slopes within the park, rendering the data especially valuable for examining the effects of climate change on distributional shifts of species along such gradients.

Grinnell's transects were resurveyed in the early twenty-first century, following a 3.7°C increase in the mean minimum monthly temperature for the region (Moritz et al. 2008). A comparison of Grinnell's data with those collected by Moritz and colleagues (2008) that took into account differences in detectability among species revealed broad changes in the distributions of species composing the small mammal community. Of the twenty-eight species surveyed, six displayed range expansions, and four of these six expansions were due to upward shifts in elevation at the upper limits of their distributions without compensatory upward shifts at the lower limits of their distributions (Moritz et al. 2008). Ten species displayed range contractions, and of these, six contractions were due solely to upward shifts at the lower limits of the species' elevational distributions, while the other four were due to downward shifts at the upper edges and upward shifts at the lower edges of their distributions (Moritz et al. 2008). The remaining twelve species displayed range stasis.

The results of the analysis by Moritz and colleagues (2008) clearly illustrate individualistic species-level responses to climate change, and suggest, furthermore, the potential for development of non-analogue communities as a result of such Gleasonian dynamics. For instance, at least two species with previously nonoverlapping elevational distributions within Yosemite National Park during Grinnell's surveys, the bushy-tailed woodrat and the pinyon mouse, are now found to overlap elevationally owing to an upward expansion of the distribution of the latter. By contrast, the elevational distribution of the bushy-tailed woodrat no longer overlaps that of the alpine chipmunk because of contractions downward by the former and upward by the latter (Moritz et al. 2008). These examples illustrate the complexity of species-specific distributional responses to climate change in laterally structured communities and their implications for community composition.

In the second example, the distributions of sixty species of birds that breed in California were compared with their projected future distributions, derived from species distribution models (Stralberg et al. 2009). The approach taken in this study differs from that in the former study in obvious ways. Primary among these differences is that the latter study is based not on a comparison of historical and contemporary data but rather on a comparison of contemporary data with inferred distributional data. As well, the latter example assumes future climatic conditions, whereas the former inferred that the observed faunal distributional changes related directly to observed changes in climate during the interval between successive sampling efforts. Both approaches have obvious disadvantages. In the case where observed changes in species' distributions are assumed to relate to climatic changes having occurred between sampling periods, other potential contributing factors may have been overlooked, including human population growth, disturbance from increased human activity, human land-use changes, or extinction of predators or competitors that influenced the distributions of focal species, to name a few. In the Yosemite example, factors related to human population growth and land use can probably be ruled out because of the protected status of the park (Moritz et al. 2008).

Species distribution models, also known as environmental niche models or bioclimate envelope models, utilize information on relationships between current presence-absence data for a given species and environmental data such as temperature and precipitation means overlapping species distributions. As we will see in the next chapter, relationships between temperature and/ or precipitation and the presence or absence of a species are applied together with predicted temperature and precipitation regimes under different climate change scenarios to project the distribution of a species under those climatic conditions. A majority of such studies to date have detected range contractions, rather than range expansions or stasis (Stralberg et al. 2009). An obvious compromise that must be accepted in using this approach is the fact that interactions among species that influence the distribution of a given focal species are not easily captured by abiotic data such as temperature and precipitation. As well, uncertainty in climate projections further complicates the prospects for accurately or meaningfully projecting species distributions based solely on climate data.

Stralberg and colleagues (2009) used species distribution models to infer changes to the breeding distributions of sixty species of birds after six decades of projected climate change. In this case, information on temperature and precipitation, as well as that on vegetation and land cover, was used to build distribution models for the focal bird species (Stralberg et al. 2009). To project future distributions, output from two climate models assuming medium to high CO_2 emissions scenarios was employed in two species distribution models. Regardless of the climate model or species distribution model used, in all cases the number of species exhibiting declining distributions exceeded the number exhibiting increasing distributions (Stralberg et al. 2009). The authors concluded that following nearly a century of expected climate change, approximately one-half of California may by occupied by non-analogue avian communities (Stralberg et al. 2009), supporting, once again, the Gleasonian model of individualistic species' responses to climate change as a factor contributing to the dynamics and composition of laterally structured communities.

The consequences of climate change for stability in laterally structured communities are likely to derive from the influences of environmental variability on interference interactions. As we saw in the chapter 4, environmental variability has the potential to promote instability at the population level if it is of sufficient magnitude to override the stabilizing influence of density dependence. At the community level, this influence may confer stability in laterally structured communities by preventing dominance of the community by one or a few competitively superior species, as we saw in the flycatcher case study earlier in this chapter. In another example of this, Adler and colleagues (2006) examined thirty-one years of continuous data on the basal cover—an index of abundance—of three species of perennial grasses within permanent livestock exclosures established during the Great Depression in the U.S. state of Kansas (Albertson and Weaver 1944a, 1944b). During the census period, the three focal species of grass, *Bouteloua curtipendula*, *B. hirsuta*, and *Schizachyrium scoparium*, which interfere with each other through competitive interactions (Adler et al. 2009), represented over 95 percent of the basal cover of the plant community, but fluctuated in relative abundance (Adler et al. 2006). As with the co-occurring species of congeneric flycatchers, the effects of climatic fluctuation on growth rates of these grass species varied widely among the three grass species. Estimates derived from empirically based simulation modeling revealed, however, that growth rates at low densities were considerably higher in climatically variable scenarios than in climatically stable scenarios (Adler et al. 2006). Moreover, the growth rates of two of the species were negative under constant climatic conditions, whereas the growth rate of the third species was positive under constant climatic conditions (Adler et al. 2006). These results bolster the hypothesis that climatic variability promotes stability in lateral communities in which interference interactions are important determinants of productivity and population growth in the constituent species (Adler et al. 2006). The opposite, as we will see, appears to be the case in vertical communities structured by exploitative species interactions.

DYNAMICS AND STABILITY IN VERTICALLY STRUCTURED COMMUNITIES

In a vertically structured community, species interact across trophic levels, and understanding the factors that convey or erode stability in them requires taking interactions among species at these different trophic levels into consideration. Originally, stability in vertical communities was described as occurring in relation to food web dynamics (MacArthur 1955). Here I would again draw a distinction between communities and food webs. Whereas a community should be viewed as the collection of species interacting predictably in space and time, a food web is a generalized depiction of the flow of energy through a system, without regard to the species in the food web. Early views of stability in vertical communities related stability to "the amount of choice which the energy has in following the paths up through the food web" (MacArthur 1955). In such a scenario, monopolization of the community by a superabundant species should erode stability; by contrast, a diversity of pathways for the flow of energy should confer stability on the community (MacArthur 1955). This is the origin of the stability-complexity debate that has found a central place in contemporary ecology (McNaughton 1977; Tilman and Downing 1994; Doak et al. 1998; Levine and D'Antonio 1999; Loreau et al. 2001; Hooper et al. 2005; Tilman et al. 2006). Almost simultaneously with MacArthur's landmark paper, Elton laid out the reasoning that supported MacArthur's theory. He observed that simple systems, in which one species preys on another, are almost always vulnerable to wild fluctuations (Elton 1958). The potentially destabilizing influence of population-extrinsic processes is recognizable in Elton's remark that such conspicuous fluctuation in simple systems was likely "even without shocks from outside like the vagaries of climate." The simplest communities, according to Elton, were inherently vulnerable to periodic outbreaks of one of the species they comprise, and almost never displayed constancy of numbers (Elton 1958).

As was demonstrated in chapter 4, there is evidence of interactions between stabilizing density dependence and potentially destabilizing density independence in population dynamics in many different types of systems. In this section, I address community-wide analyses of population dynamics in relation to climate change and attempt to place these in the context of community dynamics using a theoretical approach to estimating community stability. Predictions arising from these examples may have important implications for the persistence of communities in a changing climate. In a focal example used in this chapter, a plot of the solution of the community matrix in the complex plane for stochastic environments indicates that the community of interest in this case displays stability. This stability is likely to erode, however, with increasing abiotic contributions to dynamics at any single trophic level in the community because, as will be demonstrated later in the chapter, the solution of the community matrix falls at the boundary of stability. The loss of community stability in this case would likely owe to the deterioration of self-regulation at the trophic level displaying the weakest density dependence in the community, suggesting that the stability of vertically structured communities may be altered by climate change in two ways: directly, by disrupting self-regulation at individual trophic levels, and indirectly, by increasing the strength of exploitation interactions relative to self-regulation at one or more trophic levels.

Modeling the stability properties of populations and communities in relation to endogenous and exogenous processes has greater promise for elucidating the implications of global change for the persistence of populations and communities than does population dynamical modeling. The primary influence of environmental stochasticity on population dynamics, for example, is to move a population from deterministic variation about some equilibrium point to random variation about some equilibrium probability distribution (May 1973a, 1973b). As we saw in the previous chapter, if stochastic environmental variation exceeds the stabilizing influences of population interactions, such as density dependence, the population will tend toward instability. Climatic variability influences the dynamics of many species in a rarely consistent direction across or within species (Forchhammer, Post, et al. 1998; Post and Stenseth 1999; Root et al. 2005; Anders and Post 2006; Høye et al. 2007; Rosenzweig et al. 2008; Post, Brodie, et al. 2009). Hence, predicting the implications of climate change for vertically structured communities, in which the primary species interactions are exploitative, requires not necessarily determining whether climatic variation influences changes in density positively or negatively at any individual trophic level but rather determining the magnitude of such effects relative to the stabilizing influences operating within and between populations at adjacent trophic levels.

DEVELOPMENT OF THE PROCESS-ORIENTED MODEL FOR VERTICAL COMMUNITIES

Typically, process-oriented models are used to predict how interactions between adjacent trophic levels in two- or three-level systems will manifest in the dynamics at one of these levels as delayed density dependence (Bjørnstad et al. 1995; Stenseth, Bjørnstad, and Falck 1996; Stenseth et al.1997; Forchhammer, Stenseth, et al. 1998; Stenseth, Chan, et al. 1998; Forchhammer and Asferg 2000). Statistical process-oriented models should, therefore, be amenable to discerning the influences of climate at one trophic level on dynamics at adjacent trophic levels.

If our interest lies in quantifying the influence of climate on community dynamics and stability in vertical communities, we can begin by building a model of community dynamics incorporating vertical interactions among species and the potential influences of climate at each individual trophic level (figure 6.16). A good starting point in building the community model is mathematical models of population dynamics at each trophic level that incorporate the simplest, yet potentially important, ecological interactions among species. By focusing on trophic interactions at the most immediate time lags, we can express the dynamics of the predator (P), herbivore (H), and vegetation (V) in terms of intraspecific processes and intertrophic level (interspecific) interactions with the following mathematical equations:

$$P_t = P_{t-1} \exp(f(X_{t-1}, Y_{t-1}) + \in_t^{(X)}), \qquad (6.1a)$$

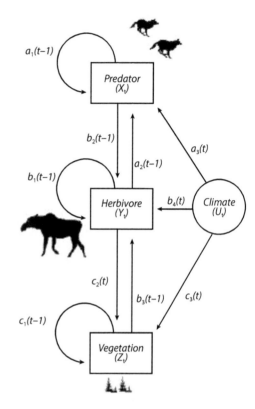

Figure 6.16. Conceptual model of the potential influence of climate change in the dynamics of a vertically structured community that also incorporates intrinsic dynamics at each individual trophic level and interactions among trophic levels. The coefficients deriving from the application of this model to time series analysis can be used to construct the community matrix for examination of its stability properties. *Adapted from Post and Forchhammer (2001).*

$$H_t = H_{t-1} \exp(g(X_{t-1}, Y_{t-1}, Z_{t-1}) + \in_t^{(Y)}), \qquad (6.1b)$$

$$V_t = V_{t-1} \exp(h(Z_{t-1}, Y_t) + \in_t^{(Z)}), \qquad (6.1c)$$

where P_i is predator density, H_i is herbivore density, V_i is vegetation growth, and X_i, Y_i, and Z_i are their respective $\log_e$ transformations. In equation (6.1c), the effect of the herbivore on vegetation is specified as a current-year effect, but is likely to operate at a lag of three to six months.

In practice, the functions $f(\bullet)$, $g(\bullet)$, and $h(\bullet)$ may assume a wide array of functional forms (May 1976). However, in estimating statistical density dependence, we may make the assumption that the functions are linear in X_t, Y_t, and Z_t. This is equivalent to the assumption that the growth rate of the population is linearly related to log-density (Bjørnstad et al. 1995; Stenseth, Bjørnstad, and Falck 1996; Stenseth, Chan, et al. 1998; Bjørnstad and Grenfell 2001; Bjørnstad et al. 2001). If we take $\log_e$ of both sides of equations (6.1), then, with Taylor expansion, we have:

$$X_t = a_0 + (1 + a_1)X_{t-1} + a_2 Y_{t-1} + \in_t. \tag{6.2a}$$

$$Y_t = b_0 + (1 + b_1)Y_{t-1} + b_2 X_{t-1} + b_3 Z_{t-1} + \in_t. \tag{6.2b}$$

$$Z_t = c_0 + (1 + c_1)Z_{t-1} + c_2 Y_t + \in_t. \tag{6.2c}$$

We can incorporate the potential direct influence of climate at each trophic level by specifying $\in_t = U_t + \varepsilon_t$, where U_t is a climate variable; then the mathematical trophic-level models can be rewritten as the following trophic-level ecological models:

$$X_t = a_0 + (1 + a_1)X_{t-1} + a_2 Y_{t-1} + a_3 U_t + \varepsilon_t. \tag{6.3a}$$

$$Y_t = b_0 + (1 + b_1)Y_{t-1} + b_2 X_{t-1} + b_3 Z_{t-1} + b_4 U_t + \varepsilon_t. \tag{6.3b}$$

$$Z_t = c_0 + (1 + c_1)Z_{t-1} + c_2 Y_t + c_3 U_t + \varepsilon_t. \tag{6.3c}$$

Though we have considered only the direct influence of climate (U) acting at each trophic level, climate may also exert an indirect influence on dynamics at the top two trophic levels through its influence on predator hunting efficiency and the susceptibility of herbivores to predation. On Isle Royale, for example, wolf pack size increases with winter snow depth, and larger packs kill more frequently (Post, Peterson, et al. 1999). In the exposition given here, the focus will remain on quantifying the direct influence of climate at each trophic level.

To develop statistical autoregressive models specifying the contributions of intrinsic (i.e., density-dependent) and extrinsic (i.e., trophic and climatic interactions) processes to the dynamics of each level, we can solve equations (6.3) in terms of the predator (X), the herbivore (Y), or the vegetation (Z).

DERIVATION OF THE PREDATOR-LEVEL STATISTICAL MODEL

Starting with the predator, it can be assumed that the dynamics at level X can be modeled in delay coordinates without including the effect of vegetation on herbivore dynamics (i.e., as a two-level system involving only predators and herbivores; note, below, that the community matrix shows an empty cell for the direct effect of vegetation on predator dynamics). Hence, for derivation of the statistical model at level X, we ignore the term $b_3 Z_{t-1}$ in equation (6.3b), and solve the herbivore and predator equations in terms of predator density. To achieve this, we start by isolating Y_{t-1} in equation (6.3a), and inserting the resulting equation into (6.3b), giving:

$$Y_t = b_0 + (1+b_1)\frac{1}{a_2}[X_t - a_0 - (1+a_1)X_{t-1} - a_3 U_t] + b_2 X_{t-1} + b_4 U_t. \quad (6.4a)$$

Let $t = t - 1$ in equation (6.4a), and insert (6.4a) back into equation (6.3a). Then

$$X_t = a_0 + (1+a_1)X_{t-1} + a_2\left[b_0 + (1+b_1)\frac{1}{a_2}[X_{t-1} - a_0 - (1+a_1)X_{t-2}\right.$$
$$\left. - a_3 U_{t-1}] + b_2 X_{t-2} + b_4 U_{t-1}\right] + a_3 U_t. \quad (6.4b)$$

By multiplying through and gathering terms, we arrive at a statistical model of predator dynamics in delay coordinates that derive from the interaction between predator and prey density, incorporating climatic effects:

$$X_t = a_0 + a_2 b_0 - a_0(1+b_1) + [(1+a_1)+(1+b_1)]X_{t-1}$$
$$+ [a_2 b_2 - (1+a_1)(1+b_1)]X_{t-2} + a_3 U_t \quad (6.4c)$$
$$+ [a_2 b_4 - (1+b_1)a_3]U_{t-1}$$

which, after re-designation of the coefficients, can be simplified to:

$$X_t = \phi_0 + (2+\phi_1)X_{t-1} + (1+\phi_2)X_{t-2} + \omega_1 U_t + \omega_2 U_{t-1} + \varepsilon_t. \quad (6.5)$$

By recalling the origin of the coefficients in equation (6.4c), we can infer the processes behind the direct and delayed density and climate terms in equation (6.5) (Post and Forchhammer 2001; Post et al. 2001). For example, the presence of significant delayed density dependence at the predator level must reflect either trophic interactions between predator and herbivore (coefficients a_2 and b_2 from equations (6.3a) and (6.3b)) or the interaction between direct density dependence at the predator and herbivore levels (coefficients $(1+a_1)$ and $(1+b_1)$ from equations (6.3a) and (6.3b)). As well, the significance of a delayed effect of climate at the predator level must reflect either the interaction between the influence of herbivore density on predator density and the influence of climate on herbivore density (coefficients a_2 and b_4 from equations (6.3a) and (6.3b)) or the interaction between direct density dependence at the herbivore level and the influence of climate on predator density (coefficients $(1+b_1)$ and a_3 from equations (6.3b) and (6.3a), respectively).

DERIVATION OF THE HERBIVORE-LEVEL
STATISTICAL MODEL

To solve the community model for the herbivore (Y) level in delay coordinates, we must include predator effects on herbivore dynamics as well as vegetation effects on herbivore dynamics (see figure 6.16; note also, as will be explained below, that the community matrix indicates that interactions both above and below the herbivore level influence herbivore dynamics). Because there is no direct interaction between predator and vegetation dynamics, however, we can solve the equations in terms of the herbivore by considering herbivore-vegetation and predator-herbivore interactions separately. Hence, we begin by solving the herbivore-vegetation equations in terms of the herbivore, temporarily ignoring the influence of the predator on herbivore dynamics (i.e., temporarily setting $b_2 X_{t-1} = 0$ in equation (6.3b)). Then we reinstate this effect and solve the herbivore and predator equations in terms of the herbivore. To pursue this approach, we first isolate Z_{t-1} in equation (6.3b), then insert this into equation (6.3c), let $t = t - 1$ in the resulting equation, and insert this back into equation (6.3b), producing an equation describing herbivore dynamics in delay coordinates that incorporates the influence of vegetation dynamics on herbivores:

$$Y_t = b_0 + b_3 c_0 - c_1 b_0 + (1 + b_1 + c_1) Y_{t-1} + (b_3 c_2 - c_1 - c_1 b_1) Y_{t-2}$$
$$+ b_4 U_t + (b_3 c_3 - c_1 b_4) U_{t-1}. \tag{6.6a}$$

To incorporate the influence of predator density, we now reintroduce the predator term to equation (6.6a), giving:

$$Y_t = b_0 + b_3 c_0 - c_1 b_0 + (1 + b_1 + c_1) Y_{t-1} + (b_3 c_2 - c_1 - c_1 b_1) Y_{t-2}$$
$$+ b_2 X_{t-1} + b_4 U_t + (b_3 c_3 - c_1 b_4) U_{t-1}. \tag{6.6b}$$

We then isolate X_{t-1} in equation (6.6b), insert this back into equation (6.3a), set $t = t - 1$ in the resulting equation, and insert this back into equation (6.6b), producing the bivariate model of herbivore dynamics in delay coordinates:

$$Y_t = b_0 + b_3 c_0 - (1 + c_1) b_0 + b_2 a_0 - (1 + a_1) b_0 - (1 + a_1) b_3 c_0 - (1 + a_1)(1 + c_1) b_0 +$$
$$+ [(1 + a_1) + (1 + b_1) + (1 + c_1) + b_3 c_2] Y_{t-1}$$
$$+ \left[\begin{array}{c} a_2 b_2 - (1 + a_1)(1 + b_1) - (1 + a_1)(1 + c_1) - \\ (1 + b_1)(1 + c_1) - (1 + a_1) b_3 c_2 \end{array} \right] Y_{t-2} + \tag{6.6c}$$
$$+ (1 + a_1)(1 + b_1)(1 + c_1) Y_{t-3} + b_4 U_t$$
$$+ [b_3 c_3 + b_2 a_3 - (1 + c_1) b_4 - (1 + a_1) b_4] U_{t-1} + [(1 + a_1)(1 + c_1) - (1 + a_1) b_3 c_3] U_{t-2}$$

By simplifying the coefficients in equation (6.6c), we can derive a statistical autoregressive model of herbivore dynamics that incorporates influences of predator and vegetation dynamics and climate (equation 6.7):

$$Y_t = \beta_0 + (3 + \beta_1) Y_{t-1} + (2 + \beta_2) Y_{t-2} + (1 + \beta_3) Y_{t-3} + \omega_1 U_t$$
$$+ \omega_2 U_{t-1} + \omega_3 U_{t-2} + \varepsilon_t. \tag{6.7}$$

The autoregressive structure of this model is not changed if we specify a one-year-delayed effect of herbivores on vegetation (Forchhammer, Stenseth, et al. 1998; Post and Forchhammer 2001), though the composition of the coefficients will be slightly different. As described above in the derivation of the predator-level model, we can, by tracking the coefficients from equations (6.3) to equation (6.6), discern the ecological processes behind the autoregressive structure in the time series analysis of the herbivore level with equation (6.7).

DERIVATION OF THE VEGETATION-LEVEL STATISTICAL MODEL

To solve the equations for the vegetation level, we can ignore the effect of predators on herbivore dynamics (see figure 6.16; note also the empty cell in the community matrix below, indicating no direct effect of predators on vegetation dynamics). Hence, we focus here on solving equations (6.3b) and (6.3c) in terms of Z_t, ignoring the term $b_2 X_{t-1}$ in equation (6.3b).

We isolate Y_t in equation (6.3c), insert this into equation (6.3b), set $t = t - 1$ in the resulting equation, and insert this back into (6.3c), giving:

$$Z_t = c_0 + c_2 b_0 - (1 + b_1)c_0 + [(1 + b_1) + (1 + c_1) + c_2 b_3]Z_{t-1}$$
$$- (1 + b_1)(1 + c_1)Z_{t-2} + (c_2 b_4 + c_3)U_t - (1 + b_1)c_3 U_{t-1} \quad (6.8)$$

Re-designating the coefficients gives:

$$Z_t = \gamma_0 + (2 + \gamma_1)Z_{t-1} + (1 + \gamma_2)Z_{t-2} + \omega_1 U_t + \omega_2 U_{t-1} + \varepsilon_t. \quad (6.9)$$

As described above for the predator- and herbivore-level models, we can track the coefficients from equation (6.3c) to equation (6.8) to draw inferences about the ecological processes behind the autoregressive patterns in the vegetation time series. As with the equation for the herbivore level, the autoregressive structure of the model at the vegetation level is not changed if the herbivore effect on vegetation dynamics occurs at a lag of one year, though the composition of the coefficients will differ (Post and Forch-hammer 2001).

Application of time series analysis using the equations derived above for the populations or species constituting the community of interest allows for the simultaneous identification of climatic influences at each trophic level in the community. Using this approach with the long-term data from Isle Royale on the dynamics of wolves, moose, and balsam fir revealed an influence of climate at each trophic level, though with important differences among them (Post and Forchhammer 2001). Climatic influences on dynamics were strongest at the top and bottom trophic levels and weakest at the middle trophic level (figure 6.17). By contrast, self-regulation, indicated by the coefficients of statistical density dependence, was strongest at the middle trophic level and weakest at the top and bottom trophic levels (figure 6.17). This suggests

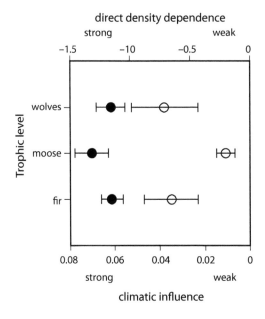

Figure 6.17. Differential strengths of the roles of climate (open circles) and self-regulation (density dependence) (solid circles) in the dynamics of each trophic level in a vertically structured community. Data were derived by applying the community model in figure 6.16 to the analysis of long-term data on wolves, moose, and balsam fir on Isle Royale, Michigan, USA. *Adapted from Post and Forchhammer (2001).*

the potential for a trade-off between the strength of density dependence and density independence in the dynamics at each trophic level in this vertically structured community that is akin to the tension between the friction (λ) and diffusion (σ^2) factors in population dynamics explored in the preceding chapter. It also suggests that any strengthening of the climatic influence at one or more trophic levels in a vertical community would have the potential to erode the stability of the population at that trophic level. To be able to examine the consequences of such an erosion for the stability of the community as a whole, it is necessary to adopt a quantitative framework that integrates population dynamics, the influences of one species on another, and a measure of community stability.

THE COMMUNITY MATRIX AND ITS STABILITY PROPERTIES

Returning momentarily to equations (6.1), we can see that the partial differentials of the functions $f(\bullet)$, $g(\bullet)$, and $h(\bullet)$ uniquely define the parameters of the log-linear autoregressive model of dynamics at each individual trophic level (equations (6.2)) and, after inclusion of the climate terms, equations (6.3). That is, we can define $(1 + a_1) = \alpha_{11}$, $a_2 = \alpha_{12}$, $(1 + b_1) = \alpha_{22}$, $b_2 = \alpha_{21}$, $b_3 = \alpha_{23}$,

$c_1 = \alpha_{33}$, and $c_2 = \alpha_{32}$ (Stenseth, Bjørnstad, and Falck 1996; Forchhammer and Asferg 2000). In other words, the coefficient $(1 + a_1)$, the statistical coefficient of direct density dependence at the predator level, represents the influence of species 1, the predator, on itself, defined in the community matrix as α_{11}. In turn, the statistical coefficient a_2 from the community model represents the effect of species 2, the herbivore, on species 1, the predator, and is defined in the community matrix as α_{12}. For dynamics at the herbivore level, the statistical coefficient of direct density dependence, $(1 + b_1)$, is represented in the community matrix as the effect of species 2, the herbivore, on itself, or α_{22}. The effect of predator abundance on herbivore abundance is quantified in the statistical community model by the term b_2 and is represented in the community matrix as the effect of species 1 on species 2, or α_{21}. The effect of vegetation abundance on herbivore abundance is represented in the community model by the statistical coefficient b_3, and in the community matrix by the term quantifying the influence of species 3 on species 2 as α_{23}. Focusing on dynamics at the vegetation level, the influence of vegetation, species 3, on itself is represented in the community model by the statistical coefficient c_1, and in the community matrix by the term quantifying the influence of species 3 on itself as α_{33}. Finally, the influence of the herbivore on vegetation abundance is accounted for in the community model by the statistical coefficient c_2 and in the community matrix as the effect of species 2 on species 3 by the term α_{32}. These interactions are summarized for the entire community in the Jacobian matrix ($\mathbf{J}$), which comprises the partial differentials of the functions in equation (6.1), and the community matrix ($\mathbf{A}$), which comprises their values around (or near) equilibrium (May 1973a, 1973b):

$$
\mathbf{J} = \begin{pmatrix} \dfrac{\partial f}{\partial X} & \dfrac{\partial f}{\partial X} & 0 \\[2mm] \dfrac{\partial g}{\partial X} & \dfrac{\partial g}{\partial Y} & \dfrac{\partial g}{\partial Z} \\[2mm] 0 & \dfrac{\partial h}{\partial Y} & \dfrac{\partial h}{\partial Z} \end{pmatrix} \quad \mathbf{A} = \begin{pmatrix} \alpha_{11} & \alpha_{12} & \alpha_{13} \\ \alpha_{21} & \alpha_{22} & \alpha_{23} \\ \alpha_{31} & \alpha_{32} & \alpha_{33} \end{pmatrix}.
$$

In this case, the zeros (or empty cells) indicate a lack of *direct* influence of vegetation on predator dynamics and of predators on vegetation dynamics; in other words, this is an example of a simple, three-level vertical community (May 1973a). To examine the stability properties of the community matrix, it is sufficient to begin with a qualitative assessment of stability in which we simply examine the signs of the elements of the matrix (May 1976). In fact, this may be the only appropriate approach to examining community stability when working with inferences from time series analysis of one trophic level, where we may be limited to conclusions about the signs of the interactions (rather than the values of the interaction coefficients).

In the qualitative assessment of the stability properties of the community matrix, we observe that if the eigenvalues (λ) of the matrix, consisting of the complex conjugate pairs of real and imaginary numbers (i.e., $\lambda(\mathbf{A}) = \zeta + i\xi$), all contain negative real parts, the community may be stable (May 1973a). The real parts of the conjugate pairs that make up the eigenvalues of the matrix represent coefficients of self-regulation, or density dependence. From this, then, the following key insight is derived: community stability may be altered by climate change, or any other extrinsic factor, through a shift in self-regulation at one or more trophic levels from negative to negligible or positive. In other words, a weakening of self-regulation, or density dependence, at any single or even multiple trophic levels in the vertically structured community may alter its stability. As was demonstrated in chapter 4, this relates to the diffusion process in population dynamics in stochastic environments. Hence, it is not necessary that climate change or any other relevant extrinsic force drive dynamics at any or all trophic levels in one direction or the other, only that the magnitude of its effect exceed that of the stabilizing influence of self-regulation at the affected trophic level. This might occur if, for example, climatic influences on vulnerability to exploitative interactions with predation or pathogens drive such interactions to exceed in magnitude and importance the influence of self-regulation at the affected trophic level, so that dynamics at that level become more strongly influenced by interspecific relations than by density dependence. Thus, this framework allows one to investigate whether stability at the population and community levels changes as a result of climatic influences on population-intrinsic or population-extrinsic processes.

While the qualitative approach to examining community stability is straightforward and valid for simple, straight-chain systems in which there are no links between nonadjacent trophic levels (May 1973a), we can also analyze the stability properties of the community matrix quantitatively through direct estimation of both the real and the imaginary parts of the eigenvalues of the matrix. Doing so requires estimation of the interspecific interaction coefficients in the ecological model through statistical analysis of the time series data (e.g., Stenseth, Bjørnstad, and Falck 1996).

TROPHIC INTERACTIONS, DYNAMIC COMPLEXITY, AND STABILITY IN VERTICAL COMMUNITIES

In the log-linear scenario, community stability is possible if there is significant density-dependent limitation at each trophic level, which is indicated if

α_{11}, α_{22}, and α_{33} are all < 1, because the coefficients of self-regulation in the log-scale statistical models include 1 (see equations (6.4c) and (6.6c)). In this approach, the real parts of the eigenvalues of the matrix can be estimated by calculating the trace of the matrix as:

$$\text{tr } \mathbf{A} = \alpha_{11} + \alpha_{22} + \alpha_{33}, \tag{6.10}$$

and we can estimate the imaginary parts of the eigenvalues of the matrix by calculating the determinant of the matrix:

$$\det \mathbf{A} = \alpha_{11}\alpha_{22}\alpha_{33} - \alpha_{11}\alpha_{23}\alpha_{32} + \alpha_{12}\alpha_{23}\alpha_{31} - \alpha_{12}\alpha_{21}\alpha_{33} \\ + \alpha_{13}\alpha_{21}\alpha_{32} - \alpha_{13}\alpha_{22}\alpha_{31}. \tag{6.11}$$

Note, however, that because of the empty cells in the matrix of the straight-chain community, equation (6.11) can be simplified to:

$$\det \mathbf{A} = \alpha_{11}\alpha_{22}\alpha_{33} - \alpha_{11}\alpha_{23}\alpha_{32} - \alpha_{12}\alpha_{21}\alpha_{33}. \tag{6.12}$$

The plot of the determinant of the matrix against the trace of the matrix in the complex plane for stochastic environments must fall to the left of the y-axis by a distance at least equal to the variability of the environment to indicate stability (May 1973b, 1976). To examine the consequences of climatic variability for community stability using this approach, then, we can calculate the community matrix under various climate scenarios, or under periods of differing climatic conditions, and then plot the eigenvalues (as $(x,y) = (\text{tr } \mathbf{A}, \det \mathbf{A})$) of the matrix in the complex plane for stochastic environments (May 1973a and Fig. 2.4 therein) for each climate scenario to determine whether the community is stable or unstable in each scenario. For instance, we might compare the stability properties of a community during a cooling phase and a warming phase, or between two periods with differing magnitudes of interannual climatic variability. In the first case, the analysis would tell us whether warming has a stabilizing or destabilizing influence on stability of the community, whereas in the latter case the analysis would indicate whether increasing climatic variability influences community stability, and if so, in which direction.

To estimate the coefficients of self-regulation and interspecific interactions composing the community matrix under different climate scenarios, the statistical models for each trophic level can be employed with observed time series data to estimate the relevant coefficients, as was demonstrated previously in an application of this approach to time series data on the dynamics of wolves, moose, and balsam fir on Isle Royale (Post and Forchhammer 2001). A plot of the solution of the community matrix in the complex plane for stochastic environments will then indicate whether the community falls within the stability

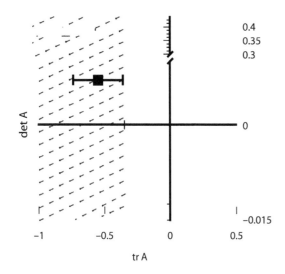

Figure 6.18. Solution of the community matrix for Isle Royale based on application of the model in figure 6.16 and derivation of the species interaction coefficients from Post and Forchhammer (2001). The community falls within the stability region (shaded) of the complex plane for community dynamics in stochastic environments but abuts the region of instability (non-shaded). Degradation of stability at any individual level of this vertically structured community may erode the stability of the entire community and is most likely to arise through a strengthening of the climatic influence on moose dynamics, which currently exhibit the weakest level of self-regulation in this community (see figure 6.17).

region of the plane, as was the case for the Isle Royale community (figure 6.18) (Post and Forchhammer 2001). Additionally, the real parts of the eigenvalues of the matrix will indicate whether the stability properties of the entire community reflect strong negative self-regulation at one or more trophic levels, and which levels these are. Furthermore, plotting the solution of the community matrix in the complex plane for stochastic environments under different climate scenarios will indicate whether the trace of the matrix moves to the right out of the stability region, indicating instability of the community under the climatic scenario of interest (figure 6.18). This will also indicate whether movement out of the stability region reflects a weakening of self-regulation at one or more trophic levels under a strengthening of the effect of climate on dynamics at other trophic levels in the community.

While caution against drawing general conclusions from attempts to model stability in relatively simple systems might be urged, the approach of deriving the simple community matrix outlined here may be applicable in systems in which climate influences trophic interactions such as predation and herbivory, as on Isle Royale (Post and Stenseth 1998; Post, Peterson, et al. 1999). Perhaps of relevance in this context, in this system the trophic level with the weakest self-regulation also displayed the greatest response to climatic fluctuation (Post and Forchhammer 2001), and may be the conduit through which climate change alters the stability of the entire community. In contrast, the examples given earlier in this chapter of dynamics in laterally structured communities suggest that climatic fluctuation may prevent interference interactions from eroding community stability in such systems. The distinction between interference and exploitation, and the role of each in the dynamics of laterally versus vertically structured communities, should aid in our attempts to infer and quantify the implications of climate change for community stability and persistence.

CHAPTER 7

Biodiversity, Distributions, and Extinction

Spawning what has since become a major subdiscipline within ecology, Elton inserted a subtle statement near the end of his landmark book on invasions in ecology (Elton 1958). In the penultimate chapter of that book, titled "The Reasons for Conservation," which sets the stage for the final chapter, "The Conservation of Variety," Elton commented, in passing, on the importance of complexity for ecosystem stability, speculating that more simple communities would likely be more vulnerable to disturbance. The relationship between species diversity and ecosystem function and stability is currently one of the most intensely studied topics in ecology. This subject is also of paramount importance in the study of the ecological consequences of climate change, most probably because of its obvious relevance to ecosystem goods and services. More classically, however, the subject of biodiversity response to climate change relates to what factors set limits to the upper and lower bounds of species diversity and how those factors might be altered by rapid climate change.

In his "Homage to Santa Rosalia," Hutchinson is famously credited for posing the question of why there are so many different kinds of animals, but he also, perhaps less famously, wondered what set the limits to the numbers of species we observe in a given system (Hutchinson 1959). Exploring a cave near Palermo, Italy, where the remains of Santa Rosalia, the patroness of Palermo, had been discovered, Hutchinson encountered two species of water bugs of the genus *Corixa* inhabiting an artificial pond within the cave. This led Hutchinson to wonder why "there should be two and not 20 or 200 species of the genus in the pond" (Hutchinson 1959). In his homage, Hutchinson reveals clues that indicate he was aware of the difference between the processes that generate and limit species diversity. For instance, he stated outright at the beginning of his remarks that he subscribed to the notion that natural selection and isolation followed by mutual invasion were the processes driving sympatric speciation. At the same time, he noted that interspecific competition sets the limits to niche boundaries.

This brings us to a central question of this chapter: how might the factors generating and limiting diversity differ in their response to climate change? In contrast to Hutchinson's perspective, Ricklefs (2004) has suggested, as we noted in chapter 6, that niche breadths of existing species in a local community should narrow to accommodate the addition of new species from the regional species pool. This is obviously at odds with the view that local interference among existing and potential members of a community limits the establishment of new species within that community. As we will explore later in this chapter, of the two processes generating diversity, speciation and immigration, the latter obviously operates at shorter time scales (Ricklefs 2009b) and is likely to respond more immediately to climate change. Of the processes reducing local diversity, extinction and emigration, the latter is, again, likely to operate at shorter time scales (Ricklefs 2009b), but both processes are likely to be influenced by climate change, although at potentially different timescales. Nonetheless, the difference in the temporal scale of response between extinction and emigration would appear to be minor relative to the difference in the temporal scale of response between speciation and immigration. As well, because changes in climate alter habitat-landscape associations and interactions among and within species across landscapes, we should expect continuous flux in the balance between the processes generating and limiting diversity, rather than a steady state in diversity with future climate change (Ricklefs 2004).

Hutchinson also wondered, in accordance with the speculations of many ecologists noted in his paper, why the Arctic was so species-poor compared to the Tropics. He thought this might have to do with the extreme seasonal variation in the Arctic, which might set a limit to the productivity of plant species and the accumulation of biomass necessary to support larger numbers of herbivores and the predators that depend on them (Hutchinson 1959). He also wondered whether it might have to do with the passage of time, and the relative youth of the Arctic compared to the Tropics, which might have limited the number of species capable of dispersing into recently exposed, high-latitude environments (Hutchinson 1959). But as we saw in chapter 2, the Arctic was much more species-rich during the Pleistocene than it is today. This should lead us to wonder why the Arctic of half a million years ago was home to so many more species than the Arctic of today. The answer appears to relate to issues on both sides of Hutchinson's musings: greater connectivity between the Old and New Worlds was afforded by land bridges during glacial maxima, when sea level was lower; and greater primary productivity characterized the mammoth steppe vegetation mosaic of the Pleistocene Arctic, which was dominated by forbs and graminoids rather than by the slow-growing, chemically defended woody shrubs that typify the region today (Guthrie 1982).

Hutchinson promoted the notion that species diversity is generated by sympatric speciation but limited by niche availability and occupancy. Hence, the generation of diversity relates to long-term processes of history and evolution. The maintenance of diversity relates to local interactions and more immediate processes over shorter timescales, some of which promote diversity and some of which limit diversity. Limits to diversity may be set, in Hutchinson's original views, by the extent of habitat complexity and primary productivity, which are viewed as determining resource availability and the potential for niche partitioning (MacArthur 1957). There is an obvious disparity of scales, both of space and time, in this view of the different processes responsible for generating, maintaining, and limiting diversity that may be more an artifact of the manner in which we think about the relationship between regional and local diversity than it is indicative of any actual differences between processes acting at regional and local scales (Ricklefs 1987, 2004).

There is considerable taxonomic and geographic variation in the extent of extinction risk in general (Pimm et al. 1995), as there likely is in the extent of extinction risk consequent on future climate change (Thomas et al. 2004). For instance, the percentage of taxa at risk of extinction is greatly skewed toward mammals, with nearly 20 percent of species facing extinction risk, while in the second most threatened group, birds, slightly over 10 percent of species are at risk of extinction (Pimm et al. 1995; Chapin et al. 2000). Because there are nearly twice as many species of birds as there are of mammals, however, the numbers of species at risk of extinction in both groups are approximately equal. A comparison of forty-two prediction-based estimates of extinction risk with thirty-two observation-based estimates concluded that the risk of extinction by the year 2100 owing to expected climate change varied between 10 percent and 14 percent across plants, vertebrates, and invertebrates (Maclean and Wilson 2011). In some cases, it may be easier to predict the risks posed to biodiversity from individual stressors than to predict their combined and synergistic effects. For instance, among the major biomes on Earth, the Mediterranean biome may be at greatest overall risk of suffering biodiversity loss from a combination of stressors if these act individually to influence biodiversity change, followed by grasslands, the boreal forest, and tropical forests (figure 7.1) (Chapin et al. 2000). The biomes at greatest risk of experiencing changes in biodiversity resulting solely from climate change are the Arctic and the boreal forest, whereas the tropical forest biome is at greatest risk of biodiversity change due to land use and at least risk of biodiversity change due to climate change (figure 7.1) (Chapin et al. 2000). Over the past five decades, geographic shifts in isotherms have occurred at a rate of 21.7 km per decade on average in the oceans and 27.3 km per decade over land, but at latitudes between 50°S and 80°N, this disparity

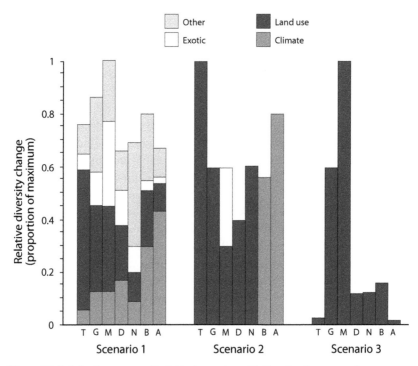

Figure 7.1. Relative importance of individual stressors to biodiversity change over the next century in the following biomes: tropical forests (T), grasslands (G), Mediterranean systems (M), deserts (D), north temperate forests (N), boreal forests (B), and arctic tundra (A). The three scenarios under consideration include (1) no interaction among the listed stressors, (2) exclusive influence of the most important factor influencing diversity change, and (3) multiplicative interaction among factors influencing diversity change. *Adapted from Chapin et al. (2000).*

is reversed, with the rate of oceanic shifts exceeding that on land (27.5 km per decade versus 27.4 km per decade, respectively) (Burrows et al. 2011). Within 15° of the Equator, the pace of geographic shifts in isotherms and seasonal temperatures in the ocean has exceeded that on land by a factor of between two and seven, suggesting that coastal equatorial biodiversity may be particularly threatened by the rapid pace of seasonal shifts due to climate change in those regions (Burrows et al. 2011).

Obviously, species gains or losses from ecosystems do not occur in a vacuum, and loss of biodiversity may, according to the facilitation/promotion hypothesis, precipitate rapid and irreversible changes to the function and stability of ecosystems. The simplest conceptualization of biodiversity change through time is a linear decline, whether due to climate change, land-use change, or a combination of these and other factors (figure 7.2a). Even under this scenario, however,

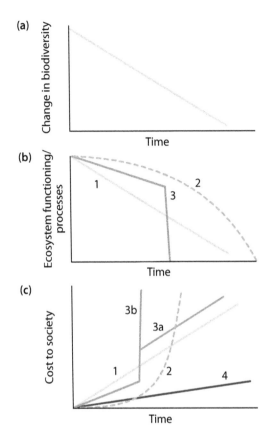

Figure 7.2. (a) The simplest scenario of a reduction in biodiversity through time is a linear decline. (b) Nonetheless, nonlinear and threshold declines in ecosystem function, as shown by lines 2 and 3, may still be precipitated by a linear decline in diversity if ecosystem function is nonlinearly related to species diversity, such as through the loss of keystone species. (c) Similarly, the costs to society of preventing or mitigating losses to ecosystem function of declining biodiversity may rise nonlinearly or through a threshold process, as depicted by lines 2 and 3. *Modified from Chapin et al. (2000).*

nonlinear and threshold changes in other variables related to biodiversity may result. For instance, a linear trend in biodiversity loss may precipitate a nonlinear decline in ecosystem function or processes through time, or even a threshold decline in function or processes through loss of a keystone species or functional group (figure 7.2b); hence, gradual changes in biodiversity under these scenarios would result in increasingly strong changes in ecosystem function (Chapin, Walker, et al. 1997; Chapin et al. 2000). The economic consequences of such changes would also rise in nonlinear and unpredictable fashion. This may be particularly troublesome in the case of threshold changes in ecosystem processes associated with abrupt loss of keystone species or functional groups. In this instance, increases in the costs associated with mitigating or compensating for the loss of ecosystem services associated with biodiversity loss could rise abruptly or almost infinitely (figure 7.2c) (Chapin et al. 2000). The interaction between land-use changes and climate change may be especially compelling in this context. The development and expansion of human habitation into the coastal wetlands that provided a natural buffer against storm surges for the land on which the city of New Orleans was built certainly contributed to the devastation of parts of that city wrought by Hurricane Katrina. Such events may see an increased probability of occurrence as climate becomes more extreme and less predictable over the next century (Tebaldi et al. 2006).

DISTRIBUTIONAL SHIFTS IN SPECIES' RANGES

An early prediction of organismal response to climate change stated that species should be expected to shift their distributions latitudinally poleward and elevationally upward as warming precipitated local extinctions at low-latitude edges of distributions and at low-elevation sites within existing distributions (Cushing 1983; Lange et al. 1990; Mooney et al. 1993). In chapter 2 it was suggested that the disproportionately severe loss of megafaunal mammalian genera during the Late Pleistocene in South America may have borne some relation to the geographic impediments to poleward range expansion on that continent. Several reports published nearly simultaneously in the mid-1990s appeared to offer mixed support for predictions of distributional shifts through local population changes related to climatic warming. Macrozooplankton in the California Current off the coast of southern California, USA, were reported, for instance, to have declined over a forty-year period by 80 percent in association with a 1.5°C increase in surface-layer water temperature (Roemmich and McGowan 1995). In an intertidal invertebrate community in Monterey Bay along the central coast of California, faunal changes between the early 1930s and early 1990s indicated that 89 percent (eight out of nine) southern species had increased in abundance, while 62.5 percent (five out of eight) northern species had declined in abundance, in association with an increase in mean coastal water temperatures of 0.75°C (Barry et al. 1995). As well, distributional shifts of up to 200 km were reported for some species of zooplankton and intertidal organisms inhabiting the English Channel in association with warming between the 1920s and early 1990s, with warm-water-associated species increasing and cold-water-associated species declining (Southward et al. 1995). Nonetheless, these studies did not provide conclusive evidence for distributional shifts associated with climate change because they did not comprise distribution-level assessments (Parmesan et al. 1999; Parmesan 2006).

A landmark study emerged on the heels of the preceding examples that met this methodological challenge by comprehensively assessing local extinctions of populations of Edith's checkerspot butterfly (*Euphydryas editha*) throughout the distribution of the species (Parmesan 1996). Among 151 known populations of Edith's checkerspot distributed along a south-north gradient in western North America from northern Mexico to southern Canada, Parmesan (1996) censused 115, excluding those that inhabited sites known to have been subject to habitat modification by human activity. Among these, a clear association was noted between local extinction and latitude, with percent extinction among southern populations exceeding that at northern populations by a factor of nearly five (figure 7.3a) (Parmesan 1996). As well, Parmesan (1996) documented a significant relationship between site elevation and percentage of local populations having undergone extinction, with extinction at low-elevation sites exceeding that at high-elevation sites by fourfold (figure 7.3b). Hence, this study provided convincing support for the prediction of increased extinction at lower latitudes and elevations in association with warming, but was unable to provide evidence for increased northward or upward expansion related to warming. Nonetheless, the strengths of this example include its assessment of local extinction at the level of the entire range of a species and its exclusion of confounding factors, such as habitat alteration by humans.

Other examples of range shifts in association with climate change have been most common to date in species with restricted ranges, particularly those in

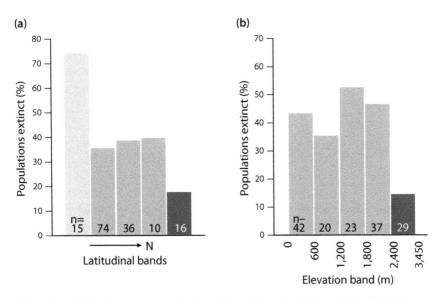

Figure 7.3. Decline in percentage of local populations of Edith's checkerspot butterfly that have gone extinct in association with (a) degrees north latitude of censused sites and (b) elevation of sites in meters above sea level. *Adpated from Parmesan (1996).*

alpine and polar environments (Parmesan and Yohe 2003). Such species have suffered the greatest range contractions from climate change, and extinction may be facilitated in such cases because of niche infringement by upward or northward expansion of competitors or simply by loss of niche space (Parmesan 2006). An interaction of this nature, in which an elevational or altitudinal specialist experiences niche loss owing to the interaction between climate change and invasion of niche space by competitors, would be predicted on the basis of the tension hypothesis driving biodiversity loss. In Europe and Scandinavia, distributions of butterflies have apparently shifted in response to climate change in manners suggestive of alteration of resource distribution along spatial gradients that was postulated in chapter 5 on niche dynamics. A comparison of historical abundance data with contemporary surveys of species' presence revealed different types of distributional changes (Parmesan et al. 1999). In the first example, *Carterocephalus palaemon* experienced a contraction at the southern edge of its distribution in the United Kingdom and Finland, but no changes at the northern edge of its distribution in either location (figure 7.4a) (Parmesan et al. 1999). This might be representative of a unidirectional contraction of niche availability along the space axis, as described in chapter 5. In the second example, *Argynnis paphia* experienced northward expansion of its distribution in southern Sweden and Finland without any changes at the southern edge of its distribution (figure 7.4b) (Parmesan et al. 1999). This appears representative of a unidirectional expansion of niche availability along the space axis, as suggested in chapter 5. Another example of such shifts from a tropical montane system, in which biodiversity loss may have resulted from the interaction between climate change and invasion of the niche space of elevational specialists by lower-elevation nonspecialists, will be detailed later in this chapter.

SCALE AND PATTERN IN DISTRIBUTION
AND ABUNDANCE

The theories of evolution by means of natural selection proposed simultaneously by Charles Darwin and Alfred Russell Wallace differed primarily in that while Darwin placed the mechanism of natural selection in the realm of competition among individuals (Darwin 1859), Wallace emphasized the role of organism-environment interactions (Wallace 1876). The concept of "Wallace's Realms of Life" grew out of Wallace's description of Earth's six regions of distinct zoogeography: the Neotropical Region, the Nearctic Region, the Palearctic Region, the Oriental Region, the Ethiopian Region, and the Australian

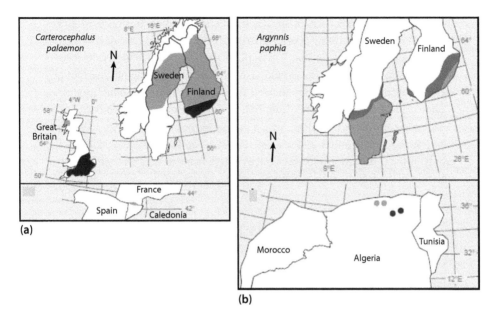

Figure 7.4. Examples of distributional changes in species that may have experienced spatial shifts in resources or niche availability in response to climate change as postulated in chapter 5. Panel (a) shows a range contraction at southern edge, in black, of the distribution of the checkered skipper butterfly as a result of local extinctions, without a concomitant northward expansion. Panel (b) shows an example of northward range expansion, in gray, without range contraction at the southern edge of the distribution of the silver washed fritillary butterfly. *Adapted from Parmesan et al. (1999).*

Region (Elton 1958). This concept was constructed to explain the roles of history and geographic separation in the evolution of species and the development of species diversity across geographic regions, subsequently grew into the discipline of zoogeography, and is considered the foundation for the theory of biogeography.

Connell's (1961) landmark research on barnacles in the intertidal zone along the coastline of western Scotland returned the field of ecology to an emphasis on the role of interspecific competition in determining the distribution and abundance of organisms, and did so, importantly, in an experimental framework that distinguished it from earlier, numerous observational studies linking patterns of distribution and abundance to competition (Elton and Miller 1954). These experiments demonstrated that local species interactions, including competition and predation, combined with different environmental tolerances for desiccation during low tide, explained the largely nonoverlapping distributions of two species of barnacles in the intertidal zone (Connell 1961). Ricklefs (2004) has suggested that the roles of species interactions and organism-environment interactions in determining species distributions should be scale invariant. That is, the same mechanisms that can explain the local distributions of species such as Connell's barnacles should also explain the geographic distributions of those same species, having allowed for the role of history in the first place (Elton and Miller 1954). The discipline of environmental niche modeling, or bioclimatic envelope modeling, which has developed out of interest in predicting species' range shifts under novel climates, appears to have returned the study of the distribution and abundance of organisms to Wallace's emphasis on the role of environment in explaining species diversity.

BIODIVERSITY CHANGES THROUGH ELEVATIONAL COLONIZATION AND EXTINCTION

Current estimates place approximately 85–90 percent of the Earth's species in the Tropics, and approximately 60 percent of all species in terrestrial ecosystems (Kier et al. 2009). As well, rates of endemism are higher in the Tropics than in all other of the Earth's biomes (Kier et al. 2009). Predicting extinction risk and biodiversity loss in the Tropics in response to future climate change should, justifiably, constitute a major conservation focus. As we saw in chapter 6, the prospects for the development of novel climates are also greater in the Tropics than elsewhere on Earth. Similarly, the probability of loss of existing climates is greatest in the Tropics, regardless of how ambitious our carbon mitigation strategies become (Williams et al. 2007). This would seem to suggest that the likelihood of loss of habitat for specialist species is greater in the Tropics than elsewhere.

Perhaps somewhat counterintuitively, Thomas and colleagues (2004) projected that the biomes with the greatest percentages of species at risk of extinction are scrubland and warm forests. In comparison, of the species inhabiting the Tropics, only approximately 4 percent are considered at risk of extinction from land use and or climate change (Thomas et al. 2004). In terms of numbers of species, however, this figure is enormous, if we consider that most of Earth's species are found in the Tropics. Because of the complexity of species interactions in the Tropics and the low warming signal there compared to higher latitudes, discerning a climatic signal in the dynamics of tropical organisms and species assemblages is particularly challenging. Further complicating the situation is that many tropical species are difficult to study, either because of the logistical challenges of conducting field research in the Tropics or because of the rarity of some species of conservation concern in the Tropics, or both.

Nonetheless, the first major extinction event ascribed to recent climate change was documented at a long-term study site in the neotropics.

The Monteverde research station is located within the cloud forest ecozone of Costa Rica's northwestern interior. The location of the site is a high-elevation, lower montane forest on the western face of the Continental Divide in the mountain range southwest of Lake Nicaragua. This area experiences regular seasonal and diurnal cycles of envelopment by clouds moving inland from the Pacific Ocean. Moisture deposited as clouds rises over the mountains and provides the most important income to the water budget of the region's biota.

According to a study by Pounds and colleagues (1999) on the changing fauna of Monteverde, since 1973, when observations began, data acquired from the use of long-term rain gauges at the study site indicate that the number of consecutive five-day periods characterized as dry days—without measurable precipitation—increased steadily over a period of more than two decades, but only during the dry season. This trend displayed a close association with both fluctuations in, and an overall increase in, sea-surface temperatures over the same period. The occurrence of dry days also, the authors noted, displayed a tendency toward increasingly lengthy runs of consecutive days without precipitation. These changes in local weather were attributed statistically to the El Niño–Southern Oscillation and increasing sea-surface temperatures (Pounds et al. 1999). Notably, the authors observed no trend in annual, seasonal, or monthly rainfall over the same period.

During the same period, observations on long-term monitoring plots revealed an overall increase in the numbers of bird species classified as premontane occurring at an elevation of 1,540 m above sea level, in the lower montane zone (figure 7.5a). These species are classified as premontane because higher-elevation conditions in the lower montane zone are unsuitable for them; that is, they are classified as cloud forest intolerant (Pounds et al. 1999). These data (figure 7.5a) suggest a trend toward colonization of the lower montane zone in Monteverde by premontane bird species at a net rate of approximately nineteen species per decade, based on the slope of the linear regression of the number of species versus year. Interannual variation in the numbers of premontane species occurring in the cloud forest zone, an index of net colonization, correlated closely with the annual number of dry days during the dry season (figure 7.5b).

Pounds and colleagues (1999) also estimated that fifteen premontane species had established breeding colonies at the study site since observations began, and had increased in numbers there since. At the same time, the number of lower montane bird species remained unchanged (figure 7.5a), and some premontane species, such as the keel-billed toucan (*Ramphastos sulfuratus*), were observed to nest in proximity to lower montane species such as resplendent

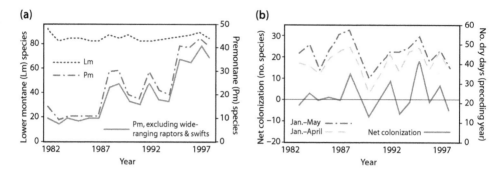

Figure 7.5. Changes in local diversity at a neotropical montane site in Monteverde, Costa Rica, between 1982 and 1997. Panel (a) shows an increase in the numbers of premontane (pm, lower elevationally breeding) species of birds in the lower montane (lm, higher-elevation) zone, and a lack of any trend in lower montane species. Panel (b) shows interannual variation in upslope colonization of premontane species in association with the number of dry days during the dry season of the previous year. *Adapted from Pounds et al. (1999).*

quetzals (*Pharomachrus moccino*) for the first time at the study site in 1995 and thereafter. This example therefore represents colonization of new habitat in response to changes in environmental conditions that have become more favorable for the colonizing species, without replacement of the original inhabitants. It remains to be seen whether continued changes in aridity in the cloud forest zone will render it inhospitable for lower montane bird species, whether competition with premontane species will displace or drive to local extinction lower montane species, or whether an interaction between continuing changes in environmental conditions and interspecific competition will prove unfavorable for lower montane species in Monteverde, as would be predicted according to the tension hypothesis.

Although the correlations between changing abiotic conditions and long-term and interannual variation in numbers of species of premontane birds occurring in lower montane cloud forest appear strong, it is nonetheless tempting to wonder what other factors may have contributed to these distributional changes. For instance, even though the number of species of lower montane birds displayed no trend or relation to changes in abiotic conditions in Monteverde, what about changes in the abundance of individuals of these species? Declines in their numbers despite overall changes in their diversity may have, according to the facilitation/promotion hypothesis, precipitated colonization of the cloud forest zone by the premontane species. As well, it is not immediately clear how changes in environmental conditions in the cloud forest zone facilitated colonization by premontane species. Subscribing to this view would appear to advocate or necessitate acceptance of the notion of niche vacancy, which, as we saw earlier, may be promoted in the Grinnellian niche concept espoused in the discipline of bioclimatic envelope modeling, but which also runs counter to the Hutchinsonian niche concept, a more realistic formulation of the niche.

Likewise, even if we were to adopt a Grinnellian perspective in this example, it seems unlikely that the bioclimate envelope of premontane bird species is determined mainly by aridity, or that conditions in the cloud forest zone are unfavorable for premontane species mainly because of the moisture regime of that region. Obviously, this is not what Pounds and colleagues (1999) have suggested, but the implication of their conclusions is that climatic changes precipitated further changes in the environment that facilitated colonization of the cloud forest zone by premontane species that previously did not or could not settle there. As for the impetus driving this elevational migration, it could be the case that conditions in the lower-elevation areas inhabited by premontane species have become less favorable with climate change, thereby pushing them along a gradient toward maintenance of their presence in favorable environmental conditions. But no data were presented by Pounds and colleagues (1999) on changes in the occurrence of premontane species in their more typical areas of occupation at lower elevations. Alternatively, human disturbance in the form of increased human presence or land use in lower elevations may have rendered those sites unfavorable for premontane species. Pounds and colleagues (1999), however, convincingly noted that land cover changes outside Monteverde occurred in the 1940s and 1950s, but not since then, which would seem to rule out any recent influence of human alteration of lower elevation habitat for premontane bird species.

Although this example illustrates how biodiversity may change as one set of species colonizes previously unoccupied habitat without replacing resident species, analysis of data on abundance of reptiles at the same site reveals a clearer and more directly negative picture of the implications of recent climate change for species diversity. Two species of anoline lizards that are endemic to Costa Rica and western Panama declined from peak abundance at Monteverde in the early 1980s to complete absence at Monteverde by the early to mid-1990s (Pounds et al. 1999). Abundance in these species was reportedly negatively associated with the number of dry days during the annual dry season (Pounds et al. 1999). Similarly, a study on Barro Colorado Island, Panama, revealed that, over a nineteen-year period, interannual variation in the abundance of a different anoline lizard species occurring on the island fluctuated in positive association with the amount of dry season rainfall (Andrews 1991). This suggests that increasing dry season aridity contributed to the declines of the two anoline species in Monteverde.

These examples of changes in the abundance and diversity of birds and liz-ards at Monteverde are important not only for illustrating the manner in which diversity may be altered by climate change but also for illustrating the inher-ent differences with which colonization versus extinction may be attributed to climate change. Neither example is without methodological difficulties. For instance, the complexity of species interactions complicates considerably the conclusions one can reach from the patterns in these data. The roles of changes in interspecific competition or pathogen-host dynamics in relation to increases or declines in diversity are impossible to discern from monitoring data, no mat-ter how exceptional such data may be. Experimental evidence indicates that, at least for some tropical species, interspecific aggression is important in deter-mining the boundaries of overlap of neotropical birds (Jankowski et al. 2010). As well, trend analysis may overlook important interactions between changes in environmental conditions and abundance of organisms. For instance, analy-sis of long-term survey data for eight species of birds in Hawaii revealed no trends in their abundance and did not detect any relation to a major environ-mental event midway through the census period (Freed and Cann 2010). In contrast, when data were divided into periods before and after the event, strong trends in opposing directions were detected on either side of the event for all eight species (Freed and Cann 2010). This analysis suggests it may have been more appropriate to apply piecewise regression to the analysis of trends in the abundance of lower montane species in Monteverde. Nonetheless, among tropical species, which characteristically display narrow thermal tolerance ranges, montane species may be at greatest risk of extinction because of the relative lack of altitudinal relief for elevational migration into favorable habi-tat zones as temperatures rise (Laurance et al. 2011). Taxonomically, tropical upper montane zone specialists tend to be biased toward plants, reptiles, and amphibians, groups that may be especially vulnerable to extinction owing to climate change in the Tropics (Laurance et al. 2011).

AMPHIBIAN EXTINCTION AND
THE CLIMATE-PATHOGEN HYPOTHESIS

More comprehensive analyses of amphibian declines across the tropics have generated debate about the potential role of climate change in pathogen-induced population crashes. Pounds and colleagues (2006) reported the results of a test of the climate-linked epidemic hypothesis. This hypothesis, although not explicitly developed to explain amphibian declines, suggests that declines in tropical amphibian populations following unusually warm years relate to

(a)

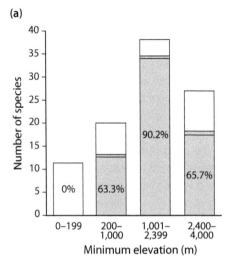

(b)

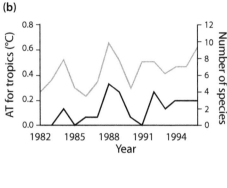

Figure 7.6. (a) Variation in the numbers of species of the harlequin frog (*Atelopus* sp.) in Central and South America by altitudinal zone, represented by total bar height, and the percentage of these that have gone extinct, represented by shading within the bars. (b) Association between the number of *Atelopus* species assumed to have gone extinct on the basis of having been observed for the last time and air temperature at the sites surveyed. *From Pounds et al. (2006).*

the increased spread or efficacy of pathogens and disease as a consequence of temperature increases (Epstein 2001; Pounds et al. 2006). Pounds and colleagues (2006) analyzed data on year of last observation for one hundred species of harlequin frogs (*Atelopus* spp.) surmised to have gone extinct in the neotropics. Extinctions in this genus have occurred most commonly at midelevations (figure 7.6a), and years of last observation of species that have gone extinct correlate closely with air temperature in the previous year (figure 7.6b) (Pounds et al. 2006).

The former of these two observations bears further consideration. Under a climate change scenario, the risk of habitat loss and niche disappearance in montane species should increase with elevation as temperatures rise because, in effect, high-elevation or mountaintop species become caught on diminishing islands (Laurance et al. 2011). But, as Pounds and colleagues (2006) emphasize, this expectation does not appear be upheld in the *Atelopus* data. Hence, there does not appear to be support for a direct role of climate change in these extinctions but rather for an indirect role (Pounds et al. 2006). Pounds and colleagues (2006) concluded that an increase in daily minimum temperatures and a decline in daily maximum temperatures at their Monteverde study site are indicative of an increase in cloudiness over the neotropical montane regions in response to warming. This change, in turn, has likely promoted the viability of chytrid fungi, which the authors believe have driven the *Atelopus* extinctions. Bolstering their argument, the authors note that most of the *Atelopus* extinctions have occurred within the elevational range where temperatures are increasing toward those most favorable for the growth of chytrid fungi (Pounds et al. 2006).

Obviously, the correlative nature of the preceding study renders it difficult to draw definitive conclusions about the role of climate change in pathogen-induced extinctions of amphibians. One of the main difficulties in drawing such a conclusion relates to the lack of information on the presence or prevalence of chytrid fungi in zones of extinction prior to the documented population crashes and species disappearances. A separate, multi-annual study in Panama attempted to take this approach. In this study, the authors did not detect the presence of chytridiomycosis, the infectious disease caused in amphibians by chytrid fungi, during the first four years of extensive monitoring, despite its documented presence and reported spread southward from Costa Rica (Lips et al. 2006). After the discovery of the first dead frog that tested positive for a fungal agent of the disease, further mortality of amphibians at the study site was rapid and widespread, culminating in infection of 70 percent of the documented anuran species at the site (Lips et al. 2006). Notably, however, Lips and colleagues (2006) reported a conspicuous absence of any climatic anomalies in 2004 that might have triggered the outbreak, although they also cautioned that it was not possible on that basis alone to rule out a role of climate in the documented declines. Most recently, a retro-analytical PCR sampling technique employed to examine museum specimens for the presence of *Batrachochytrium dendrobatidis* in samples collected before, during, and after amphibian declines provided evidence of an outbreak wave of the fungus deriving from Mexico in the 1970s and expanding southward through Central America to Monteverde in Costa Rica by the late 1980s (Cheng et al. 2011).

Amphibian extinctions have not, however, been confined to the neotropics, and the genesis of the pathogen hypothesis explaining the rapid disappearance of hundreds of high-elevation species derived from a study of widespread declines in anuran species in Australian rain forest (Laurance et al. 1996; Berger et al. 1998). A subsequent study (Laurance 2008), revisiting the disappearance of fourteen species of frogs from the rain forests of eastern Australia, was undertaken to test the warming-plus-pathogen hypothesis presented by Pounds and colleagues (2006) to explain the loss of neotropical amphibians. Laurance (2008) did not find direct support for this hypothesis but did conclude that frog population declines in tropical Australia followed three consecutive years of anomalously warm temperatures. While urging caution because of the correlative nature of his study, Laurance (2008) suggested that his results were consistent with the notion that a run of years characterized by increasing mean minimum annual temperatures could leave montane amphibians vulnerable to disease outbreaks resulting from improved conditions for the development of chytrid fungi.

In my opinion, one of the most fascinating aspects of the global decline in amphibians is the fact that, even where chytrid fungi occur, there appears to be variability in the extent to which local populations and species are vulnerable to it. A ripe avenue for research in this area would be to examine the role of variation in breeding phenology in the variable susceptibility of individuals, populations, and species to this pathogen. Because the chytrid fungus pathogen is waterborne, its spread among amphibian populations is closely linked to their congregation in water bodies during the mating season (Berger et al. 1998). If climate change has altered the hydroperiod of ephemeral ponds, streams, and other bodies of water utilized by tropical amphibians during the breeding season, it seems likely that this could influence the prevalence of diseases spread through contact during this period. To my way of thinking, this would be particularly relevant in areas or for species experiencing loss of some bodies of water due to drying associated with rising temperatures.

BIODIVERSITY AND STABILITY

Perhaps the earliest thoughts on diversity and stability, or the tendency of a community or ecosystem to return to its prior state following a pulse perturbation or disturbance, were expressed in MacArthur's famous essay on population fluctuation and community stability (MacArthur 1955). Although MacArthur's ideas were theoretical and concerned mainly with community stability, they are relevant to a discussion about the origins of the relation between biodiversity and stability. Despite observing that populations in nature often fluctuate, MacArthur subsequently observed that some communities, obviously comprising populations of multiple species, appear to remain constant. He suggested that this community-level stability may relate to two factors, namely, interactions among the species within the community or an intrinsic tendency toward stability within the species themselves. This latter notion was subsequently developed further in the idea of the so-called balance of nature, which enjoyed considerable favor in population and community ecology (Mech 1966) for at least a decade following MacArthur's initial conjecture.

The relation between diversity and stability in MacArthur's theoretical system focused on the more readily generalized role of species interactions. The focus here was on trophic interactions, or primarily vertical interactions among species in food webs. MacArthur supposed, as did Elton (1958) after him or at roughly the same time, that as the number of connections among species increased along with diversity within the system, the consequent increase in the number of pathways through which energy could flow in the system (Odum 1953) would lend stability to the system as a whole. This might even be the case despite the loss of one or more component species in the system. These ideas have since become central to the development of the concept of functional redundancy, according to which one species belonging to the same functional group or performing the same ecosystem services as another species may assume or approximate the role of that species if it goes extinct within the system (Walker 1992). At the biome scale, functional redundancy is assumed to be lower in, for example, the Arctic than in the Tropics, and it is the former in which loss of species due to climate change may have greater consequences for community and ecosystem stability (Post, Forchhammer, et al. 2009).

The diversity-stability subdiscipline in ecology has produced numerous hypotheses to explain or predict the nature of this relationship, four of which have been most prominent in the progression of its development (figure 7.7) (Johnson et al. 1996). MacArthur's original hypothesis is, perhaps not surprisingly, the simplest and most intuitive of these, with some generalized ecosystem process increasing linearly with diversity (figure 7.7a). The so-called rivet hypothesis (Ehrlich and Ehrlich 1981; Ehrlich and Walker 1998), close in conceptualization to the functional redundancy hypothesis (Naeem 1998), draws on the analogy in which species in a system are likened to rivets holding together the parts of an airplane. As the number of rivets, or species, increases, so does the structural integrity of the system held together by them, until at some number of species further additions improve stability only marginally more than that characterized by fewer species (figure 7.7b). Similarly, the redundancy hypothesis predicts that ecosystem processes and stability will increase rapidly with diversity, but then reach a plateau after which further increases in diversity exert no further influence on these ecosystem properties because of functional redundancy among the species. Conversely, erosion of species diversity may precipitate no changes in the stability of the system until a threshold in diversity is crossed, after which a precipitous crash in ecosystem stability results (figure 7.7c) (Scheffer et al. 2001, 2009; Scheffer and Carpenter 2003). Hence, in the rivet/functional redundancy conceptualization, it is the effect of loss of species on stability that is of interest. Moving from right to left on the x-axis of figure 7.7b, for instance, we can observe that at the highest levels of diversity, loss of species exerts only a gradual effect on stability until some critical level of diversity is reached, after which further loss of species will lead to more rapid declines in stability. Finally, the idiosyncratic

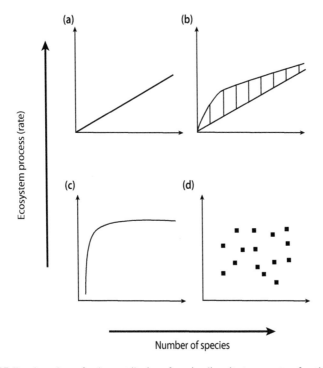

Figure 7.7. Four hypotheses for the contribution of species diversity to ecosystem function or processes. (a) The diversity-stability hypothesis. According to this hypothesis, ecosystem productivity and resilience should increase linearly with the number of species present. Conversely, removal of a species through extinction results in a linear decrease in productivity or resilience. (b) The rivet hypothesis. According to this hypothesis, the nonlinear association between ecosystem processes and species diversity may render ecosystems relatively insensitive to species losses, primarily because of functional redundancy among species, until enough species are removed to result in a dramatic decline in productivity or resilience. (c) Expanding on this notion, the redundancy hypothesis similarly predicts that loss of some species from ecosystems may be compensated for by the expansion of their functional role by other species until a threshold of diversity is crossed, at which point ecosystem function collapses. (d) Finally, the idiosyncratic hypothesis predicts a lack of any discernible relation between number of species and ecosystem function. *Adapted from Johnson et al. (1996).*

hypothesis predicts no inherent relation between diversity and ecosystem function or stability (figure 7.7d).

More recently, the four preceding conceptualizations of the stability-diversity relationship have given way to explanations based on niche complementarity and trait selection (Loreau 2000; Schmid et al. 2001), which in turn derive from studies investigating the relationship between diversity and ecosystem productivity (Naeem et al. 1994, 1995, 1996; Tilman and Downing

1994; Tilman 1996, 1999; Naeem and Li 1997; Tilman et al. 1998, 2006, 2007; Naeem 2002). According to the niche complementarity hypothesis, the stability of community or ecosystem productivity through time derives from enhanced resource exploitation with increasing diversity of species in the system (Tilman et al. 1997; Loreau 2000). In contrast, according to the trait selection hypothesis, the stability of a community or ecosystem will increase with the number of species composing it because as this number increases, it becomes more likely that the system will become dominated through trait selection by a species or group of species with extreme trait values that are characterized by reduced temporal variability (Loreau 2000; Nijs and Roy 2000).

The obvious relevance of these hypothesized relations between diversity and properties such as ecosystem function and stability to climate change extends from the implications of rapid climate change for niche dynamics (McKane et al. 2002), as explored in chapter 5, and for species losses or gains (Fonseca and Ganade 2001). As was discussed earlier in this chapter, the processes by which species may be removed from communities or ecosystems by climate change include emigration of species following the development of unsuitable conditions in portions of the currently occupied ranges to areas of more suitable conditions and extinction. Conversely, the processes by which species may be added to communities or ecosystems include immigration to new locations and the colonization of sites or communities formerly unoccupied by those species, or speciation. Whereas emigration, immigration, and extinction are processes that occur over relatively short timescales and can even occur abruptly, diversification through speciation is the outlier in terms of the temporal scale over which this process unfolds in comparison to the others (figure 7.8a). I will deliberately avoid ascribing exact temporal scales to these processes in this example because, aside from the fact that doing so is inherently arbitrary and contentious, my intent here is merely to compare the relative scales at which these processes occur. Immigration and emigration can obviously occur suddenly and over the briefest of periods, from years to decades, although they can also occur much more slowly and are inherently dependent on the species of interest. Likewise, extinction can occur very suddenly and over brief periods, but also over much longer periods, over decades to tens of centuries. Diversification through speciation, on the other hand, occurs, at least in vertebrates and higher plants, over much longer time periods, on the order of mega-anna, or millions to hundreds of millions of years (Hedges et al. 1996; Kumar and Hedges 1998; Hedges and Kumar 2009).

The effect climate change may have on the rates at which, and the temporal scales over which, these processes occur is also likely to vary among them (figure 7.8b). For instance, climate change may increase rates of immigration and

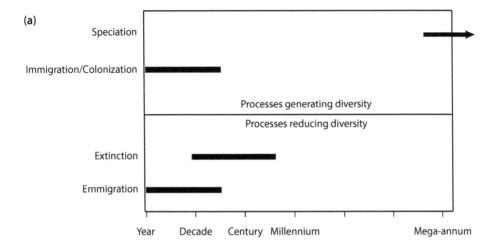

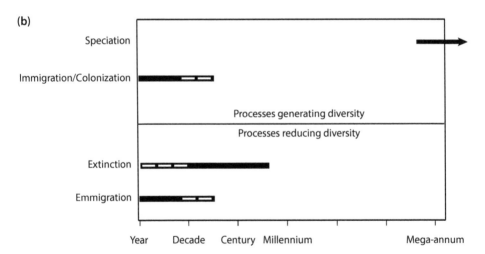

Figure 7.8. Relative temporal scales of the processes generating and reducing species diversity (a), and the direction and relative magnitude by which they may be altered by climate change (b).

emigration, thereby shortening the temporal scale over which these processes occur, but is unlikely to reduce their rates or extend the scales over which they occur. Similarly, as we saw in chapter 2 on the Pleistocene extinctions, rapid climate change and consequent extreme events may hasten extinction in some cases, thereby extending the temporal scale over which this process occurs. Rates of speciation, on the other hand, are less likely to be altered by climate change, although climatic shifts may precipitate radiations, as noted earlier in this chapter. Hence, the net effect on biodiversity of climate change may very well be loss of species, either locally or globally, because the balance appears to lie in favor of those processes that erode diversity.

The remaining question, then, in terms of the implications of climate change for biodiversity changes resulting from climate change, is whether ecosystem processes will respond gradually and linearly, as predicted by the full complementarity hypothesis (figure 7.7a), gradually and then suddenly, as predicted by the partial complementarity hypothesis (figure 7.7b), or imperceptibly and then increasingly abruptly, as in the facilitation/promotion hypothesis (figure 7.7c) (Schmid et al. 2001). Although Schmid and colleagues (2001) used the term "positive interactions hypothesis" for what I refer to in this case as the facilitation/promotion hypothesis, they are essentially the same. In my estimation, the choice really comes down to the facilitation/promotion hypothesis. As Schmid and colleagues (2001) described these hypotheses, relating ecosystem function to biodiversity, the main feature distinguishing the full and partial niche complementarity hypotheses is that whereas ecosystem function decreases linearly with the number of species lost according to the full complementarity hypothesis, much as MacArthur (1957) had suggested, it declines slowly at first and then at an accelerating pace with species loss under the partial complementarity hypothesis.

By contrast, according to the positive interactions, or facilitation/promotion, hypothesis, species losses will precipitate an initially gradual decline in ecosystem function, which then declines more rapidly with further species losses (Schmid et al. 2001). Because of the low functional redundancy of species in the Arctic, I view this system as lying at one end of a continuum of response most likely to follow the facilitation/promotion hypothesis relation between diversity loss and ecosystem processes (figure 7.9a). At the other end of this continuum I would place the Tropics, where high species diversity and presumed high functional redundancy among species will render detection of declines in ecosystem processes difficult until a threshold of diversity is crossed with loss of further species (figure 7.9b). The difficulty in detecting such changes in the Tropics is further compounded by our lack of full understanding of the extent of species diversity in the Tropics and our consequent lack of complete knowledge about the current rates of species loss there.

At the study site near Kangerlussuaq, Greenland, I began tracking species diversity of forbs on my long-term experimental plots in 2005 (Post 2013). As mentioned in chapter 1 in the description of this site and of the ongoing experiments and observations I have been conducting there since 1993, an unexpected outbreak of caterpillar larvae of a noctuid moth occurred at the

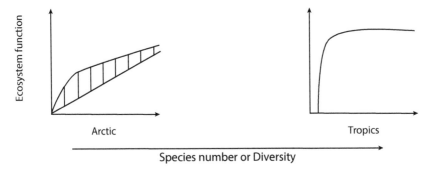

Figure 7.9. The ecosystem function–species diversity hypothetical relationships as they apply in arctic and tropical systems. Species losses due to climate change are likely to incur greater ecosystem consequences in arctic than in tropical systems for two reasons. First, species diversity is lower in the Arctic than in the Tropics, so ecosystems in the Arctic lie further to the left along the *x*-axis, where the relation between ecosystem function and species number is steeper, than do those in the Tropics. Second, functional redundancy is likely lower in the Arctic than in the Tropics, suggesting that the rivet hypothesis is more appropriate in the former and the redundancy hypothesis more appropriate in the latter.

site in 2004 and peaked in 2005. This represented a strong pulse perturbation to the system, and may be representative of a phenomenon that will occur with increasing frequency as the climate continues to warm (Post, Forchhammer, et al. 2009).

In 2006, very few caterpillars were detected at the study site, and it was evident that the outbreak had ended. Data on forb species diversity collected at the site since the last year of the outbreak reveal a strong recovery of forbs since 2005 (figure 7.10). A lack of species-level forb data before the outbreak precludes a comparison with levels of diversity preceding the disturbance. Nonetheless, we can examine the data for evidence of an effect of the warming treatment on forb diversity during this recovery period. These data reveal that forb abundance was exceedingly low on all plots at the peak of the caterpillar outbreak but has since increased on all plots. Incidentally, a second outbreak of caterpillars began in 2009 and persisted through the growing season of 2010, reducing, once again, mean numbers of forbs on all plots, though temporarily. Ignoring the grazed plots momentarily and focusing solely on the exclosed plots not exposed to herbivory by large mammals, we can see that prior to this second outbreak, the species diversity of forbs had in 2008 reached an apparent peak and was approximately 50 percent greater on ambient plots than on warmed plots. It would be tempting to conclude on this basis that experimental warming had reduced the species diversity of forbs in this system. However, this conclusion depends entirely on whether

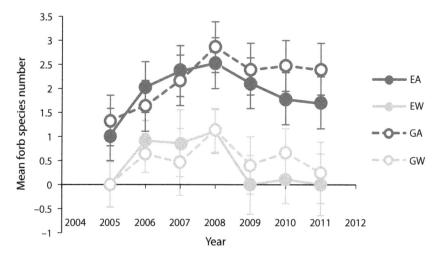

Figure 7.10. Variation in mean number of forb species by treatment at the study site near Kangerlussuaq, Greenland, following the peak of a major caterpillar outbreak in 2005. The line plots represent annual means under the following treatments: ambient plots exclosed from herbivory (EA), warmed plots exclosed from herbivory (EW), ambient plots exposed to herbivory (GA), and warmed plots exposed to herbivory (GW). *From Post (2013).*

or not we take into consideration exploitative species interactions between trophic levels in the system (Post 2013).

Turning now to the data from plots that were exposed to herbivory by vertebrates as well as those that were protected from it by the exclosure treatment, we can see that as the recovery from the caterpillar outbreak progressed, forb species diversity plateaued and was sustained on grazed ambient plots, but declined subsequently on exclosed ambient plots (figure 7.10). In addition, on examining data from the grazed plots only, it is appears that experimental warming substantially reduced the diversity of forbs (Post 2013). If we further consider only warmed plots, we can see that the lowest diversity of forb species occurred on plots that were exclosed from vertebrate herbivory in addition to receiving the warming treatment (figure 7.10). Thus, in this instance, exploitation interactions played a role in the recovery response of forb diversity to warming. Further complicating matters is the second caterpillar outbreak. In this case, it appears that herbivory by large mammals influenced the recovery response of forb diversity to the second outbreak, with warming having little to no effect. In 2010, at the peak of the second caterpillar outbreak, diversity was lowest on warmed exclosed plots, that is, those on which large mammals were prohibited from grazing (figure 7.10). In contrast, forb diversity was greater by a factor of nearly 50 to 300 percent on grazed plots, whether they were warmed or not, though that warming reduced substantially the diversity of forbs on grazed plots (Post 2013).

In such species-poor systems as this arctic tundra site, where forb diversity averages approximately four to eight species per square meter, it is tempting to speculate that random events play a more prominent role in the presence or absence of species than they do in more species-rich systems such as the Tropics. If this is the case, indirect, climatically driven events such as may be exemplified by this caterpillar outbreak may have a greater influence on biodiversity and subsequent trajectories of recovery in the Arctic than they would have in the Tropics. On the other hand, although the Arctic may be species-poor, the numbers of individuals per species per unit area may be greater there than in the species-rich Tropics, where numbers of individuals per species per unit area are characteristically lower. This, too, undoubtedly shapes the importance of random events in determining the trajectory of biodiversity recovery from disturbance.

Tilman's landmark investigations of the diversity-stability relationship in a Minnesota grassland (Tilman and Downing 1994; Tilman 1996, 1999) catalyzed a number of subsequent studies and debate (Givnish 1994) concerning the nature and biological meaning of this relationship. The central relationship uncovered by Tilman was one of increasing community stability with increasing species richness (Tilman and Downing 1994). In this case, stability was defined as the ratio of the community-level biomass following drought, a pulse perturbation, to that preceding the drought. The tendency of the community to return to predisturbance biomass, an indication of stability, increased with species richness on the experimental plots on which biomass was monitored (figure 7.11). Examining longer-term temporal stability in the dynamics of this community, Tilman and colleagues subsequently utilized a statistical quantification of stability as the inverse of the coefficient of variation of community biomass through time (Tilman et al. 2006). This metric likewise displays a positive association with diversity (figure 7.11b), indicating that more species-rich plant communities are rendered temporally stable by increasing diversity of the species they comprise (Tilman et al. 2006).

This latter observation has direct implications for the role of species diversity in community stability under climate change. As we saw in chapter 6, the temporal stability of communities in a climate change context relates to the stability of the populations of which they are composed and to the nature of the species interactions that characterize them as vertically or laterally structured. Population stability and species interactions, in turn, are vulnerable to the potentially perturbing effects of climatic fluctuation (Levin 1970; Post and Forchhammer 2001; Ellis and Post 2004). It appears that the interactions

(a)   **(b)**

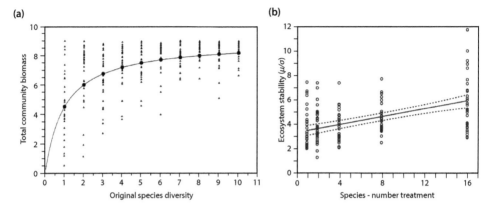

Figure 7.11. Increase in ecosystem stability, defined as drought resistance (a), and defined as the inverse of variability in aboveground biomass through time (b), in association with plant species richness in a north temperate grassland. *Modified from Tilman and Downing (1994), Tilman (1999), and Tilman et al. (2006).*

between climate change and species interactions in the diversity-stability relationship have been understudied, if at all.

The multi-annual data on species diversity of arctic forbs under warmed and ambient and grazed and exclosed conditions at my long-term study site may shed some light on this subject. As mentioned above, I began monitoring species composition among forbs on my long-term plots in 2005. Biomass has been quantified annually on these plots since 2003, but prior to 2005 the biomass of forbs was quantified as a composite estimate only. Resolution of the forb data to species level allowed for an assessment of the relationship between species diversity and stability according to four treatments: exclosed, ambient; exclosed, warmed; grazed, ambient; and grazed, warmed. In this experiment, the plots exposed to herbivory but not warmed (grazed, ambient) represent the double control plots, or the natural state of the system. As described in chapter 1, warming was achieved passively using cone-shaped, open-top chambers that elevate near-surface temperature on the plots by about 1–3°C on average (Post et al. 2008a), and exclosure of resident large herbivores, caribou and musk-oxen, was achieved by fencing off three 800 m² treatment sites.

Across treatments, there is a clear increase in the stability of the forb community with species diversity (figure 7.12). In this example, I quantified stability as the inverse of the coefficient of variation in plot-scale community biomass among years (Tilman et al. 2006), from 2005 through 2011 (Post 2013). Diversity was quantified as the mean number of forb species per plot over the same period. Although this relationship is not as strong as that reported originally

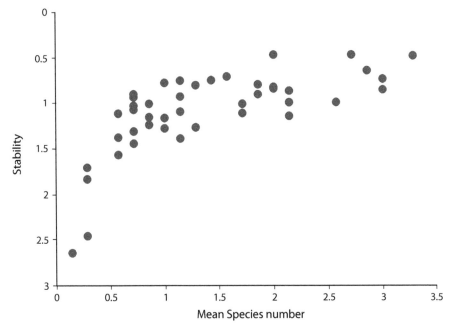

Figure 7.12. Stability of the forb community at the study site near Kangerlussuaq, Greenland, from 2005 through 2011, defined as the inverse of the coefficient of variation of community biomass versus the mean number of species per plot on which biomass was measured over that seven-year period. The y-axis is oriented to depict an increase in stability (i.e., decrease in variability) with increasing mean species number. *From Post (2013).*

for the Minnesota grassland studied by Tilman (Tilman and Downing 1994; Tilman 1996), it is still highly statistically significant (Post 2013).

However, both removal of large herbivores and experimental warming appeared to reduce, if very modestly, forb diversity following recovery from the caterpillar outbreak mentioned earlier (figure 7.10). The consequences of this for the strength of the stability-diversity relationship may be examined by parsing the relationship in figure 7.12 into the individual correlations by treatment. This exercise reveals that the exclosure treatment reduced the strength of the relationship between forb species diversity and the long-term stability of biomass production by the forb community (figure 7.13). In other words, exploitation by large herbivores appears to have been integral to maintenance of the stability-diversity relation. A comparison of the Spearman correlation coefficients for this relationship reveals two further, interesting aspects. First, the strongest relationship between forb diversity and community stability occurred on grazed plots that were also warmed, and second, removal of large

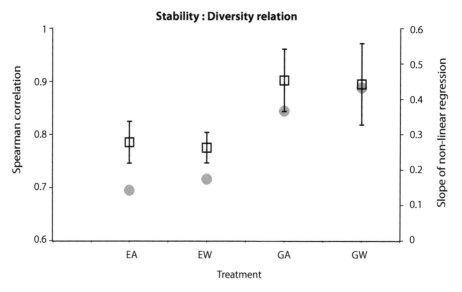

Figure 7.13. Variation in the strength of the relationship between forb species diversity and forb community stability across experimental treatments at the study site near Kangerlussuaq, Greenland, 2005–11. Treatments are exclosed ambient (EA), exclosed warmed (EW), grazed ambient (GA), and grazed warmed (GW). Spearman correlation coefficients are plotted as solid circles, and nonlinear regression slopes (± 1 SE) are plotted as open squares. *From Post (2013).*

herbivores reduced the strength of this relationship on warmed plots by approximately 20 percent (figure 7.13) (Post 2013).

What explains the erosion of the stability-diversity relationship under the exclosure and warming treatments? There are many potential explanations that are beyond the scope of the data presented here, including those relating to the roles of warming and herbivory in competition among forb species and consequent changes in productivity. However, it is interesting to note that the variation among treatments in the strength of the correlation between diversity and stability does not appear to mirror variation in mean stability among the treatments (figure 7.13). In fact, the greatest plot-scale stability, in terms of the lowest interannual variability in biomass throughout the study period, was observed on grazed, warmed plots (Post 2013). Rather, it appears to be the case that variation in the strength of this relationship among treatments is more closely mirrored by variation in the mean number of species per plot among treatments (Post 2013). Hence, although herbivory appears to promote plant species diversity, likely by suppressing dominance by the plant community of a single species, warming tends to work in opposition to this, seemingly because it facilitates domination of the plant community by one or a few species (Post and Pedersen 2008), with consequences for community stability.

TROPICAL DEFORESTATION
AND CLIMATE CHANGE

A potentially catastrophic nexus exists between climate change and deforestation in the Tropics. The most urgent points of concern over the tropical deforestation crisis center on the loss of biodiversity and the loss of carbon sequestration potential for the entire biosphere. The former is the focus of this section, while the latter will be addressed at greater length in the next chapter. In the first case, rates of species loss and extinction risk are closely related to loss of total forest area and forest fragmentation (Lovejoy et al. 1986). Although most of the Earth's biodiversity lies in the Tropics, many species likely remain as yet unidentified by Western science; the potential for biodiversity loss with tropical deforestation is, hence, likely underestimated. In the second case, tropical forests contribute substantially to the global carbon budget (Grace et al. 1995), and quantifying rates of deforestation is therefore of central importance to estimating net carbon gains and losses. For instance, although the terrestrial biosphere transitioned from a neutral carbon balance during the 1980s toward becoming a net carbon sink in the 1990s, the Tropics remained largely carbon neutral, suggesting that an increase in carbon uptake there was offset by carbon losses through deforestation (Schimel et al. 2001). Uncertainties estimating rates of deforestation and net carbon gain or loss from Tropical forests and changes in these rates through time are discussed in the next chapter.

The aggregate absolute rate of tropical deforestation in the Americas, Asia, and Africa totaled approximately 15.5 million hectares per year through the 1980s (Laurance 1998, 1999). Analysis of satellite imagery indicates that the rate of deforestation of the humid tropical forest biome on the same continents between 1990 and 1997, and from 2000 to 2005, averaged 5.4–5.8 million hectares of forest loss per year, with the total loss during the latter period equaling 2.36 percent of the existing biome area (Achard et al. 2002, Hansen, Stehman, et al. 2008b). In the Brazilian Amazon alone, the annual rate of deforestation increased on average through the 1990s to approximately 1.5 million hectares per year (Laurance 1999), and reached, between 2000 and 2005, nearly 50 percent of total humid tropical forest loss (Hansen, Stehman, et al. 2008). Although the neotropics are subject to the greatest absolute rate of deforestation, a rate approximately twice that seen in the Asian and African tropics, the Asian Tropics experience the greatest relative rate of deforestation—that is, the amount of deforestation relative to the amount of forest in total (Laurance 1999). The single greatest factor contributing to variation in the rate of deforestation among regions is the size of the human population in those areas: forests are cleared for development, agriculture, and timber use in some of the fastest-growing parts of the world (Laurance 1999). Growing human populations throughout the Tropics will continue to exert enormous pressure on the integrity of tropical forests through direct removal of forest tracts for agricultural use of land and for timber products associated with expansion of existing urban populations. For instance, satellite-based estimates of tropical deforestation for the period 2000–2005 indicate that the rate of forest loss was related primarily to the annual growth rate of human populations in associated tropical urban centers in Africa, Asia, and Latin America, but most strongly so in Latin America (DeFries et al. 2010). By contrast, during the 1980s and 1990s, deforestation was primarily associated with expanding rural populations in the Tropics (Ehrhardt-Martinez et al. 2002; Jorgenson and Burns 2007).

As we saw in chapter 1, climate change in the Tropics is expected to alter rainfall patterns spatially and temporally, but drying is expected to accompany warming across broad regions currently harboring extensive tracts of tropical rain forest. Deforestation is expected to interact with climate change in the Tropics by rendering forest fragments more susceptible to degradation. For instance, climate change is likely to increase drying, fire frequency, and wind disturbance at forest fragment edges, thereby increasing tree mortality (Laurance 2004) and exacerbating forest fragmentation (Brodie et al. 2010).

Tropical forest dynamics associated with climate change are, however, far from straightforward. In Amazonian forests, tree mortality on long-term observational plots has increased, but so have rates of growth and recruitment (Laurance et al. 2004). This apparent paradox can best be understood as an overall increase in the complexity of tropical forest dynamics. Rapidly growing canopy and emergent species have increased in abundance, while subcanopy species with slower growth rates have declined (Laurance et al. 2004). These dynamics are apparently a response to increasing atmospheric CO_2 concentrations and greater rates of carbon assimilation in faster-growing than in slower-growing species; no concurrent trends in precipitation, which might also have contributed to the recent dynamics, were observed over the study period (Laurance et al. 2004).

An especially troubling aspect of tropical climate change is the potential for forest drying to increase human access to remote and disconnected forest fragments. Such a development could lead, for example, to more effective harvest of economically valuable tree species as well as more efficient use of fire as a means of converting forested areas to tillable land or livestock pastures (Brodie et al. 2010). Drying associated with warming in the Tropics would thereby increase forest fragmentation, as well as the efficacy of anthropogenic forest burning (Brodie, Post, and Laurance 2012), which in turn would exacerbate extinction risk as predicted by island biogeography theory (MacArthur and Wilson 1967; MacArthur 1972). The association between forest fragmentation or deforestation and extinction risk derives from the species-area relationship predicted by the theory of island biogeography (MacArthur and Wilson 1967), which, when applied in reverse, predicts species loss on the basis of diminishing habitat availability (Simberloff 1972; Connor and McCoy 1979; Gaston 2000). Forest clearing in the eastern United States has, for example, precipitated the loss of four species of birds in that region, and may have put a total of twenty-eight bird species at risk of extinction (Pimm and Askins 1995). By comparison, experimental deforestation in the Brazilian Amazon indicated increased extinction risk with total area cleared in thirty-six of fifty-four focal bird species (Ferraz et al. 2007). Ongoing deforestation in Southeast Asia has led to the extinction of between 34 percent and 87 percent of species in some taxonomic groups, with further extinctions of endemic species expected within the next century (Brook et al. 2003). The lag time between habitat destruction and realized extinction, known as the extinction debt (Tilman et al. 1994), relates to the persistence of subpopulations, and in some cases individuals, in relict habitat pockets, and presents the possibility of threshold dynamics in extinction (Brooks and Balmford 1996; Brooks et al. 1999; Hanski and Ovaskainen 2002). Typically, species display a negative exponential relationship between time since isolation owing to habitat destruction and expected persistence; the so-called half-life of species is the time required for 50 percent of the number of species originally present in a continuous tract of habitat to go extinct (Brooks et al. 1999). Approximately one-half of Earth's species of vascular plants and one-third of all species of terrestrial vertebrates occur exclusively in twenty-five biodiversity hotspots (Brooks et al. 2002), and the locations of these hotspots overlap with regions at greatest risk of developing novel climates or experiencing loss of existing climates over the next century (Williams et al. 2007). Although resilience to climatic fluctuation is prevalent among species that have persisted through millennia of natural climatic variability (Dawson et al. 2011), in species with low adaptive capacity, such as habitat specialists characteristic of the tropics and polar regions, climate change may, in fact, hasten the arrival of the threshold in the relationship between habitat loss and extinction debt. As mentioned in chapter 1, the difference between prehistoric climate change and present, ongoing climate change is the role of modern humans in habitat destruction and direct exploitation of natural systems.

BIODIVERSITY, CLIMATE CHANGE,
AND HUMAN EXPLOITATION

Just as human exploitation of tropical forests for wood products and conversion to agriculturally arable land may interact with climate change to exacerbate deforestation and biodiversity loss, human exploitation of wildlife may interact with climate change to hasten the demise of species already at risk of suffering declines due to hunting. Almost paradoxically, pressure on wildlife populations is actually greater at the margins of protected areas than in rural regions away from protected areas because it is at the margins of protected areas where growth rates of human populations are greatest and increasing (Wittemyer et al. 2008). A recently developed, individual-based model of population dynamics of red howler monkeys indicates that increased access to hunted populations increases local extinction risk through disproportionate increases in unit-area specific population losses and increased infanticide resulting from turnover of males in hunted troops (Wiederholt et al. 2010). If climate change improves access by humans to currently remote or difficult to access regions of the Tropics, it is likely also to increase extinction risk in exploited wildlife populations. The interaction among climate change, human access, and exploitation could be even more devastating for such species because populations of ateline primates in the neotropics also covary negatively with the El Niño–Southern Oscillation, suggesting that elevated temperatures and reduced precipitation are detrimental to these species (Wiederholt and Post 2010).

Similar interactions may accrue in the Arctic, where climate-induced restrictions to human access to hunted populations and areas may be relieved by warming. In northeastern Greenland, for example, recent increases in the harvest of narwhals appear to have been related to increased hunter access to the animals in Smith Sound resulting from temperature-related changes in sea ice conditions (Nielsen 2009). Similarly, any increase in resource extraction, mineral exploration, and associated infrastructure development in the Arctic associated with increased access to remote areas fostered by climate change is likely to increase the disturbance of wildlife species of preeminent cultural and economic importance (Johnson et al. 2005). Such synergisms do not appear to be under consideration currently by management, conservation, or policy groups (Post and Brodie 2012).

CHAPTER 8

Ecosystem Function and Dynamics

Through their interactions with each other and with the abiotic environment, organisms transport energy and mineral nutrients within and among communities, across landscapes and the surface-subsurface boundary, between terrestrial and aquatic or marine environments, and between the biosphere and the atmosphere. At any stage along the multiple pathways of energy and nutrient exchange, climate change may act to alter the strengths of these interactions. The ecosystem is the most comprehensive interactive framework in ecology and, perhaps for this reason, difficult to conceptualize in a consistent manner. Consider, for instance, the two heuristic representations in figure 8.1. The first depicts the structure of a simplified ecosystem linking plant-based and detritus-based food webs through arrows denoting consumption, the reverse of the direction of energy flow through the system (figure 8.1a) (Schmitz 2010). The second similarly depicts inorganic nutrient and energy flow through compartments of a food web that is itself a compartment of an abstracted ecosystem (figure 8.1b) (Loreau 2010). At the surface, there are multiple subtle differences between these idealized representations of ecosystem structure. One contains fungivores, detritivores, decomposers, and organic litter (figure 8.1a), arguably the engines and the source, respectively, of inorganic nutrient flow through the system, none of which are represented in the other (figure 8.1b), probably because they are collected in the further abstracted pool of inorganic nutrients.

At a coarser scale, however, these two models are actually quite similar in comparison to the type of conceptualization of ecosystem structure and function that is more typically considered a tractable and useful depiction by ecosystem modelers (figure 8.2). In this representation, the key components and fluxes are represented without regard to the actual species, functional groups, or trophic positions of the organisms involved. Instead, temperature, primary production, and heterotrophic respiration interact with primary producers, nutrient availability, moisture, and litter to determine net ecosystem production (NEP), or the capture of carbon from the atmosphere. As is abundantly evident, there are almost as many ways in which to conceptualize ecosystems as there are ecosystems themselves.

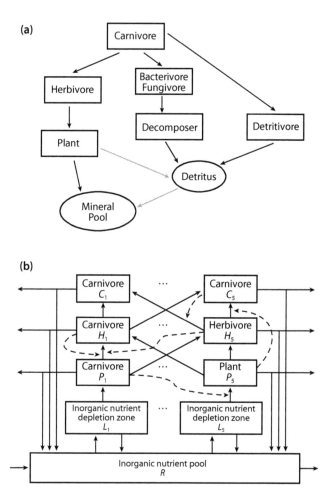

Figure 8.1. Two representations of the ecosystem concept. In panel (a), arrows represent pathways of consumption among organisms and between organisms and the detritus and inorganic nutrient pools. Modified from Schmitz (2010). In panel (b), arrows represent the flow of energy among and away from compartments that are organized according to their trophic positions. *Modified from Loreau (2010).*

Yet, while others have suggested that there is an inherent conceptual difficulty in merging individual- and population-based processes and dynamics with those occurring at the ecosystem scale, arguing instead for a unified theoretical approach (Loreau 2010), this chain of hierarchical interactions could not, in my opinion, be more apparent and intuitive. So far in the book we have transitioned from individual-based life history strategies to population dynamics, and from there to community dynamics and their biodiversity implications,

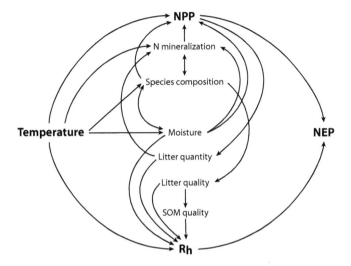

Figure 8.2. A more generalized conceptualization of an ecosystem in which abiotic conditions, such as temperature, influence net primary productivity (NPP) and heterotrophic respiration (R_h), and thereby influence net ecosystem production (NEP), through processes involving nutrient mineralization, plant community composition, and soil organic and inorganic properties. SOM, soil organic matter. *Modified from Shaver et al. (2000).*

arriving finally at a consideration of the ecosystem consequences of variation in all of these combined.

I would, however, agree with Loreau (2010) in that perspectives on ecosystem dynamics appear to represent two endpoints of a gradient from abiotic control over nutrient dynamics and primary productivity to organism-environment interactions that drive energy and nutrient flow, though without much overly obvious middle ground between these endpoints. Characterizations of ecosystem dynamics fall generally into two broad conceptualizations: those that place the emphasis on environmental parameters and nutrient and energy flow and those that place the emphasis on species interactions and nutrient and energy flow (Loreau 2010; Schmitz 2010). In both conceptualizations there is considerable potential for climate change to alter the strength of interactions determining the flow of energy and nutrients through ecosystems. Most prominent among the candidate features of ecosystems vulnerable to alteration by climate change is carbon exchange between the atmosphere and biosphere. This is because carbon uptake by ecosystems has the potential to ameliorate or even offset anthropogenic fertilization of the atmosphere with CO_2, while carbon emission by ecosystems can exacerbate it. At this point, it is important to distinguish between ecosystem function and ecosystem dynamics. The former term refers to the manner in which the ecosystem of interest works, and interactions among its component parts and fluxes, including biotic and abiotic compartments. The latter refers to variation in ecosystem function through time in response to perturbations that are continuous or stochastic in nature, or in relation to changes in ecosystem components. Hence, the study of ecosystem dynamics derives from an understanding of ecosystem function, and this, in turn, depends critically on successful identification of the important drivers within the ecosystem.

Inevitably, a discussion of ecosystem function and dynamics boils down to the factors that influence and contribute to variation in NEP. Net ecosystem production is the result of net primary productivity (NPP) and ecosystem respiration (ER) and, according to the Woodwell-Whittaker model, which has represented the state of the science for more than forty years (Chapin et al. 2008), can be expressed simply as

$$NEP = GPP - ER = NPP - HR, \qquad (8.1)$$

in which GPP represents gross ecosystem photosynthesis and HR represents ecosystem heterotrophic respiration, and ER can be further decomposed to autotrophic respiration (AR) plus HR (Woodwell and Whittaker 1968). In terms of ecosystem processes, NPP represents the uptake of carbon by plants during photosynthesis, which can ultimately result in carbon sequestration if atmospheric carbon is locked away in long-lived or recalcitrant dead plant tissue. ER, by contrast, represents the release of carbon back to the atmosphere by plants and through decomposition or trace gas release by soil microorganisms. Whereas at high latitudes, NPP occurs mainly during summer, ER occurs year round, though with reduced rates during winter. Many of the ecosystem components that contribute to the processes of NPP and ER are temperature dependent (figure 8.2). In addition to influencing rates and magnitudes of NPP and ER directly, temperature can also influence rates of soil nutrient mineralization, the species composition of plant communities, and soil moisture content (Shaver et al. 2000). The rates and magnitude of NPP are in turn influenced by, aside from temperature directly, soil nutrient mineralization rates, species composition of the plant community, and soil moisture content. ER is influenced by, aside from temperature, the quality of soil organic matter and litter, litter quantity, and soil moisture (Shaver et al. 2000). As we will explore later in this chapter, there is considerable potential for species interactions such as herbivory to influence many of these pathways, and to interact with temperature in doing so.

The global distribution of seasonal and annual patterns of temperature and precipitation regimes is a key determinant of the distribution and primary productivity characteristics of Earth's biomes (figure 8.3) (Reich et al. 1997).

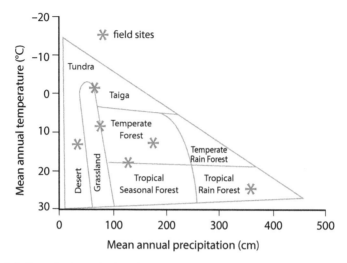

Figure 8.3. Generalized organization of Earth's biomes along gradients of mean annual temperature (*y*-axis) and mean annual total precipitation (*x*-axis). Asterisks indicate positions of field sites within the parameter plane from which the zonation of biomes has been derived. *From Reich et al. (1997).*

Although simplistic, the distribution of biomes within the parameter space defined by variation in mean annual precipitation and mean annual temperature in figure 8.3 is useful for drawing generalized expectations of the potential for long-term increases in mean annual temperature and coincident long-term trends in mean annual precipitation at regional scales for biome shifts. For instance, the tundra biome falls at one extreme, with the lowest annual temperatures and precipitation; warming coincident with increases in mean annual precipitation would be expected, therefore, to lead to a shift toward a boreal forest or taiga biome (figure 8.3). In comparison, the drying of tropical regions coincident with expected increases in mean annual temperature (Rind 1998) has the potential to shift the tropical rain forest biome toward a tropical seasonal forest (figure 8.3). Indeed, both large-scale rainfall exclusion experiments (Nepstad et al. 2002) and empirical analyses of interannual variability in precipitation and primary productivity (Brando et al. 2010) have demonstrated declines in aboveground NPP associated with reduced precipitation in tropical forests.

In the latter study, however, which linked a reduction in remotely sensed primary production with declines in wet-season rainfall and plant-available water over a nine-year period in the Amazon basin, the results applied only in open-canopy forest; in closed-canopy forest an increase in leaf flush and, by extension, photosynthesis accompanied the observed regional-scale drying (Brando

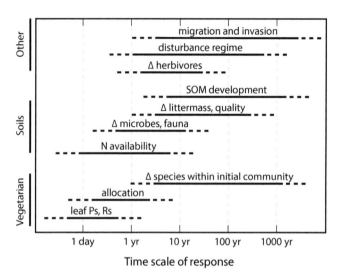

Figure 8.4. Scales of temporal response of factors influencing ecosystem function. SOM, soil organic matter; P$_s$, photosynthesis; R$_s$, respiration. *Adapted from Shaver et al. (2000).*

et al. 2010). This dichotomy illustrates the complexity of predicting ecosystem response to climate change deriving from the myriad scales of processes involved in ecosystem function. The processes contributing to ecosystem function depicted in figure 8.1 are likely to display widely divergent temporal scales of response to warming. For instance, leaf-level rates of photosynthesis, patterns of carbon allocation among plant tissues, and changes in availability and turnover rates of soil nutrients are likely to respond to rising temperatures on scales of days to months or years (figure 8.4) (Ryan 1991; Shaver et al. 2000). By comparison, changes in plant community composition, the evolution of soil organic matter characteristics, the invasion of plant communities by new species, and the concomitant effects of all of these processes on ecosystem function are more likely to unfold over decades to centuries (figure 8.4) (Shaver et al. 2000; Ryan and Law 2005). The inherently disparate scales of space and time over which the processes influencing ecosystem function are likely to respond to climate change are reflected in the myriad approaches to studying ecosystem response to climate change.

STABILITY, DIVERSITY, AND ECOSYSTEM RESILIENCE

Before examining the wide variation in methodology used in studying ecosystem response to climate change, it is worth taking time to revisit the relationship between diversity and stability introduced in the previous chapter. Doing so may reveal broad patterns that bear on the variation in response of ecosystems to climate change across biomes. As we saw, community stability tends to increase with species diversity, and we should expect the same relationship to accrue for ecosystem stability. By way of a brief reiteration, this relationship owes to the operation of either niche complementarity, trait selection and the sampling effect, or the portfolio effect (Chapin, Walker, et al. 1997). According to the portfolio effect, a broad diversity of species implies that response heterogeneity to environmental disturbance will result in overall stability of the community or its resilience. This effect likely also captures functional redundancy among species, in which the contribution of one species to, say, community-level biomass production may be compensated for in case of the loss of that species by persistence of another species. In its inverse, the portfolio effect also predicts that regions or biomes with inherently low functional redundancy among species may be particularly vulnerable to environmental change (Chapin, Walker, et al. 1997). Regardless of the mechanism, however, the relationship between species diversity and stability appears robust.

In general terms, the diversity-stability relationship can be depicted as log-linear in form (figure 8.5a) (Tilman 1996). Along the x-axis in figure 8.5 we can place the Arctic and the Tropics at two ends of the continuum of species diversity, with very low diversity in the Arctic and very high diversity in the Tropics. Hence, even though the diversity-stability relationship should be expected to apply in both regions, the y-axis intercept must differ widely between them. Similarly, the relationship between functional redundancy and stability can be depicted as approximately linear, with the Arctic and the Tropics again depicting extremes along a continuum represented by variation along the x-axis (figure 8.5b). In general, we might expect to encounter very low functional redundancy among species—defined as the number of redundant species—in arctic communities, but much higher functional redundancy among tropical species.

We can justify the log-linear assumption for a relationship between diversity and stability (Tilman and Downing 1994; Tilman 1996, 1999) by realizing that stability must approach some threshold beyond which only very minor increases in stability accrue with further addition of species. If, then, the relationships in figure 8.5 are in general effective representations of what we might expect in natural systems, two important insights emerge. The log-linear relationship between stability and species diversity suggests that the slope of this relationship should be greater in species-poor than in species-rich systems. Hence, in systems in which net species loss is an expected outcome of climate change, such as in the Arctic (Yurtsev 1997), there should be a risk of erosion of stability, as we have already seen in the previous chapter. Such a loss of stability should occur, however, more rapidly or with greater probability for

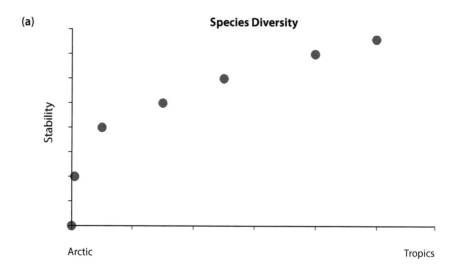

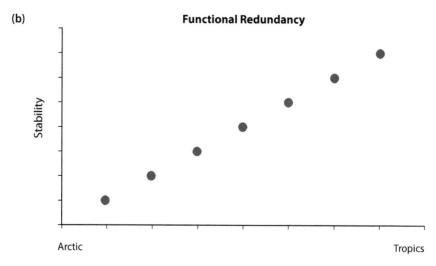

Figure 8.5. Adaptation of relationships between ecosystem stability and species diversity (a) and functional redundancy among species (b) according to a generalized gradient in species diversity from the Arctic (species-poor) to the Tropics (species-rich).

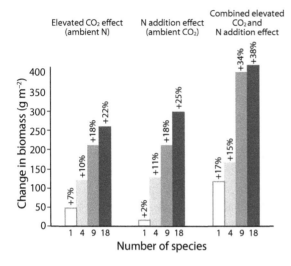

Figure 8.6. Diversity dependence of community-scale biomass response to elevated CO$_2$ only (left), nitrogen (N) fertilization only (middle), and elevated CO$_2$ and nitrogen combined (right). Responses to each manipulation individually, and to their combined effect, increased with the number of species on the experimental plots. *Modified from Reich et al. (2001).*

each species lost in species-poor systems, such as the Arctic, than in species-rich ones, such as the Tropics. We might expect, then, that ecosystem dynamics are more vulnerable to rapid and pronounced environmental change, such as climate change, in the Arctic than in the Tropics.

Experimental evidence from a study in Minnesota grasslands that utilized a combination of the treatments simulating climate change (explored below) provides evidence of diversity dependence in community-scale biomass response to simulated global change (Reich et al. 2001). The increase in biomass documented in this study in response to experimentally elevated atmospheric CO$_2$ concentration, soil nitrogen availability, and their combination increased with plant species diversity on experimental plots (figure 8.6) (Reich et al. 2001). We might expect, on the basis of such studies, that species losses will have consequences for ecosystem response to continued climate change, and that such consequences will be greater in the Arctic than in the Tropics because of lower species diversity in the Arctic. Although there is evidence to support the first of these expectations, the differences in approaches to investigating ecosystem-scale responses to expected climate change among studies in different systems makes a direct comparison problematic.

NUTRIENT, TEMPERATURE, AND CO$_2$ MANIPULATIONS

Early investigations of the anticipated effects of climate change on ecosystem function and dynamics focused, in arctic systems, on manipulating nutrient availability. The premise behind this approach was that northern ecosystems are inherently nutrient limited (Chapin 1978; McCown 1978; McKendrick et al. 1978; Ulrich and Gersper 1978), and rising temperatures were likely to influence ecosystem function by increasing soil microbial activity and, thereby, soil nutrient dynamics and availability (Kielland and Chapin 1992; Nadelhoffer et al. 1992; Chapin, Hobbie, et al. 1997; Nadelhoffer et al. 1997).

The passive warming manipulation utilized at my field site and described in the opening chapter of this book has been used across the Arctic and other high-latitude systems, including in the Antarctic, to simulate temperature increases expected to occur over the next century as a result of climate change (Henry and Molau 1997). Meta-analyses of the results of multi-annual experiments following the protocol of the International Tundra Experiment, for which the use of open-top warming chambers was designed, have demonstrated increases in plant stature and biomass in response to warming across several sites, but have also identified substantial variation among plant functional groups in the direction and magnitude of response (Arft et al. 1999; Walker et al. 2006). There also appears to be a general tendency at high latitudes for woody plants to become increasingly dominant under experimental warming (Chapin et al. 1995; Walker et al. 2006; Post and Pedersen 2008; Elmendorf et al. 2012), which is expected to pose further implications for the carbon uptake potential of the tundra biome. Most notably, on moist tussock tundra, dwarf birch accounted for 25 percent of total plant biomass on control sites but for 65 percent of biomass on fertilized plots after nine years of treatment (Chapin et al. 1995; Shaver and Jonasson 1999). Increases in dwarf birch in fertilized and warmed plots nearly eliminated ericaceous shrubs such as *Ledum palustre* and *Vaccinium vitis-idaea* after nine years (Chapin and Shaver 1996).

Temperature, and by extension warming, are not, however, the sole drivers of primary production and ecosystem dynamics at high latitudes. The additional importance of nutrient turnover and availability to ecosystem function at high latitudes is highlighted by the fact that long-term experiments in Alaska and Sweden have revealed that warming alone does not consistently increase net aboveground primary productivity (NAPP) and plant biomass (Shaver and Jonasson 1999). In both Alaskan wet sedge and tussock tundra, the same experiments indicate that nitrogen fertilization alone, at the rate of 10 grams per square meter per year, resulted in significant increases in NAPP and aboveground biomass, while significant increases in biomass were obtained in heath and fellfield in arctic Sweden in response to nitrogen fertilization at the rate of 4.9 grams per square meter per year (Shaver and Jonasson 1999). In combination, warming and nitrogen fertilization increased NAPP on tussock and wet sedge tundra in Alaska, but increased total biomass only on wet sedge tundra

(Shaver and Jonasson 1999). In Alaskan tussock tundra, the addition of nitrogen and phosphorus increased biomass and the production of deciduous shrubs but reduced the growth of evergreen shrubs (Chapin et al. 1995). In arctic Sweden, warming plus fertilization increased biomass more than just fertilization alone (Shaver and Jonasson 1999). Across the five ecosystems studied in these experiments in Alaska and Sweden, the clearest consistency was the association of the greatest changes in biomass with fertilization treatments; temperature treatments matched fertilization treatments only in evergreen shrubs in one of the Swedish ecosystems (Shaver and Jonasson 1999).

At the community level, field experiments in the Arctic have revealed that warming or the combination of warming and fertilization leads to a decline in species diversity, driven primarily by a loss of forbs (Shaver and Jonasson 1999). A nine-year warming and nutrient-addition experiment at Toolik Lake in Alaska resulted in declines in the abundance of evergreen shrubs and the complete loss of forbs from experimental plots (Chapin et al. 1995). Evidence from subarctic systems indicates as well that shifts in species composition of plant communities in response to increases in temperature may also have consequences for soil nutrient availability. For example, long-term warming of soil and vegetation in a subalpine meadow produced changes in litter input to the soil as a result of a shift in community composition from forbs (which produce labile litter) to shrubs (which produce more recalcitrant litter) (Harte and Shaw 1995). This community shift subsequently translated into altered carbon cycle dynamics that promoted substantial losses of carbon from the soil (Harte et al. 2006).

As we will examine later in this chapter, such community-level responses to warming may be mediated to some extent by herbivory. For instance, herbivory by caribou and reindeer has been associated with a decline in forb species cover in northern Norway (Bråthen and Oksanen 2001) and with diebacks of dwarf birch in Canada (Manseau et al. 1996) and at my study site in West Greenland (Thing 1984). The community-level effects of herbivory might also become a potentially important component of biospheric feedbacks to climate in the Arctic because colonization by deciduous shrubs would allow greater biomass accumulation and carbon uptake than is possible in a graminoid-dominated community (Shaver et al. 1998; Cahoon et al. 2012).

The effects of climatic warming on the overall carbon balance of ecosystems (NEP) are difficult to extrapolate from the responses of plants or communities to warming because warming can directly increase both NPP, which is important in the uptake of carbon by the ecosystem, and ER, the efflux of carbon from the ecosystem. Moreover, both NPP and ER may be influenced indirectly by the effects of warming on, for example, nitrogen availability and plant species composition (Oechel et al. 1995, 1997). As well, a response to warming or increased nutrient availability may be minor at the ecosystem level because the responses of individual plant species may effectively cancel each other out if the biomass and productivity of some are enhanced while those of others are reduced (Chapin and Shaver 1985). Hence, the most meaningful insights into the effects of warming on ecosystem carbon balance that derive from warming experiments are those based on multi-annual studies. A soil warming experiment conducted in a mixed deciduous, north temperate site within the Harvard Forest, for example, revealed a cumulative net loss of carbon from the experimental site compared with an adjacent control site over the seven-year duration of the experiment (Melillo et al. 2011). This net loss was attributed to stimulation of soil microbial activity by the warming treatment. Microbial respiration on the warmed site accounted for 82 percent of total soil respiration and exceeded that on the control site, where it accounted for 74 percent of total soil respiration, for the duration of the experiment, resulting in net carbon losses from the system even though the aboveground vegetation storage of carbon was also greater on the warmed site than on the control site (Melillo et al. 2011).

In addition to several experiments that have utilized temperature manipulations to investigate ecosystem response to projected climate change, others have utilized CO_2 fertilization experiments. Interest in the ecosystem consequences of CO_2 enrichment is linked to the recognition that CO_2 emissions from human activity are a major contributing factor to recent, ongoing, and potentially future climate change. Aside from providing insights into the effects of elevated atmospheric CO_2 on primary productivity, species turnover, and ecosystem function and dynamics, experiments employing CO_2 enrichment may also help inform us of the potential for carbon uptake by plants and the capacity of ecosystem carbon uptake to mediate the effects of anthropogenic CO_2 emissions.

The terrestrial biosphere contains approximately 600 petagrams of carbon locked in plant biomass, mostly in trees in tropical regions (Olson et al. 1983). The capacity for the terrestrial biosphere to sequester carbon is expected to increase with rising atmospheric CO_2 concentrations, especially in conjunction with rising temperatures and increased nitrogen availability, but the capacity for natural systems to absorb human emissions of carbon may be limited to approximately 20 percent of the carbon expected to be emitted by humans over the next century (Körner 2000). Currently, it is estimated that terrestrial and marine components combined take up approximately 50 percent of the CO_2 emitted by humans, with oceans comprising an annual carbon sink of approximately 2.2 petagrams, and the land carbon sink amounting to approximately 2.8 petagrams of carbon annually, for the period 2000–2006 (Canadell et al. 2007). While the terrestrial biosphere was in a state of neutral carbon balance during the 1980s, it shifted to a state of net carbon uptake in the 1990s, likely as a result of large-scale patterns of changing land use in the Northern Hemisphere, including abandonment of agricultural areas (Schimel et al. 2001; Houghton 2003). At the end of this period, the northern mid-latitudes were estimated to operate, as a result of land-use changes, as a net carbon sink amounting to approximately 2 petagrams of carbon taken up annually from the atmosphere (Houghton 2003). In contrast, estimates of carbon balance for the Tropics are variable and range from the region being a net carbon source to approximately carbon neutral (Houghton 2003). Uncertainty in estimating the carbon balance of tropical regions will be examined subsequently in this chapter, but it appears that massive carbon losses from the neotropics, as a result of deforestation related to urban population growth and the agricultural exports of forest products (DeFries et al. 2010), are nearly balanced by large terrestrial carbon sinks in the Northern Hemisphere, on the order of between 0.6–0.7 petagrams and 0.6–2.7 petagrams of carbon per year, depending on the method of estimation (Goodale and Davidson 2002; Goodale et al. 2002). It is important at this stage in the chapter to emphasize that responses to experimental CO_2 fertilization may not be entirely useful for estimating, even grossly, the carbon sequestration potential of the biomes, ecosystems, or communities in which such studies are conducted because these responses are short term, transient, and not necessarily indicative of longer-term processes occurring at larger scales (Körner 2000).

The free-air carbon exchange (FACE) plot approach employs a ring of CO_2 fertilization towers to elevate atmospheric CO_2 concentration by approximately 200 ppm above ambient concentration over fairly large forest plots of several hundred square meters (Hendrey 1992; Hendrey et al. 1999). This design allows the manipulation of CO_2 levels at and above the canopy of experimental plots without the unwanted side effects that might arise from the use of open-top greenhouses infiltrated with CO_2 (Körner 2000), and allows manipulations over large areas to test for ecosystem responses to expected future atmospheric conditions, independent of temperature changes (Hendrey et al. 1999). CO_2 enrichment experiments have revealed that productivity and biomass responses of vegetation to CO_2 fertilization are rarely consistent among plant tissues, plant species or functional groups, or ecosystems (Körner 2000) and in some cases are closely dependent on nitrogen availability (Crous et al. 2010).

For instance, during the first three years of a long-term FACE plot experiment at the Oak Ridge National Laboratory experimental facility in the U.S. state of Tennessee, treatment plots initially displayed an increase in carbon allocation to woody stems, but this response attenuated after the first year, beyond which carbon allocation to more labile tissues such as leaves and fine

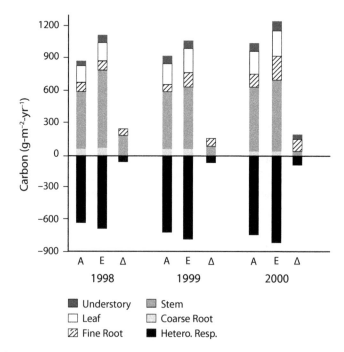

Figure 8.7. Decline in increased carbon allocation to woody stem tissue (dark gray shaded portions of bars) from the first year of CO_2 fertilization (1998) through the third year of CO_2 fertilization (2000), suggesting an attenuation of the fertilization effect that may be characteristic of woody plant response to elevated atmospheric CO_2 over longer periods. Within years, A indicates ambient conditions, E indicates elevated CO_2, and Δ represents the increase in carbon allocation under elevated CO_2 conditions. *Modified from Norby et al. (2002).*

roots increased (figure 8.7) (Norby et al. 2002). This suggests that transient responses to elevated CO_2 may reflect an initial investment of carbon in structural tissues with high carbon sequestration potential but subsequent investment in photosynthetic tissues that also turn over more quickly. This raises a key issue in the study of ecosystem response to elevated CO_2, namely, whether rising concentrations of atmospheric CO_2 will elicit greater ecosystem carbon uptake and biomass accumulation or whether they will result merely in increased rates of carbon cycling through ecosystems. The latter seems to be indicated by additional data emerging from the Oak Ridge FACE study. After six years of CO_2 fertilization, forest plots displayed a doubling of fine-root production by trees under 200 ppm atmospheric CO_2 concentration above ambient compared to those on control plots (Norby et al. 2004). Of note, investment in fine roots accounted for the greatest single fraction of the observed increase in NPP on experimental plots (Norby et al. 2004). Moreover, there appears to be a consistent level of NPP stimulation across deciduous forest FACE plots on the order of a 21–25 percent increase above that on control plots, and most of this increase in NPP appears to be attributable to an increase in leaf area in response to rising CO_2 concentrations (Norby et al. 2005).

A comprehensive meta-analysis of 120 publications stemming from twelve separate FACE experiments revealed that the greatest responses to CO_2 enrichment seem to occur in deciduous trees and shrubs, and the strongest of these responses appears to be in leaf area (Ainsworth and Long 2005). That such responses may indicate a transient response to elevated CO_2 that may not be indicative of longer-term responses (Körner 2000) is also suggested by a decline after six years of CO_2 enrichment in the NPP response observed at the Oak Ridge site (figure 8.8a), although this decline appears to be related to the exacerbating effects of increasing limitation by nitrogen availability, which declined as the treatment forest stand matured (Norby et al. 2010). This latter notion appears to be supported by maintenance of carbon allocation to woody tissue in response to CO_2 fertilization on plots that also received a nitrogen fertilization treatment, and a decline in carbon allocation to woody tissue on plots that received no such nitrogen addition but remained under CO_2 fertilization (figure 8.8b) (Norby et al. 2010).

Early microcosm experiments with cores taken from coastal wet tundra revealed that increased CO_2 alone had no effect on the biomass of vascular plants, but that nitrogen addition coupled with CO_2 enrichment enhanced net CO_2 uptake (Billings et al. 1977, 1978, 1984). Subsequent in situ experiments on upland tussock tundra revealed that elevated CO_2 led to a tripling of carbon gain by tundra, but this was followed by relatively rapid homeostatic adjustment, suggesting little potential for long-term carbon sequestration (Grulke et al. 1990). Similarly, a four-year CO_2-enrichment field experiment in alpine grassland resulted in no long-term increase in aboveground biomass (Körner et al. 1997), though increased biomass was observed in the initial years of the experiment (Niklaus et al. 2001). This initial biomass response to CO_2 enrichment was greatest in communities with the highest number of species (Niklaus et al. 2001), echoing the results mentioned above of experiments in temperate grassland and forest ecosystems indicating that species-rich plant communities show a greater NPP response to CO_2 enrichment and nitrogen fertilization than do species-poor communities (DeLucia et al. 1999; Oren et al. 2001; Reich et al. 2001).

Field experiments with CO_2 enrichment have shown that such manipulations may also elicit shifts in plant species composition. In alpine grassland, for example, five years of CO_2 enrichment produced a shift in species composition, with a significant increase in cover of graminoids, including *Festuca* species

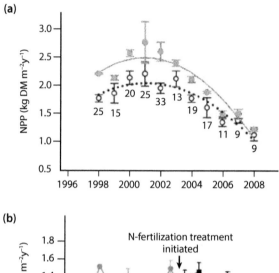

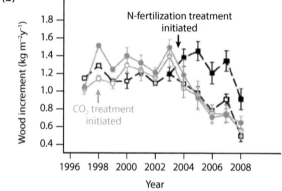

Figure 8.8. (a) Decline in tree biomass growth response to elevated atmospheric CO_2 at the Oak Ridge forest FACE plot, as indicated by both the trend in the treatment means (solid circles) and their difference from control means (open circles) over the course of the experiment. (b) Nitrogen (N) fertilization (shaded squares) promoted a sustained allocation to woody growth that otherwise diminished in unfertilized trees (open squares) on plots under ambient (open circles) and elevated (solid circles) CO_2 regimes. *From Norby et al. (2010)*.

(Niklaus et al. 2001). Similarly, in a sandstone grassland community, Joel and colleagues (2001) documented an increase in proportional biomass of the grass *Bromus* species in response to a combination of elevated CO_2 and nutrient addition; the same species declined, however, in response to the same treatment in serpentine grassland. Important insights emerging from these studies are that the biomass response to CO_2 enrichment appears to depend on nutrient availability in a range of vegetation and ecosystem types, and the productivity response to CO_2 enrichment appears to increase with plant species diversity. A further important consideration in this context is that biomass carbon sequestration is linked to the residence time of carbon, not just to the rate of carbon fixation (Körner 2000), and the residence time of carbon can be greatly reduced by the foraging activities of large herbivores (Kielland et al. 1997).

CARBON DYNAMICS AND PROJECTED RESPONSES TO GLOBAL CLIMATE CHANGE

A major focus of studies of the ecosystem consequences of observed and expected changes in climate is carbon dynamics, or the exchange of carbon between the biosphere and the atmosphere. Long-term monitoring stations on Mauna Loa, Hawaii, and near Point Barrow, Alaska, both in the United States, have recorded seasonal and annual CO_2 dynamics since 1958 and 1961, respectively. Although these records demonstrate an increase in atmospheric CO_2 concentration since measurements began, they reveal little to no trend in the *amplitude* of the annual CO_2 cycle prior to the mid-1970s, but a steadily increasing trend since then (figure 8.9) (Keeling et al. 1996). The amplitude of the

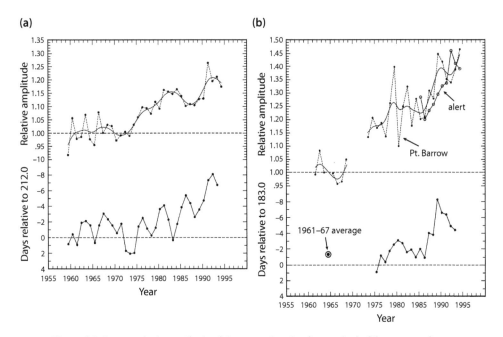

Figure 8.9. Increases in the amplitude of the seasonal cycle of atmospheric CO_2 concentration since the mid-1970s at Mauna Loa, Hawaii (a) and Point Barrow, Alaska (b), and associated advances in the timing of the annual downward zero-crossing, or CO_2 uptake by vegetation, at both locations. *Modified from Keeling et al. (1996).*

annual CO_2 cycle quantifies the difference in atmospheric CO_2 concentration during Northern Hemispheric winter, when atmospheric CO_2 concentration increases as a result of reduced seasonal NPP, and Northern Hemispheric summer, when it declines because of increased seasonal NPP (Fung et al. 1987). In general, high-latitude systems display the greatest amplitude in the annual atmospheric CO_2 cycle because of the pronounced seasonality of primary production at high latitudes (Fung et al. 1987). An increase in the amplitude of the annual cycle at both Mauna Loa and Point Barrow has been interpreted to suggest that the annual drawdown of CO_2 from the atmosphere has increased in the Northern Hemisphere since the mid-1970s (Keeling et al. 1996). Other potentially contributing factors might include warming-associated increases in plant respiration during winter or heterotrophic respiration by soil microorganisms during winter (Chapin et al. 1996). Nonetheless, the increase in the amplitude of the CO_2 cycle coincides with an advance in the timing of the onset of the annual period of downward zero crossing, the point during the annual cycle when it begins to decline, indicating uptake of CO_2 by vegetation (figure 8.9, lower panels—"Days relative") (Keeling et al. 1996). This relationship suggests that the increase in the amplitude of the CO_2 cycle relates not necessarily to increased NPP during a fixed period of the year but rather to a lengthening of the annual growing season in both regions (Keeling et al. 1996).

The increase in the amplitude of the annual CO_2 cycle reported by Keeling and colleagues (1996) measured 20 percent at the Hawaiian site and 40 percent at the Arctic site, and correlated with regional surface temperatures at both sites. This suggests that temperature increases, which, as we saw in chapter 1, have been twice as great in the Arctic as elsewhere, may elicit substantially greater effects on biosphere-atmosphere carbon dynamics in the Arctic than in other regions. This is not only interesting academically but should be cause for concern as well because of the enormous potential for exacerbation of atmospheric CO_2 concentrations if carbon emissions from arctic ecosystems increase. Despite having been a sink for atmospheric CO_2 during the 1980s and 1990s, some arctic ecosystems may have begun functioning since then as a net source of carbon to the atmosphere owing to a warming-related increase in soil heterotrophic respiration (Oechel et al. 1993), although the extent to which the Arctic functions overall as a net carbon source or sink is likely to depend on changes to soil moisture regimes across the tundra biome (McGuire et al. 2009). Adding complexity to this issue, the carbon dynamics of arctic ecosystems have been characterized as displaying net carbon uptake during the growing season and net carbon efflux during winter (Oechel et al. 1997; Vourlitis and Oechel 1997, 1999; Mastepanov et al. 2008; Zulueta et al. 2011). Nonetheless, an analysis of long-term (1960–98) carbon flux data from several tundra sites in the Alaskan Arctic revealed, despite summer carbon sink activity, a net release of CO_2 to the atmosphere on the order of 40 grams of carbon per square meter annually (Oechel et al. 2000). Moreover, arctic ecosystems constitute a major source of atmospheric methane (McGuire et al. 2009), a much more powerful greenhouse gas than CO_2. As well, arctic tundra constitutes the largest single soil carbon pool on Earth, exceeding the sum of four tropical forest pools, despite accounting for only 28.6 percent of the total surface area covered by tropical forests (Post et al. 1982). The most recent estimate of the magnitude of the soil organic carbon pool in the circum-Arctic tundra region represents a doubling of earlier estimates (Billings 1987), placing it at approximately 1,672 petagrams of carbon, or half the total global soil organic carbon pool (Tarnocai et al. 2009). As will be examined in detail in a subsequent section in this chapter, animals may have the potential to substantially influence the carbon dynamics of arctic ecosystems in opposition to the effects of warming. Recent evidence also indicates that plant pathogens may similarly have the potential to reduce carbon uptake of plants in arctic ecosystems. An outbreak of a parasitic fungus of plants increased shoot mortality of the dominant species, *Empetrum hermaphroditum*, in the focal community, ultimately reducing net ecosystem exchange significantly below that on unaffected plots (Olofsson et al. 2011).

The nutrient-addition and warming manipulations described above revealed trends toward increasing availability of soil nitrogen and phosphorus with warming; increased growth and abundance of deciduous shrubs; reduced growth and abundance of evergreen shrubs and mosses; and reduced plant species diversity, primarily of evergreen shrubs and forbs (Chapin et al. 1995; Chapin and Shaver 1996). Although these responses canceled each other out at the ecosystem level, in terms of total biomass production (Chapin et al. 1995) they nonetheless indicate the potential of such changes to influence important ecosystem dynamics because of the influence of plant community composition on litter quality, soil chemistry, and nutrient turnover rates (Hobbie 1992). The warming response of ecosystem CO_2 flux, for instance, may be highly sensitive to species-specific responses to expected effects of climate change.

A long-term, multi-annual field experiment at Toolik Lake, Alaska, in which nitrogen and phosphorus fertilization treatments were applied singly and in combination with a warming treatment on wet sedge tundra, illustrates this point (Shaver et al. 1998). Both GEP and NEP at the ecosystem scale increased with warming over the course of this experiment, but NEP actually declined in response to the warming plus nutrient-addition treatment over the first nine years of this experiment (Shaver et al. 1998). This was explained as being due to a greater increase in ER, which included soil and plant respiration, in response to the warming plus nutrient-addition treatment than the increase elicited in GEP (Shaver et al. 1998). As this experiment continued, eventually spanning two decades, nutrient addition simulating the increased nutrient availability and mineralization expected to occur in response to warming resulted in a net ecosystem loss of approximately 2 kg of carbon per square meter over the duration of the experiment (Mack et al. 2004). This net ecosystem carbon export occurred primarily from deep soil layers, presumably as a result of increased soil microbial activity, and more than offset carbon gains to the system through primary productivity, which doubled in response to the nutrient-addition treatment (Mack et al. 2004).

I recall here that NEP is the difference between GEP and ER. Hence, increases in aboveground NPP in response to warming or nutrient additions should not be expected to translate directly into increased carbon uptake by the system because of the complicating influence of the response of ER, and in particular heterotrophic respiration, to these expected features of climate change. Moreover, a further interesting detail emerging from this experiment is that species-level dynamics at the community scale play an important role in ecosystem response to simulated and, by extension, expected effects of climate change. In this experiment, control plots were dominated by cotton grass (*Eriophorum angustifolium*), a common wet-sedge tundra species, but treatment plots displayed an increasing domination by another sedge, *Carex cordorrhiza*. This latter species exhibited greater rates of photosynthesis per unit mass, which apparently contributed to differences in ecosystem CO_2 flux between treatment and control plots, independent of changes in aboveground biomass (Shaver et al. 1998).

A closer look at plant community composition and ecosystem characteristics highlights further important differences in the response of carbon flux to warming and nutrient addition among community types. The study site at Toolik Lake, Alaska, contains both moist nonacidic tundra and moist acidic tundra, and there is evidence to suggest that these two communities display divergent responses to fertilization simulating an expected increase in soil nutrient availability with warming. In both community types, the addition of nitrogen at a level of 10 grams per square meter and phosphorus at half that level increased biomass and aboveground NPP, but in different ways (Hobbie et al. 2005). On moist acidic tundra, the increases were due primarily to the response of dwarf birch (*Betula nana*), but on moist nonacidic tundra, where dwarf birch is virtually nonexistent, community-level increases in biomass and aboveground NPP reflected responses by the sedge *Carex* species, the shrub *Rhododendron* species, and several species of forbs (Hobbie et al. 2005). As well, canopy-level Normalized Differential Vegetation Index (NDVI) measurements made on experimental plots at the Toolik Lake site revealed differences between sedge and tussock tundra in the relationship between NDVI and aboveground biomass, and in the fit of NDVI as a predictor of NEP, GEP, and ER that relate to heterogeneities in plant functional group composition of the local communities (Boelman et al. 2003, 2005).

Hence, the response to nutrient addition in one community type was due almost exclusively to the dominant species occurring there, whereas on the other site it was due to responses by several species. These differential responses likely hold implications for future carbon flux dynamics in the two community types. Because dwarf birch is a woody species with a strong biomass response to warming, its dominance of plant community composition on acidic tundra suggests that increased nutrient availability in association with climate change will improve the carbon sequestration potential of this community type (Chapin, Hobbie, et al. 1997; Hobbie and Chapin 1998; Hobbie et al. 2005). In contrast, there is little potential for increased carbon sequestration in response to increasing nutrient availability in nonacidic tundra, but rather greater potential for carbon turnover because of enhanced production by nonwoody species with higher rates of tissue turnover and more labile litter (Hobbie et al. 2000, 2005).

Considering the effects of experimental and observed warming on plant community composition in these relatively simple arctic systems, we may wonder what the consequences of warming-induced changes in primary productivity, species turnover, or shifts in species dominance might be for short-term CO_2 flux. An early soil warming experiment conducted at a north temperate deciduous forest site in Harvard Forest in the U.S. state of Massachusetts used heating cables buried at a depth of 10 cm to elevate soil temperature by 5°C and examined the consequences of this manipulation for carbon flux (Peterjohn et al. 1994). Compared to ambient plots, the efflux of CO_2 from soil nearly doubled in response to the warming treatment, and was estimated to have increased by 538 grams of carbon per square meter (Peterjohn et al. 1994). As Peterjohn and colleagues (1994) noted, the relationship between soil temperature and CO_2 efflux was exponential within their study site, a relationship that appears to be common among sites distributed across the Northern Hemisphere (figure 8.10a) and globally (figure 8.10b) (Raich and Potter 1995). This suggests that soil warming has the potential to increase rates of carbon emission from soils nearly universally, at least until limited by soil drying. The net effect of climate change on global soil carbon stocks is, however, an issue of considerable complexity, depending not only on the influences of warming and drying on heterotrophic respiration but also on potential increases in carbon input to soils through increases in primary productivity (Davidson and Janssens 2006).

At Toolik Lake, Alaska, a warming treatment using passive open-roofed greenhouses to elevate surface temperature by 4°C resulted in increases of approximately 50 percent in gross ecosystem photosynthesis and ER (Hobbie

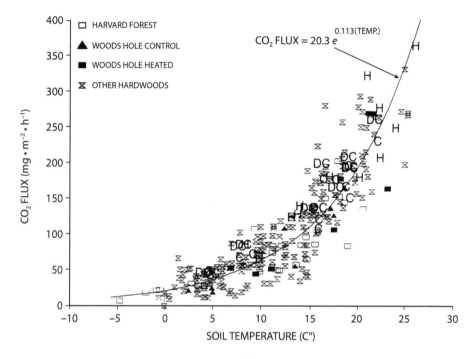

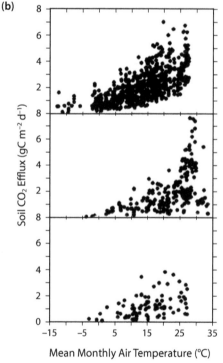

Figure 8.10. (a) Nonlinear increase in soil carbon efflux with increasing temperature from an analysis of deciduous hardwood forests in Massachusetts (Harvard Forest and Woods Hole), Minnesota, Missouri, and Tennessee in the United States; and from the United Kingdom and Italy, indicating the generality of the form of this relationship. (b) Similar relationships, but with mean monthly air temperature from studies representing moist biomes without a dry season, biomes with a dry season, and wetland biomes. *Panel (a) adapted from Peterjohn et al. (1994). Panel (b) adapted from Raich and Potter (1995).*

and Chapin 1998). However, as a consequence of increases in both uptake and release of carbon, there was no change in NEP (Hobbie and Chapin 1998). Notably, however, the increases in GEP and ER attenuated across the growing season, and were mainly apparent only during the early part of the growing season. This suggests the response may have been at least partly driven by phenological advancement in response to the warming treatment.

Overhead electrical heaters were used to elevate near-surface temperature on experimental plots in two types of alpine plant communities, dry and moist meadows, at the Rocky Mountain Biological Laboratory in the U.S. state of Colorado during a long-term warming experiment initiated in 1991 to test for the effects of warming on plant community dynamics (Harte and Shaw 1995; Harte et al. 1995) and, subsequently, ecosystem CO_2 dynamics (Saleska et al. 1999). The early results from this experiment indicated that ecosystem carbon uptake was significantly reduced by warming, by approximately 133 grams of carbon per square meter over the course of the growing season, but only on dry meadow plots, and only during mid-summer (Saleska et al. 1999). The early reduction in carbon uptake on warmed plots appears to have been driven by a shift in the timing of diurnal carbon uptake that resulted in a shortening of the length of the daily period when the community constituted a carbon sink, but not by changes in soil respiration (Saleska et al. 1999). Concomitantly, soil nitrogen mineralization rates initially doubled in response to warming on xeric plots, but not on mesic plots, and then declined to baseline levels as soil organic matter diminished (Shaw and Harte 2001). Over the next ten years of experimental warming at this alpine meadow site, soil organic carbon declined to levels 22 percent lower in warmed than in control plots, while the biomass of sagebrush increased and that of shallow-rooting forbs declined (Harte et al. 2006; Cross and Harte 2007). The reduction in soil organic carbon on artificially warmed plots was eventually mirrored on control plots during a multi-annual drought from 2000 to 2004 (Harte et al. 2006).

Studies of wintertime ecosystem function in cold environments typified by prolonged snow cover have recently revealed additional important components of the carbon cycle. The timing of snowmelt in spring at high latitudes and at high elevations reflects the interaction between total snow accumulation and temperatures, and can be an important driver of the onset of the annual growing season and other ecosystem properties (Forchhammer et al. 2005, 2008; Høye et al. 2007). As was noted in chapter 1, observational data indicate that snow cover is diminishing at high latitudes in response to rising global temperatures (Post, Forchhammer, et al. 2009; Brodie and Post 2010). Experimental manipulation of snow cover at high-latitude and high-elevation sites is performed out of interest in examining the ecosystem consequences of altered snowfall and snow persistence regimes and is accomplished using fences that accumulate snow on the windward side. Such manipulations can elevate subnivean temperature by, in some cases, up to 25°C above temperatures under ambient snow, and delay the snowmelt date by two to three weeks compared to the timing of snowmelt on ambient snow plots (Walker et al. 1999).

In response to elevated subnivean wintertime temperatures, CO_2 efflux from soils can increase up to twofold during the depth of winter (Walker et al. 1999). Observational studies have noted a similar increase in efflux of CO_2 from subnivean soils during winter in response to rising temperature. For instance, at a High Arctic site and a northern boreal forest site in Alaska, soil carbon emissions increased with soil temperature in the same nonlinear manner as did rates of carbon efflux from soils in the other systems reviewed herein (figures 8.11a and b) (Sullivan et al. 2008). Note that the temperature gradient along the x-axis of each of the panels in figure 8.11 is observational and derives from seasonal increases in temperature rather than experimental manipulation. Nonetheless, the relationships in figure 8.11 suggest that the rate of carbon efflux may increase with each 1°C of soil warming by 0.29 grams of carbon per square meter per day in High Arctic soils and by nearly double that rate, 0.44 grams of carbon per square meter per day, in upland boreal forest soils. The difference between the rates of carbon efflux from soils at these two sites may relate to the different depths at which soil temperature was monitored on each of them, 10 cm and 5 cm, respectively (Sullivan et al. 2008). However, the estimate of carbon flux from subnivean soil during winter at the High Arctic site was consistent with that reported for other tussock tundra sites (Sullivan et al. 2008) and suggests that wintertime warming is likely to play a considerable role in changes to the annual carbon budget of high-latitude systems.

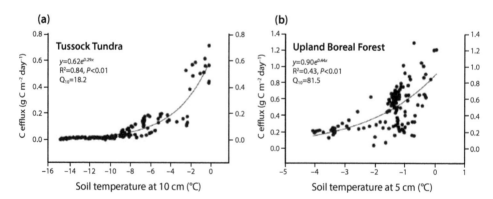

Figure 8.11. Relations between carbon efflux from soils and soil temperature in tussock tundra at Toolik Lake, Alaska (a), and a boreal forest site in south-central Alaska (b). Note the different soil depths at which temperature was measured at the two sites. Nonetheless, soil carbon emissions increased nonlinearly with temperature in both systems. Q_{10} indicates the increase in efflux associated with a 10 degree Celsius increase in temperature. *Modified from Sullivan et al. (2008).*

Northern fen and bog ecosystems in the U.S. state of Minnesota were the focus of a mesocosm experiment designed to test for the effects on CO_2 and CH_4 emissions of elevated temperature and changes in water-table depth (Updegraff et al. 2001). Excised soil and vegetation monoliths from both ecosystems were placed in holding tanks in the laboratory, where water level was manipulated and soil temperature was elevated using overhead infrared radiators. In response to warmer soil temperatures, at the highest treatment level of approximately +4°C, CO_2 emissions increased by 20–30 percent in both fen and bog types across all water-table levels, but the increases appeared greater in fen than in bog samples (Updegraff et al. 2001). Methane emissions were universally higher in bog than in fen, and increased most in response to warming—by almost 50 percent—in the high-water-table treatment (Updegraff et al. 2001). Hence, soil warming has considerable potential to increase carbon emissions to the atmosphere in bog and wet fen ecosystems, which make up nearly one-third of Earth's soil carbon stores (Gorham 1991; Wardle 2002), but the magnitude of this response may be mediated, in the case of CH_4 emissions, to some extent by soil drying.

The results of a similar experiment conducted in the field at a wet sedge tundra site near Barrow, Alaska, reveal, however, partially opposing patterns to those reported above from the mesocosm experiment in Minnesota. Huemmrich and colleagues (2010) elevated soil temperatures by 5°C above ambient temperature using subsurface heaters attached to the inside of cylinders embedded in the soil, and measured CO_2 flux on warmed and ambient plots and in relation to water-table manipulations. ER increased with soil temperature, but did so differently for each level of water-table manipulation, with the strongest relationship observed in the reduced-water-table treatment (Huemmrich et al. 2010). In fact, for every 1°C increase in soil temperature, ER increased at a twofold greater rate in the reduced-water-table treatment plots than in the elevated-water-table treatment plots, where the relationship between soil temperature and CO_2 efflux all but disappeared (figure 8.12). Water-table level appears to have exerted a primary influence over carbon flux at this site because, by the end of the growing season, sites exposed to the reduced-water-table treatment were all net sources of carbon regardless of whether or not they simultaneously experienced experimental warming, but all other sites were net carbon sinks (Huemmrich et al. 2010).

At a montane site in the Swiss Alps, a soil warming manipulation was performed using heating cables laid on the surface of experimental plots, achieving a warming effect of +4°K, and during which soil carbon efflux and plant carbon uptake were measured (Hagedorn et al. 2010). The focal plant community in this experiment was alpine treeline, consisting of several conifer

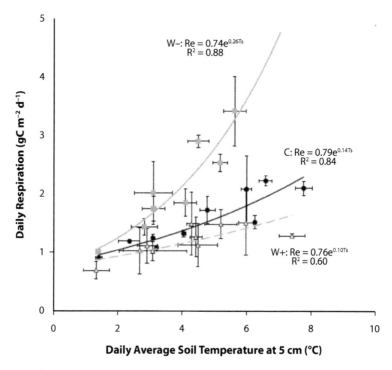

Figure 8.12. Effect of water-table depth on the relationship between soil temperature and soil respiration rate (grams of carbon emitted per square meter per day) at a moist sedge tundra site near Barrow, Alaska, USA. Black symbols represent mean rates of respiration at various experimentally manipulated soil temperatures under ambient conditions, while gray and white symbols represent means on plots experiencing lowered and elevated water-table levels, respectively. *Adapted from Huemmrich et al. (2010).*

species with an understory of ericaceous shrubs and forbs. In response to soil-surface heating, soil CO_2 efflux increased exponentially, as in the other studies described herein, and achieved a rate exceeding the premanipulation efflux by approximately 120 grams of carbon per square meter per year, a 45 percent increase above ambient conditions (Hagedorn et al. 2010). This increase in soil carbon emissions apparently resulted from stimulation by warming of decomposition of existing soil organic matter rather than from increased root respiration (Hagedorn et al. 2010). In every case examined herein, rates of carbon efflux from soils bear an exponential relationship to soil temperatures. Globally, soil carbon emissions average, at a minimum, approximately 68 petagrams of carbon annually, and are expected to increase with future warming (Raich and Schlesinger 1992).

It is tempting at this point to run a simplistic comparison of rates of change in photosynthesis versus heterotrophic respiration in response to rising air and

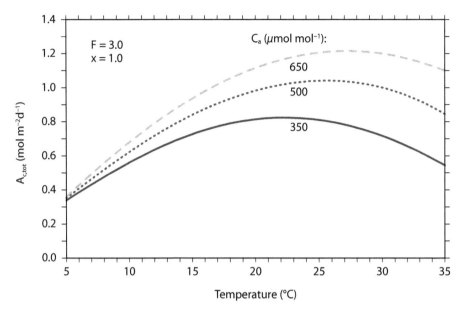

Figure 8.13. The apparently robust nature of the relationship between net canopy CO_2 uptake and air temperature across three different levels of atmospheric CO_2 concentration. The initial increase in CO_2 uptake with temperature is inevitably followed by a decline as temperatures continue to rise. The point at which this curve crosses the curve describing the increase in heterotrophic respiration, or release of carbon from soils, with increasing temperature in the preceding examples will determine the transition from net sink to net source of atmospheric carbon in northern biomes. *Adapted from Long (1991).*

soil temperatures. Generally, in C3 plants, which dominate at northern latitudes, rates of photosynthesis increase with temperature as rubisco production and efficiency initially increase, but then decline with additional warming as a result of stomatal closure as plants attempt to conserve moisture (figure 8.13). A higher concentration of carbon in the atmosphere may render photosynthesis more efficient despite the trade-off inherent to the need to conserve water, but the general, inverse parabolic shape of the relationship between photosynthetic rates and temperature appears robust across varying levels of atmospheric CO_2 concentration (figure 8.13) (Long 1991). Consider, now, the relationship between rates of heterotrophic respiration—carbon emission from soils—and temperature that has been documented in multiple examples in this chapter, one of exponentially increasing soil carbon emissions with soil temperature owing to increasing soil microbial activity with warming. The transition of ecosystems from net carbon sinks to net carbon sources depends on the threshold at which carbon efflux due to respiration exceeds carbon uptake due to photosynthesis, as the examples herein have illustrated. Might northern ecosystems, then, become on balance net sources of carbon under the continued warming expected with rising atmospheric CO_2 levels over the next century?

A canopy CO_2 flux model developed recently for Low Arctic ecosystems (Shaver et al. 2007; Street et al. 2007) suggests that the consequences for short-term ecosystem-atmosphere carbon exchange of vegetation response to warming may have little to do with the species composition of arctic plant communities and more to do with standing live biomass and canopy leaf area. Recognizing the tremendous degree of variation among ecosystems across the Arctic in terms of their capacities for primary productivity, rates of nutrient cycling, and carbon stores, all of which can differ by up to three orders of magnitude across the region (Callaghan et al. 2005), Shaver and colleagues (2007) developed their canopy flux model as a means of simplifying the quantification of CO_2 flux rates. Application of the model to measurements taken across six sites in arctic Alaska and Sweden revealed the considerable predictive capacity of canopy leaf area for CO_2 flux. Approximately 80 percent of the variance in net ecosystem exchange was explained by the combination of an index of canopy leaf area based on NDVI measurements taken at the plot level, surface air temperature, and photosynthetically active photon flux density (Shaver et al. 2007).

TROPICAL DEFORESTATION, CARBON TURNOVER, AND MODEL PROJECTIONS OF CHANGES IN CARBON DYNAMICS

The discussion of carbon dynamics to this point has focused mainly on extratropical, especially high-latitude, systems. A biosphere-level understanding of carbon dynamics in relation to climate change requires taking into account carbon dynamics in tropical forests. Whereas I will argue later in this chapter that animal influences may represent an important biotic factor in ecosystem response to climate change in northern ecosystems, in this section the importance of human land use, especially deforestation, in the tropical ecosystem response to climate change will be emphasized as the biotic factor of overriding importance.

The total amount of carbon released to the atmosphere through deforestation since the Industrial Revolution amounts to approximately one-half of the amount released through burning of fossil fuels, and approximately two-thirds of that resulting from deforestation derives from the Tropics (Houghton 1999; Malhi et al. 1999). Among the world's forests, the greatest standing pool of carbon in vegetation occurs at low latitudes, in tropical forests, whereas the largest pool of soil carbon is found at high latitudes, especially in the boreal

(a)

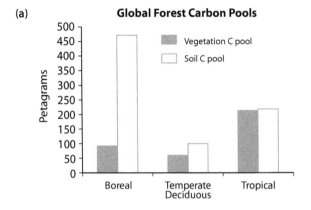

(b)

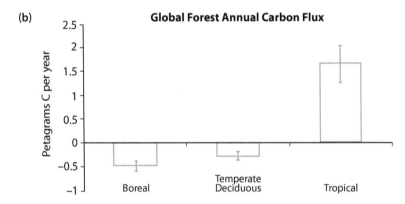

Figure 8.14. (a) Magnitude of global forest carbon pools according to forest type and vegetation and soil pools. (b) Magnitude of the total annual global carbon flux in petagrams (10^{15} g) of carbon by forest types. Negative values in panel (b) indicate carbon draw-down from the atmosphere, while positive values indicate carbon release to the atmosphere. Tropical forests are a net carbon source of greater total magnitude than the combined carbon sinks of boreal and temperate deciduous forests. *Adapted from Dixon et al. (1994).*

forest (figure 8.14) (Dixon et al. 1994). Indeed, soil carbon stocks in high-latitude forests are estimated to exceed the vegetation carbon pool of all global forests combined by approximately 30 percent, or more than 100×10^{15} grams (Dixon et al. 1994). The highest rates of carbon flux among the world's forests, however, currently occur in tropical forests, and these fluxes indicate substantial carbon release to the atmosphere (figure 8.14b) (Dixon et al. 1994). In contrast, carbon fluxes in temperate deciduous and boreal forests, although substantially smaller than those occurring in tropical forests, indicate annual net carbon uptake from the atmosphere (figure 8.14b) (Dixon et al. 1994).

Quantifying the magnitude of carbon flux on regional and global scales is, however, problematic and characterized by variation in estimates of the extent of carbon uptake and release by the Earth's major biomes (Achard et al. 2007; Ramankutty et al. 2007). The values in figure 8.14 derived from Dixon and colleagues (1994), for example, estimated total annual carbon efflux from low-latitude tropical forests at 1.65 petagrams and total annual carbon uptake by boreal and temperate forests at 0.74 petagrams. More recent assessments have noted, however, that regional- and global-scale estimates of forest carbon balance vary according to whether they are based on so-called inverse modeling and inversion calculations or on analyses of forest inventories and land-use change (Houghton 2003; Ramankutty et al. 2007). According to the former method, tropical forests act as a net source of carbon to the atmosphere of 1.2 petagrams annually, while boreal and temperate forests act as a net carbon sink from the atmosphere of 2.4 petagrams annually (Houghton 2003). By contrast, according to the latter method, tropical forests release approximately 2.2 petagrams of carbon to the atmosphere annually, while boreal and deciduous forests take up only 0.03 petagrams of carbon annually from the atmosphere.

Generally, rates of carbon turnover should be higher in tropical forests than in temperate forests or high-latitude and tundra alpine systems because of higher rates of decomposition and soil nutrient turnover owing to warmer soil temperatures and high soil moisture regimes. Hence, while soil carbon reserves may be lower in the Tropics than in colder, higher-latitude systems such as the Arctic, the tremendous amount of standing biomass in the form of mature forest trees in the Tropics renders them among the largest standing stocks of aboveground carbon on the planet, amounting to three to four times the amount of carbon in vegetation compared to that in extratropical regions (Dixon et al. 1994). For this reason, deforestation through logging and burning is a major source of carbon release to the atmosphere from tropical systems (Houghton 2003), and the management of tropical forest clearing has the potential to become a powerful tool for climate change mitigation (Canadell and Raupach 2008).

It has been estimated that almost 50 percent of mature tropical forest has been removed to date (Wright 2005) as land has been cleared for agriculture, livestock grazing, and development. In addition to the intentional use of fire as an agent of forest clearing, unintentional burning may have accounted for a greater proportional loss of mature forest in the eastern Amazon basin during the late 1990s than intentional deforestation itself (Cochrane et al. 1999). While deforestation is the single largest source of carbon loss from tropical forests attributable to human land use, accounting for up to 71–96 percent of carbon efflux in those regions (Houghton 2003; Achard et al. 2004), forest harvest for timber products is the single greatest source of carbon loss from temperate and boreal forests (Houghton 2003). Across all tropical regions, more than half of the total land area put into use for agriculture derives from cleared forests (Gibbs et al. 2010). In humid tropical forests, typified by, for example, the Brazilian Amazon, the rate of deforestation during the period 1990–97 averaged nearly six million hectares of forest loss annually, with an additional degradation of nearly three million hectares annually (Achard et al. 2002). By comparison, between 2000 and 2005, humid tropical forests were cleared at an average rate of 4.5–5.4 million hectares annually, despite efforts to curb deforestation, with nearly half of the total clearing occurring in Brazil alone (Hansen, Stehman, et al. 2008).

Drought conditions such as those deriving from El Niño events can drive carbon loss from tropical forests directly, as well as interact with logging to increase the risk of further loss of tropical forest to fires, and have the potential to double regional carbon emissions under such conditions (Nepstad et al. 1999). The 2005 drought that affected the Amazon rain forest, for example, reversed a local trend toward carbon uptake and resulted in a net loss of between 1.2 and 1.6 petagrams of carbon from the forest basin that year (Phillips et al. 2009). Forest fragmentation exacerbates tropical deforestation by increasing access to previously remote forest regions and, worldwide, facilitated further degradation of tropical forests at a rate in excess of two million hectares annually through the 1990s (Achard et al. 2002), as well as releasing enormous quantities of carbon to the atmosphere. Approximately 3×10^9 tons of tropical forest biomass is burned annually worldwide, one-third of which is released to the atmosphere (Fearnside 2000). During the 1990s, tropical deforestation worldwide amounted to a release of an estimated 3 petagrams of carbon annually (Pan et al. 2011). Over the ensuing decade, this number declined to 2.82 petagrams of carbon released annually by tropical deforestation, due to greater tropical forest regrowth during the period 2000–2007 compared to 1990–99 (Pan et al. 2011). By contrast, carbon uptake by high-latitude boreal forests accounted for only 0.5 petagrams annually, offsetting only one-sixth of the carbon emitted annually by tropical deforestation, over both periods (Pan et al. 2011). Of note, however, intact tropical forests (i.e., those not altered directly by human activity) act as net carbon sinks (Grace et al. 1995), and their uptake of carbon combined with that of boreal and temperate forests amounts to a net global forest carbon sink of slightly more than 1 petagram of carbon annually (Pan et al. 2011).

The recovery of tropical forests following abandonment of cleared land results in carbon uptake through secondary forest succession (Bonan 2008), but regrowth forests are not strictly comparable to old-growth forests in species numbers, diversity, or composition of trees (Wright 2005). It is not pushing the point to compare secondary successional tropical forests to annual sea ice in the Arctic: the surface area covered by regrowth forest may approach the original area covered by old-growth forest in abandoned or recovering clearings, but this does not necessitate similarity in ecological characteristics such as carbon uptake and sequestration potential, habitat and niche quality and diversity, or species abundance and diversity. Similarly, annual sea ice may reform and cover areas previously covered by multi-annual ice, but it lacks the characteristics, such as stability, important to the species dependent on or associated with sea ice for habitat, nor does it provide the lasting albedo effects of thicker, multi-annual sea ice.

Estimating or projecting ecosystem- or biome-scale carbon balance in a changing climate requires modeling built from empirical estimates of the important drivers of carbon dynamics, which can quickly become a lengthy list of variables even with a simple heuristic model such as that in figure 8.2 as a starting point. Ecosystem process models have played an essential role in this effort. The Century model, for instance, was used to quantify the sensitivity of the terrestrial soil carbon cycle to temperature, estimating the effect of increasing temperature on soil carbon loss at the rate of −11.1 petagrams per 1°C increase in temperature, and was essential in identifying the importance of soil nitrogen turnover rates in this sensitivity (Schimel et al. 1994). Similarly, the Hybrid v3.0 terrestrial biosphere model, an individual tree–based model of ecosystem dynamics, was important in identifying the role of phenology in ecosystem productivity response to rising temperature and atmospheric CO_2 concentrations (Friend et al. 1997). The Terrestrial Ecosystem Model (TEM) was important in identifying the role of soil nutrient response to elevated temperature in ecosystem productivity response to elevated atmospheric CO_2 in far northern ecosystems, as well as in quantifying differences among regions in expected carbon uptake potential (Melillo et al. 1993). According to TEM output, the greatest increases in NPP in response to a doubling of atmospheric CO_2 concentrations are expected to occur in tropical deciduous forests under all climate change scenarios explored, whereas the weakest responses to a doubling of CO_2 are expected to occur in polar deserts, polar tundra, and boreal woodlands (Melillo et al. 1993).

The TEM has also been applied to investigate the implications of interannual climate variability for short-term variation in carbon storage in undisturbed Amazonian ecosystems (Tian et al. 1998). This exercise revealed the effects of extreme climatic fluctuation on carbon balance in a region of the world that represents approximately 10 percent of annual terrestrial NPP (Phillips et al. 1998). During droughts occurring during El Niño years, the Amazon basin became a net source of carbon to the atmosphere, emitting 0.2 petagrams of carbon to the atmosphere during the driest years. In contrast, during nondrought years, the region behaved as a net carbon sink, taking up on the order of up to 0.7 petagrams of carbon annually under such conditions (Tian et al. 1998). This nearly 1 petagram swing from carbon source to carbon sink in the Amazon basin between drought and nondrought years is remarkable in light of the fact that logging during the 1990s resulted in carbon emissions of approximately 0.3 petagrams per year (Tian et al. 1998). A coordinated, multi-annual experimental and observational effort to understand the ecological consequences of tropical deforestation, the Biological Dynamics of Forest Fragments Project, further highlighted the negative effects of forest fragmentation for biomass accumulation in the Amazon basin (Lovejoy et al. 1983). This experiment involved creating several forest fragments of varying sizes within an expanse of intact forest in Brazil, and then monitoring a suite of ecological responses, including abundance and diversity of plants and animals (Lovejoy et al. 1986). Tree mortality and wind disturbance increased along forest fragment edges, with consequences for longer-term changes in the biomass of forest vegetation (Laurance 1997). As proximity to forest fragment edges increases, forest biomass change becomes negative (figure 8.15) as a result of the increased mortality of large overstory trees (Laurance et al. 2002). Decomposition of dead trees from forest fragmentation accelerates the rate of carbon turnover in tropical forests by orders of magnitude in comparison to the long-term residence of

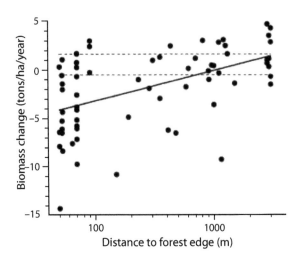

Figure 8.15. One of the more notable results from the long-term Biological Dynamics of Forest Fragments Project in the Amazon basin of Brazil, illustrating the influence of forest fragmentation on biomass loss from tropical forests. Closer to fragment edges, biomass loss increases owing to mortality of mature trees, possibly from exposure to wind damage. *Modified from Laurance et al. (2002)*.

carbon in living, mature trees (Laurance et al. 2002) and may contribute millions of tons of carbon to the atmosphere in tropical systems worldwide (Laurance et al. 1998).

Not surprisingly, the different dynamic vegetation models and ecosystem process models currently in use produce varying estimates of the magnitude of carbon sources and sinks and the overall carbon balance of the terrestrial biosphere. In part, this variability among model projections owes to the different climate scenarios they employ, but it can be attributed as well to the manner in which the component flux compartments of the models are constructed and calibrated. For this reason, comparisons of model projections to common climate-forcing scenarios are particularly useful for evaluating the potential sources of variability in estimates of the responsiveness of the global carbon cycle to changes in atmospheric CO_2 concentration and climate. A synthesis of results from six dynamic global vegetation models illustrates this point. Dynamic global vegetation models (DGVMs), in contrast to biogeochemical models and ecosystem process models, which assume equilibrium vegetation and invariant vegetation distributions, allow for important ecological processes to develop through time, and incorporate processes occurring at multiple temporal and spatial scales. For instance, DGVMs incorporate the rapid, nearly immediate responses of leaf-scale physiology to variation in temperature, moisture, and CO_2 concentration; daily to weekly processes such as timing of the onset and progression of the plant growing season captured in plant phenological dynamics; processes of vegetation growth dynamics and succession that unfold over months to years; and changes in soil nutrient content, availability, and turnover that accompany successional changes, as well as annual biomass production occurring over periods of months to years (figure 8.16).

Cramer and colleagues (2001) undertook a comparison of the projections of six DGVMs, each of which linked changes in atmospheric CO_2 concentration and climate according to midrange projections of the HadCM2-SUL general circulation model to changes in the structure and dynamics of vegetation and ecosystem function. By focusing on a single climate change scenario, the authors were able to limit variability in model projections to the structure of the models themselves, while at the same time ascribing generality to common features of the various model projections. A brief exploration of the model components common to the DGVMs compared by Cramer and colleagues (2001) is warranted. The vegetation composition in these DGVMs is commonly characterized according to the fractional representation of each biome by plant functional types, while vegetation dynamics consist of NPP and the accumulation of biomass on an annual basis. Additionally, the roles of competition, disturbance, and succession are included in the temporal dynamics of

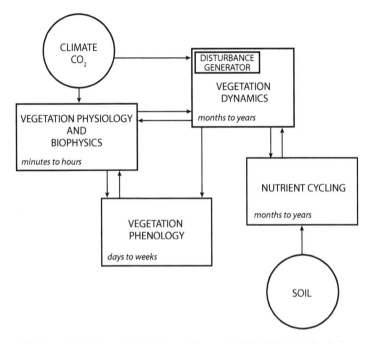

Figure 8.16. Example of a dynamic global vegetation model (DGVM) illustrating its incorporation of components representing processes that occur over widely disparate scales of time. *From Cramer et al. (2001)*.

vegetation types. The phenology component of DGVMs consists of the timing of seasonal onset of leaf bud opening, termination of leaf production, and leaf shedding, and the nutrient cycling component incorporates decomposition of litter produced in the vegetation component, heterotrophic respiration, and soil nitrogen mineralization. Finally, the soil component accounts for the physical properties of soil, including moisture content and texture. Although not depicted as such in figure 8.16, quite obviously all of these components are themselves potentially directly responsive to climatic variation (Cramer et al. 2001).

Under current ambient CO_2 and climatic conditions, the six DGVMs arrived at estimates of total annual NPP ranging between 45 and 60 petagrams of carbon per year, a standing vegetation biomass of 500–950 petagrams of carbon, and a global soil carbon pool of 850–1,200 petagrams of carbon (Cramer et al. 2001). Simulations running to the end of the twentieth century produced an estimate of global NEP, averaged across all six models, of 2.4 petagrams of carbon per year in response to the observed increase in rising atmospheric CO_2 concentration alone. In contrast, simulations over the same period in response to rising temperatures alone revealed a net release of carbon to the atmosphere of 1 petagram of carbon annually, the result mainly of net carbon loss in the Tropics (Cramer et al. 2001). When the models were run with coincident changes in climate and elevated CO_2 observed through the end of the twentieth century, global NEP was estimated at an average of 1.6 petagrams of carbon annually, that is, approximately 35 percent less than the estimated NEP resulting from simulation runs including only the observed increase in atmospheric CO_2 concentration. Hence, the past century of elevated temperature has suppressed carbon uptake globally and reduced the magnitude of the terrestrial carbon sink, despite a rise in atmospheric CO_2 concentration, and Cramer and colleagues (2001) attribute this to the responses of tropical ecosystems to recent warming.

A similar intermodel comparison was undertaken more recently using eleven climate–carbon cycle coupled C^4MIP models (Friedlingstein et al. 2006). In addition to several more minor differences among them and the six DGVMs utilized in the preceding comparison, the eleven C^4MIP models incorporate both terrestrial and marine carbon dynamics, and comprise both Earth System Models of Intermediate Complexity and the first generation of coupled Atmosphere-Ocean General Circulation Models (Denman et al. 2007). The suite of C^4MIP models was forced with the so-called A2 CO_2 emissions scenario for the period 1850–2100 (Denman et al. 2007). The average increase in atmospheric CO_2 concentration across all eleven models resulting from a positive climate–carbon cycle feedback was 87 ppm by the year 2100 (Denman et al. 2007). This intermodel comparison resulted in an estimate of net terrestrial CO_2 flux by the year 2100 of an uptake of 11 gigatons of carbon annually to a release of 6 gigatons of carbon annually, and an estimate of net oceanic CO_2 uptake by the year 2100 of 3.8–10 gigatons of carbon annually (Friedlingstein et al. 2006). Under a scenario of a doubling of atmospheric CO_2 concentration alone over the next century, nine of eleven of the C^4MIP models examined projected a potential carbon uptake by the terrestrial biosphere by the year 2100 of 0.9–1.6 gigatons of carbon, the range reflecting variability in the sensitivity of NPP to CO_2 fertilization among models (Denman et al. 2007). In response to a temperature increase of 4°C over the next century, the C^4MIP models projected a net carbon release by the terrestrial biosphere to the atmosphere of anywhere from just under 100 gigatons up to 800 gigatons of carbon, owing mainly to a decline in NEP with warming in the Tropics (Friedlingstein et al. 2006). All of the eleven models considered projected an increase in soil heterotrophic respiration in response to expected warming over the next century (Denman et al. 2007). A general conclusion deriving from these modeling exercises is that the combined influences of climate change and rising CO_2 concentrations reduce the NEP potential of the terrestrial biosphere in comparison to its potential response to CO_2 fertilization alone, primarily because of strong expected increases in heterotrophic respiration with warming, which averaged 6.2 percent per 1°C increase in global temperature across the eleven C^4 MIP models (Denman et al. 2007).

As is evident from the preceding review of case studies, there is little, if any, overlap between observational, experimental, and synthesis modeling efforts to quantify and project the consequences for ecosystem function and dynamics of climate change. Although these approaches are conceptually coordinated, in the sense that observation informs and shapes experimentation and vice versa, and the results from both approaches feed efforts to parameterize ecosystem process models, biogeochemical models, and dynamic vegetation models, there is a greater need for integration of all these approaches to improve the empirical basis of our understanding of the ecological consequences of global climate change (Luo et al. 2011).

Chapin and colleagues (2009) presented a thorough and critical evaluation of the state of the science of modeling global carbon dynamics in relation to climate change and identified a series of shortcomings inherent to current biogeochemical, ecosystem process, and linked climate–carbon cycle modeling efforts. Chief among the shortcomings were a general failure to consider potential differential responsiveness of NPP and organic matter decomposition to changes in climate, CO_2 concentration, and nutrient inputs, and the omission of several aspects of soil biology not currently considered explicitly in any modeling framework, such as allocation of GPP to belowground biomass and partitioning of the responses of root and mycorrhizal respiration from heterotrophic respiration, and changes in microbial community composition with warming (Chapin et al. 2009).

Less obvious by dint of its omission, however, is any mention of animal influences on ecosystem structure and function in the preceding examples. As we touched on briefly in chapter 2, the extinction of Pleistocene megafauna from northern biota during the transition to the Holocene may have played an important role in the transition from productive grassland steppe to the tundra biome of today. There are numerous reasons to expect that animals may play an important role in ecosystem response to contemporary and future climate change. In the 1980s and 1990s, ecology witnessed an explosion of research on the influences of vertebrate herbivores on plant biomass, soil nutrient dynamics, and the species composition of plant communities (Pastor et al. 1988, 1993; McInnes et al. 1992; Molvar et al. 1993; Frank and Groffman 1998; Kielland and Bryant 1998). We should expect that, either alone or in combination, such influences can and will mediate vegetation productivity response to climate change, nutrient cycling rates, and even vegetation response to increasing levels of atmospheric CO_2 because this response itself is limited by leaf area and nutrient availability (Billings et al. 1984; Shaver and Jonasson 1999; Shaver et al. 2000).

ROLE OF ANIMALS IN ECOSYSTEMS
OF RELEVANCE TO CLIMATE CHANGE

Animal influences on ecosystem structure and function provide compelling examples of the importance of biotic interactions in ecological response to climate change. A recent series of experiments in old field meadows in northeastern North America illustrates this clearly. For instance, data from two years of experimental warming at a field site at the Yale-Myers Research Forest in the U.S. state of Connecticut largely corroborated patterns evident in data from fifteen years of experimental manipulation of a spider-grasshopper-graminoid/forb system conducted against a natural background variation in temperature and precipitation of 2°C and 2.5 cm, respectively (Barton et al. 2009). Experimental warming caused spiders to seek refuge by occupying lower layers in the vegetation canopy, and this in turn increased encounters between spiders and their grasshopper prey (Barton and Schmitz 2009). As a consequence, the indirect effects of spider predation on grasshoppers promoted the production of grasses but reduced that of forbs, though neither plant functional group responded directly to changes in temperature or precipitation (Barton et al. 2009). A latitudinal comparision among spiders from three sites in the northeastern United States spanning a 4.8°C gradient subsequently demonstrated that such indirect effects on vegetation of predator response to warming declined markedly from northern to southern populations (Barton 2011).

Such cascading effects of interactions between climate and trophic dynamics influencing vegetation productivity are also evident in mammalian communities. Declining snowfall in the western mountainous zone of the United States over the past two decades has concorded, in Arizona, with a decline in the density of deciduous woody plant stems, presumably because of improved access to woody stems by elk during the winter (Martin and Maron 2012). As woody stem densities have declined, the breeding densities of five of six species of migratory songbirds have also declined at the study site (Martin and Maron 2012). The erection of exclosures to prevent elk browsing on experimental plots promoted the regrowth of aspen, maple, and locust species and reversed declines in all five songbird species within just five years, despite continued declines in snowfall (Martin and Maron 2012).

Referring briefly back to chapter 2, we will recall that the Zimov model of megafaunal extinction at the Pleistocene-Holocene transition allowed for the possibility that the biome shift at far northern latitudes from a productive, graminoid-forb complex to one of chemically defended, less productive woody plants may have been driven by the interaction between climatic warming and the extinction of large mammalian grazers. Such a perspective, along with that deriving from the study by Martin and Maron (2012), suggests a tension between the effects of large herbivores and climate change on ecosystem structure and function that will be explored throughout the rest of this section. To place this in the context of the tension versus facilitation/promotion hypotheses of the role of biotic interactions in ecological response to climate change, the influences of foraging by herbivores may in some cases, especially in northern environments, oppose vegetation and ecosystem responses to warming by reducing plant biomass and constraining carbon uptake potential. Before examining whether and in what capacity herbivores may influence ongoing ecosystem response to climate change, I first review briefly some prominent examples of the role of animals in ecosystem processes according to their short- and long-term effects on primary productivity and plant communities. This brief review will focus on some of the landmark studies of herbivore ecology because the influences of herbivores are of immediate relevance to ecosystem response to climate change. Nonetheless, we must recognize that predators also exhibit well-documented influences on plant productivity and community composition and dynamics that are of relevance to ecosystem response to climate change, both by removing herbivores from ecosystems through predation and by influencing where and how herbivores forage through predation risk (McLaren and Peterson 1994; Schmitz 1998, 2008; Post, Peterson, et al. 1999; Schmitz et al. 2003; Wilmers et al. 2006; Wilmers et al. 2007a; Barton and Schmitz 2009).

The most prominent short-term effects of herbivores in northern environments include influences on NAPP, standing biomass, and rates of soil nitrogen mineralization, all of which can be dramatic. In subarctic salt marshes, moderate levels of grazing by lesser snow geese (*Chen caerulescens caerulescens*) can increase the NAPP of graminoids threefold; this effect reflects a combined response to tissue removal and nitrogen fertilization through fecal output (Hik and Jefferies 1990). In temperate grasslands, experiments with movable exclosures have revealed that grazing by bison and elk increases the NAPP of graminoids by up to 45 percent above that in fenced, ungrazed plots (Frank and Groffman 1998). The standing biomass of vegetation may be reduced dramatically by large herbivores over the course of a single growing season. A comparison of standing crop outside and inside temporary exclosures erected on summer ranges of reindeer in northern Norway, for example, revealed a 33 percent reduction in the standing biomass of all vascular plant species after one grazing season (Bråthen and Oksanen 2001; Bråthen et al. 2007). Among preferred forage species, reductions varied among vegetation types, with cryptogams displaying 75 percent less, herbs 65 percent less, and graminoids 47 percent less standing biomass outside compared to inside exclosures (Bråthen and Oksanen 2001). Similar variation in the effects of tissue removal by caribou according to vegetation type have been noted in subarctic Canada, where the standing biomass of dwarf birch (*Betula glandulosa*) was reduced on birch heath, but no differences in standing biomass were noted between grazed versus nongrazed sites on shrub tundra (Manseau et al. 1996). The increase in cover of *Betula glandulosa* in the Low Arctic of Eastern Nunavik, Canada, over the past forty years (Tremblay et al. 2012) suggests that this chemically defended species of dwarf birch may possess a greater capacity for expansion in response to warming than does *Betula nana*, which, as described below, is constrained in its response to warming by herbivory.

These observations indicate that the NAPP of graminoids may be dramatically increased, and the standing biomass of woody species and forbs may be dramatically reduced, by large herbivores over the course of a single plant growing season. Clipping experiments, exclosure experiments, and observational studies indicate that intense use of summer ranges by caribou and reindeer can reduce the cover of woody, browsed species such as willows (*Salix lanata*) and dwarf birch (*Betula nana*) and promote the establishment of graminoids (Thing 1984; Ouellet et al. 1994; Manseau et al. 1996; Bråthen and Oksanen 2001; Olofsson et al. 2001). Such biomass off-take may play a role in the prolongation of juvenile growth stages of plants and constrain enhancement of net ecosystem production by CO_2 enrichment (Körner 2000). As well, such off-take may constrain potential increases in productivity associated with global change.

Prominent examples of the long-term effects of herbivores on plant community composition derive from exclosure experiments. Comparisons of vegetation and soil characteristics inside and outside long-term (i.e., thirty or more years) exclosures have been a major focus of research on effects of browsing by moose, and grazing by elk and bison, on biogeochemical processes, plant community composition, and succession. A major influence of browsing in boreal forest and taiga systems is the suppression of early successional competitive dominants and promotion of the establishment of late successional species (Kielland and Chapin 1992; McInnes et al. 1992). Comparisons outside and inside forty-plus-year-old exclosures on Isle Royale have also indicated that browsing by moose reduces tree biomass, increases shrub and herb biomass, and increases species diversity of (mainly) herbs (McInnes et al. 1992). As well, on Isle Royale moose browsing suppresses the recruitment of saplings into the canopy, resulting in a more open canopy with a well-developed understory of shrubs and herbs (McInnes et al. 1992). Similar effects are evident on Alaskan taiga, where moose browsing reduces the canopy height of preferred willow species by 12–50 percent, resulting in higher light intensity, lower humidity, and warmer and drier soils on browsed sites (Kielland and Bryant 1998).

Over the long term, browsing also alters soil chemistry through its effects on litter quality and decomposition (Kielland et al. 1997). For instance, browsing on Alaskan taiga increases the rate of litter decomposition and the pool of mineralizable carbon in litter, ultimately accelerating carbon turnover (Kielland et al. 1997). These effects may result in an increase in the average daily flux of carbon (i.e., CO_2 production) from the litter of browsed plants by up to 50 percent above that of nonbrowsed plants (Kielland et al. 1997). The long-term effects of grazing are in some respects at variance with those of browsing. In Yellowstone National Park, comparisons inside and outside thirty-five-year-old exclosures revealed higher rates of soil nitrogen mineralization (by, on average, a factor of two) on grazed versus nongrazed sites, apparently reflecting herbivore-mediated enhancement of the quality of soil organic matter (Frank and Groffman 1998). Grazing also apparently promotes the development of productive, steppe-like communities dominated by graminoids: thirty-year-old exclosures on reindeer ranges in Norwegian tundra reveal a dominance of graminoids, higher NAPP, and higher rates of nitrogen cycling on highly grazed sites compared to fenced sites (Olofsson et al. 2001). Similarly, Thing (1984) attributed the dieback of dwarf birch (*Betula nana*) and the spread of *Poa* species–dominated grazing lawns in West Greenland to long-term use by caribou there.

These observations indicate that the prominent long-term effects of large herbivores include suppression of growth of woody forage species, enhancement of growth of graminoid forage species, reduction of rates of soil nitrogen mineralization in woody plant communities, enhancement of soil nitrogen mineralization in graminoid communities, and alteration of species composition of plant communities. The long-term influences of grazing on standing biomass and plant community composition—away from shrub-dominated communities and toward graminoid, steppe-like communities—could substantially alter the carbon sequestration potential of vegetation because woody plants have much greater carbon sequestration capacity than do graminoids (Shaver et al. 1998). Moreover, herbivores exert considerable control over plant establishment (Kielland and Bryant 1998), plant community composition (McInnes et al. 1992), and forest succession (Kielland and Bryant 1998).

HERBIVORES, WARMING, AND ECOSYSTEM CARBON DYNAMICS

Insect outbreaks such as those of bark beetles and defoliating caterpillars can devastate vegetation and have enormous potential to influence ecosystem dynamics (Maclean 1980; Royama 1984; Veblen et al. 1991; Paine et al. 1997), in addition to exhibiting the potential to increase in extent and magnitude with warming (Chapin et al. 2009; Dukes et al. 2009; Post, Forchhammer, et al. 2009; Duehl et al. 2011; Valtonen et al. 2011). Such outbreaks can reduce carbon uptake by, and potentially increase carbon emissions from, forests at the landscape scale through defoliation and killing of trees in the first case and through decomposition of dead trees or increased susceptibility to fire in the second case. A recent, and still ongoing at the time of this writing, outbreak of mountain pine beetles in western North America that began in 2000 is the most geographically extensive outbreak to date by an order of magnitude (Taylor et al. 2006). The vast magnitude of this outbreak is believed to reflect the combined influences of increased extent of mature coniferous trees and warm conditions favorable to the spread of mountain pine beetles. This outbreak is expected, over the course of two decades, to reduce the carbon stock of the forest by 270 megatons, reducing net biome production, at the peak of the outbreak, by between 20 and 32 megatons of carbon annually (figure 8.17) (Kurz et al. 2008). Moreover, effects of the outbreak, including defoliation, tree mortality, and increased susceptibility to fire, have the capacity to transform the forest from a net carbon sink of 0.59 megatons of carbon annually to a net carbon source of 17.6 megatons of carbon annually (Kurz et al. 2008), although it should be noted that there is some suggestion that beetle outbreaks

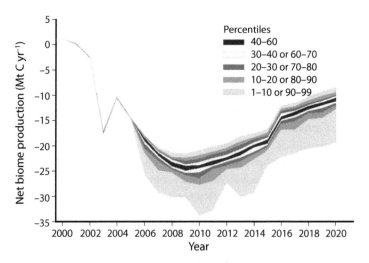

Figure 8.17. Estimated variation in net biome production during and following the mountain pine beetle outbreak centered in British Columbia, Canada, based on Monte Carlo simulations using the carbon budget model of the Canadian Forest Sector that accounts for direct tree mortality and susceptibility to fire resulting from the outbreak. Even under minimal estimates of forest area burned, the pine beetle outbreak is expected to result in net carbon emissions to the atmosphere of several megatons annually. *Modified from Kurz et al. (2008).*

may reduce the risk of fire by reducing canopy bulk density (Turner 2010). Nonetheless, such a transformation would translate to an impact similar in carbon loss from this forest of combined carbon emissions over all of Canada's forest fires during the four decades from 1959 to 1999, and would amount to the CO_2 equivalent of five years of output by the entire transportation sector of Canada (Kurz et al. 2008).

In North American spruce and fir forests, the spruce budworm (*Choristoneura fumiferana*) undergoes periodic, multi-annual outbreaks that unleash extensive tree mortality over millions of square kilometers of forest through defoliation (Maclean 1980; Dymond et al. 2010). The effects on ecosystem carbon dynamics may consequently be of similarly enormous magnitude. An empirically parameterized simulation of a spruce budworm outbreak lasting thirteen years in an eastern Canadian mixed conifer and deciduous forest resulted in a transformation of the forest from a net carbon sink of approximately 13 grams of carbon per square meter annually to a net carbon source of nearly 17 grams of carbon per square meter annually as repeated defoliation throughout the course of the outbreak precipitated, in some model runs, up to 90 percent cumulative mortality of softwood trees (Dymond et al. 2010). At the landscape scale, over the entire study forest, comprising approximately 106,000 km², net biome production was reduced by 3.4 teragrams of carbon per year at the peak of the outbreak (Dymond et al. 2010). Over the entire outbreak period of thirteen years, landscape-scale NPP was reduced by a total of 25 teragrams of carbon, while heterotrophic respiration increased by 6.5 teragrams of carbon (Dymond et al. 2010).

Outbreaks of moth larvae may have extensive ecosystem consequences at higher latitudes in subarctic mountain birch andarctic tundra biomes. The autumn moth (*Epirrita autumnata*) and winter moth (*Operophtera brumata*) undergo cyclic outbreaks approximately every decade in mountain birch forests of northern Fennoscandia but can reach extreme outbreak densities at much longer intervals of up to once every century, the effects of which are noticeable in the biomass recovery of forests for several decades afterward (Tenow and Bylund 2000). Some evidence suggests that such outbreaks may be moving northward or expanding in association with recent warming trends at high latitudes (Jepsen et al. 2008, 2009).

A recent outbreak of the autumn moth in the Lake Torneträsk catchment of subarctic Sweden in 2004 (Heliasz et al. 2011) coincided with the onset of a two-year outbreak of larvae of another cyclic moth, *Eurois occulta*, at the Low Arctic study site near Kangerlussuaq, Greenland (Pedersen and Post 2008; Post and Pedersen 2008). In the latter case, we reported a reduction in aboveground biomass of all functional groups of plants at the study site by up to 90 percent

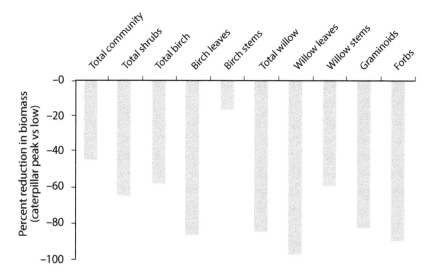

Figure 8.18. Reduction of plant aboveground biomass response to experimental warming by an outbreak of larvae of a noctuid moth during the second and third years of an ongoing experiment currently in its eleventh year at the study site near Kangerlussuaq, Greenland. *Adapted from Post and Pedersen (2008).*

as a result of intense defoliation (Post and Pedersen 2008). The most extensive defoliation was experience by birch and willow leaves and by graminoids and forbs (figure 8.18). The simultaneous outbreak at the Lake Torneträsk site in Sweden, which covered an estimated area of 3,300 km² (Jepsen et al. 2009), was monitored for its effects on ecosystem carbon dynamics. After defoliation reached its peak during the 2004 growing season, the magnitude of the carbon sink was estimated at 12 grams of carbon per square meter, whereas two years later, after the vegetation had recovered from the outbreak, it was estimated at 105 grams of carbon per square meter (Heliasz et al. 2011). The regional reduction in carbon uptake is evident in the estimated sink in 2004 of 4,000 tons of carbon compared to that in 2006 of 33,000 tons of carbon (Heliasz et al. 2011). Hence, the caterpillar outbreak reduced the carbon sink in this catchment by approximately 89 percent.

Vertebrate herbivores may similarly influence ecosystem response to climate change, especially, as argued above, in northern, nutrient-limited systems such as the Arctic. Management and conservation practices in temperate regions, together with changing patterns of land use, have spurred increases in numbers of migratory ducks and geese, some of which breed in the Arctic. A field study conducted on the High Arctic island of Spitzbergen, part of Norway's Svalbard

(a)   Net ecosystem exchange (NEE)

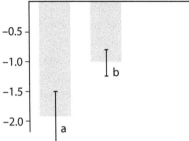

(b)   Gross ecosystem photosynthesis (GEP)

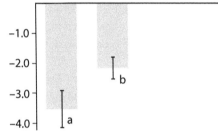

Carbon flux (µmol C m⁻² s⁻¹)

(c)   Ecosystem respiration (R_e)

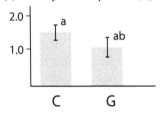

C   G

Figure 8.19. Alteration of net ecosystem exchange (a) by the grubbing action of geese on Svalbard, Norway, through the influence of grubbing on gross ecosystem photosynthesis (b) but not on ecosystem respiration (c). Letters "a" and "b" indicate statistically significant differences between bars within panels. *Adapted from Van der Wal et al. (2007).*

archipelago, compared carbon flux measurements on sites grubbed by pink-footed geese (*Anser brachyrhyncus*), one such arctic breeding anserine, to carbon flux measurements on adjacent nongrubbed sites (Van der Wal et al. 2007). The authors hypothesized that grubbing activity—during which geese penetrate and disturb the moss layer of the tundra to access belowground plant tissues—might alter the carbon dynamics of the system because of its effect on subsurface plant biomass and productivity. Their measurements indicated that grubbing by pink-footed geese significantly reduced the strength of net eco-system exchange by reducing gross ecosystem photosynthesis without altering ecosystem respiration (figure 8.19) (Van der Wal et al. 2007). This effect de-rived from a substantial reduction in the biomass of mosses and vascular plants across wet, moist, and mesic tundra types and a reduction in total plant cover, which in turn reduced ecosystem carbon uptake (Van der Wal et al. 2007). In a related study, the same authors utilized captive barnacle geese (*Branta leucopsis*) in a warming experiment that employed open-top chambers to elevate surface temperatures at the same study site, and investigated the interactive effects of geese grazing and warming on ecosystem CO_2 exchange. Warming alone had a tendency to increase the carbon sink strength on wet tundra, but grazing eliminated this response (Sjogersten et al. 2008). At the highest level of grazing, herbivory by geese reduced net ecosystem exchange, or the rate of carbon uptake on wet tundra, by reducing gross ecosystem photosynthesis without influencing ecosystem respiration, effectively eliminating the carbon sink on this tundra type (Sjogersten et al. 2008). The effect of herbivory by geese on carbon uptake in response to warming operated through a reduction of the aboveground biomass and the effect of this reduction on ecosystem photosynthesis (Sjogersten et al. 2008).

Large mammalian herbivores appear to exert similar influences. Data collected during the fourth year of an ongoing experiment at the study site near Kangerlussuaq, Greenland, indicated that herbivory by caribou and muskoxen reduced the aboveground biomass response to warming of the entire plant community, primarily through the negative effects of browsing on deciduous shrubs, the most abundant functional group at the study site (Post and Pedersen 2008). We suggested at the time of publication of these results that large herbivores had the potential to reduce the carbon uptake capacity of arctic tundra in response to warming because their foraging behavior tended to negate the warming-induced shift toward dominance of the plant community by shrubs while maintaining it in a graminoid-dominated state (figure 8.20) (Post and Pedersen 2008).

Woody plants, such as dwarf birch and gray willow at the site, should have a much greater carbon uptake and sequestration potential than graminoids because of the extensive production of leaf biomass and the accumulation of carbon in woody stems of the former. Data from carbon flux measurements taken weekly during two subsequent field seasons on all of our experimental plots, both grazed and exclosed and warmed and ambient, support this hypothesis. Gross ecosystem carbon uptake increased in association with the biomass of deciduous shrubs on our plots but declined in association with the biomass of graminoids (figure 8.20) (Cahoon et al. 2012). The negative association between graminoid biomass and gross ecosystem carbon uptake may be an artifact of the negative association between shrub biomass and graminoid biomass, but the important consideration here is that herbivory tends to push the system away from one of increasing carbon uptake by reducing the aboveground biomass of shrubs (figure 8.20) (Cahoon et al. 2012). Warming, by

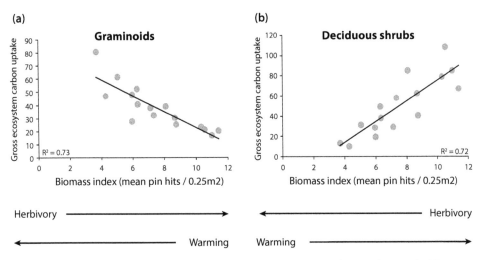

Figure 8.20. Relationship between gross ecosystem carbon uptake and biomass of (a) graminoids and (b) deciduous shrubs, as well as the directionally opposing influences of herbivory and experimental warming on the biomass of these two functional groups. *Data from the ongoing experiment at the study site near Kangerlussuaq, Greenland, and modified from Cahoon et al. (2012).*

contrast, pushes the system toward biomass accumulation by, and increasing dominance by, deciduous shrubs (figure 8.20) (Post and Pedersen 2008). As a consequence of the positive association between leaf area and net ecosystem photosynthesis (Shaver et al. 2007; Street et al. 2007), the reduction in leaf area of shrubs, and its response to experimental warming, by caribou and muskoxen at the Kangerlussuaq study site results in a reduction in net ecosystem carbon uptake (Cahoon et al. 2012). Over the course of the eight-year (to date) experiment, we estimate that large herbivores have increasingly reduced the carbon uptake potential of the system by up to 50 percent, or by up to 100 grams per square meter per year (figure 8.21) (Cahoon et al. 2012).

A similar influence of large herbivores on ecosystem carbon dynamics has been inferred in the Serengeti ecosystem, but in this case through a cascading effect of a pathogen. Analysis of a forty-four-year time series on wildebeest population dynamics in the Serengeti-Mara ecosystem indicated that increases in the population following eradication of the rinderpest virus resulted in fire suppression through grazing, which reduced accumulation of vegetative fuel, and promoted the expansion of trees in the system (Holdo et al. 2009). As a result, the system shifted from being an estimated net source of carbon to a carbon sink on the order of 10 million megagrams of carbon annually (Holdo et al. 2009). Observations such as those deriving from the studies of invertebrate,

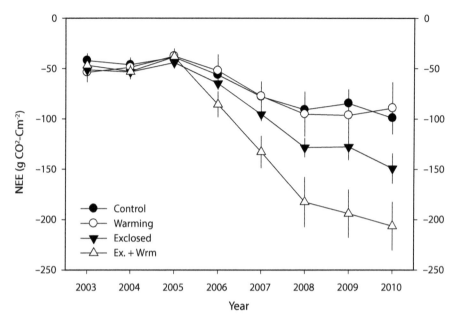

Figure 8.21. Estimated net ecosystem exchange at the study site near Kangerlussuaq, Greenland, derived from the relationship between leaf area index and NEE, according to the four treatments composing the experiment. Control plots receive no experimental warming and are exposed to natural herbivory; warming plots receive experimental warming and are exposed to natural herbivory; exclosed plots receive no experimental warming and are exclosed from natural herbivory; and finally exclosed + warming plots receive experimental warming and are exclosed from natural herbivory. *Modified from Cahoon et al. (2012).*

avian, and mammalian herbivory reviewed here strongly suggest a need to account for animal influences in ecosystem process models and models of linked climate-ecosystem carbon dynamics to better account for their effects on ecosystem response to climate change.

Brief Remarks on Some Especially Important Considerations

Earth's climate is warming at a pace that may very well be unprecedented, and it is doing so from a higher baseline average temperature than that which was the starting point for the most recent episode of rapid warming, which signaled the end of the Pleistocene and the demise of most of its large mammals. That most recent warming episode also coincided with geographically widespread biome shifts. Perhaps more tellingly, current warming, still in its early stages, has already heralded similarly geographically widespread and taxonomically broad shifts in phenological dynamics, population dynamics, species distributions, and ecosystem carbon dynamics. We may be, as some have asserted, on the threshold of the sixth major extinction event on Earth (Barnosky et al. 2012). The message that is at risk of getting lost in discussions of extinction risk from climate change is the obvious one that extinctions do not occur in vacuums, may be symptomatic of unforeseen and major changes in ecosystem function across Earth's biomes, and, much like those changes, are irreversible. In this chapter I revisit, and in a case or two present a slightly different perspective on, the themes of major importance emerging from the literature presented thus far.

TRENDS AND VARIABILITY REVISITED

The trends in abiotic conditions overviewed in chapter 1, especially trends in temperature and precipitation, are the engines of species redistributions, biome shifts, and ecosystem evolution. The variability about those trends, by contrast, may be viewed as the brakes or speed bumps encountered by those engines. Climatic trends may drive comparatively gradual changes in habitat availability, reducing it for some species but increasing it for others, thereby altering over longer terms, on the order of centuries, the composition and distributions of biomes. Some species will adapt to such changes and keep pace with them; many others will not. Fluctuations in climatic conditions, on the other hand, especially when increasing in magnitude over short temporal scales, such as from year to year, render adaptation and adjustment to climatic trends by individuals and populations increasingly difficult. Variability in climatic conditions has the potential to drive populations to local extinction and species to global extinction, and we, as ecologists, must improve our thinking about and approaches to incorporating variability explicitly into analytical and predictive models of population dynamics and species distributions. I would argue that the contribution of changes in short-term climatic variability to extinction risk, and in particular to extinctions during the Middle to Late Pleistocene, deserves greater attention. Attempts to assign attribution to climate change in Pleistocene extinctions have been hampered by the confounding though important contributions of human exploitation, but these attempts have also focused largely on climatic shifts and trends. Analyzing the contribution of changes in interannual and longer-term climatic variability to extinction dynamics in the Pleistocene and more recently should prove fruitful.

In general, more emphasis needs to be placed on climatic variability in both time and space, perhaps the more urgently on such variability in time. It is my conviction that the magnitude of population response to climatic variability must be at least as important as the direction of such response to climatic trends in determining the persistence of populations and, by extension, of species in a changing climate. The nexus of human land use or other modifications of the natural landscape and climate change is of particular relevance in this context. To understand the importance of this nexus, we must acknowledge the role of density dependence in stabilizing population dynamics and the role of habitat availability and quality in setting limits to density, and recognize that as climate change intensifies with increasing interannual variability in climatic conditions, any reduction in natural habitat has the potential to tip the balance between population-stabilizing and population-destabilizing forces in favor of the latter.

Of primary importance in this subtheme may be the occurrence of extreme events. Events such as extreme droughts, floods, storms, or heat waves are inherently stochastic in both time and space, and are exceedingly difficult to predict or incorporate into climate projections. It is similarly challenging to try to incorporate them into analyses of observed dynamics, to say nothing of projections of expected ecological dynamics. Yet they may occur with greater frequency in the near future. Incorporating variability into our way of thinking about and approaches to studying population- and species-level responses to climate change must increasingly take into account unpredictable elements such as stochasticity and extreme events.

COMMUNITY RESPONSE TO CLIMATE CHANGE: FURTHER CONSIDERATIONS

The preceding thoughts on trends versus variability relate as well to considerations of community response to climate change. This response, and its study, cannot be generalized under the umbrella of community ecology. We should try to think of species interactions at what we conveniently refer to as the scale of the local community in terms of exploitation and interference, for these broad categories will help us understand the factors influencing the structure, dynamics, and stability of vertically versus laterally structured communities and the differential roles of climate change in them. By now we should be comfortable with devoting attention to the role of climatic variability, in addition to climatic trends, in thinking about and analyzing population dynamics and stability. Likewise, we should be prepared for the potentially disparate contributions of climatic variability to stability in vertically and laterally structured communities. Although this variability may destabilize population dynamics, the consequences of such destabilization will differ between primarily vertically structured and laterally structured communities. In the former we may expect similarly destabilizing outcomes, whereas in the latter we may expect stability to result or persist if, as in the case study of sympatric *Ficedula* flycatchers, it prevents competitive exclusion (Sætre et al. 1999).

Thinking along such lines may also help us to refine, in general, our notions of the role of the primary interference interaction, competition, in the dynamics of laterally structured communities. In such communities, interference interactions are more likely to involve conspecifics than they are members of other species in inherently species-poor systems such as the Arctic. In contrast, interference interactions are more likely to involve members of other species than they are conspecifics in species-rich systems such as those found in the Tropics. Hence, the composition and dynamics of primarily laterally structured communities may reflect predominantly the influence of intraspecific competition (i.e., density dependence) in species-poor communities, where individuals are most likely to encounter and compete with members of their own species, and the influence of interspecific competition in species-rich communities, in which individuals are most likely to encounter and compete with members of another species. It is in the latter rather than the former that null models of community composition and dynamics (Hubbell 2001) may be more appropriate.

THE SCALE-INVARIANT NATURE
OF NON-ANALOGUES

The concept of non-analogues that figured prominently in our assessment of community-scale responses to past (Graham et al. 1996) and future (Williams and Jackson 2007) climate change should be viewed as scale invariant. By this I mean that the composition of populations, communities, ecosystems, and biomes should all be regarded as possessing the potential for non-analogue characteristics under future climatic conditions. Hence, the known contributions of climatic variability to demography, and demographic contributions to population dynamics, may apply differently under future climatic conditions. Similarly, the contribution of phenology to community dynamics and composition may change with future warming. As well, the distribution of ecosystem types among biomes, and the distribution and relationships among biomes across the Earth, may differ under future climatic conditions in ways that are unfamiliar based on current relationships. The heuristic potency of the non-analogue concept should not be limited to discussions of communities and climatic regimes; we should apply this thinking as widely as possible in our study of the ecological consequences of climate change.

LACK OF DETECTION DOES NOT ALWAYS
MEAN LACK OF RESPONSE

Statistical approaches to quantifying responses to climate change have made considerable progress in the past decade alone, and state-space modeling appears to hold additional promise for dissecting deterministic, stochastic, and error-laden contributions to observed dynamics (Clark and Bjørnstad 2004; Clark 2007). Nonetheless, the documentation of weak or null responses to measured indices of climate change should not be accepted with conviction as equivalent to a lack of such response. Many data sets employed in retrospective analyses of ecological responses to recent climate change are not the result of foresighted efforts to assess or identify such responses. Hence, data collected for one purpose are often applied post hoc in studies with a distinctly different focus. This approach is not without merit, and indeed has figured prominently in multiple influential publications cited in this book and in even more publications not cited herein. It is important to bear in mind, however, the limitations of such data sets and the conclusions that can be drawn from analyses based on them. Similarly, experimental and observational protocols rarely change midway through what eventually turns out to be a long-term project, even though limitations to study design may become apparent during the course of the first few years of the study. Ecology is an inherently shifting-target sort of discipline. The more a population, species, community, or system is studied, the more the honest investigator realizes how little he or she truly understands of it. It is immensely wise not only, as is common, to urge caution when presenting results that indicate a response to climate change but also, as is far less common, to be prudent when presenting results that appear to indicate no such response.

A GREATER EMPHASIS ON PHENOLOGY

As was emphasized in chapter 3, our understanding of the contribution of the timing of life history events to demography and population dynamics is still in its infancy. As ecologists interested in climate change, we need—urgently, I would argue—to improve our appreciation for the myriad ways in which the timing of reproduction influences offspring production, survival, and recruitment, and thereby population dynamics. As well, we should not shy from investigating the contribution of phenological dynamics to offspring production and survival in comparatively aseasonal environments such as the Tropics. These points are likely obvious at this stage of the book.

Less obvious, perhaps, is the need for further consideration of the role of phenology in community dynamics. As many of the studies highlighted in chapter 3 made clear, species are highly individualistic in the extent to which, and even in the direction in which, they respond phenologically to climate change. We still understand little, however, of the role of such differences in the variable success of species in capitalizing on seasonal resource pulses and the consequences this may pose for the outcomes of interference interactions and shifts in species' dominance within local communities in the near future and over the longer term.

On a related note, greater consideration should be given to the direct contribution of phenology to seasonal carbon dynamics between the atmosphere and biosphere. As mentioned in the preceding chapter, dynamic global vegetation models may include phenology components, but these are largely blind to species differences in phenological response to climate change, and so are static to a certain extent. Even within functional groups, plant species display a wide array of responsiveness to warming in the timing of their emergence

and leaf-out. This variable response suggests the possibility that rates of phenological response to warming may similarly vary through time, with some
species becoming increasingly or decreasingly responsive to warming. In addition to continuing to incorporate the influences of leaf area and standing
biomass into models of ecosystem carbon dynamics, we should also begin to
think in terms of the *phenological community* and its influence on seasonal
carbon exchange. For instance, warm years may be characterized by a different
sequence of emergence, growth, and senescence of plant species in the local
community than holds in cold years. If so, this difference may have important
consequences for patterns of the timing and duration of carbon draw-down and
release by the ecosystem, as well as for the strength of both exploitation and
interference interactions.

DIRECT VERSUS INDIRECT ECOLOGICAL
RESPONSES AND THE THIEF IN THE NIGHT

It is not uncommon for laymen, or even for students and scientists, to imagine
direct responses to changes in temperature and precipitation when thinking
of ecological responses to climate change or when hearing, more specifically,
about extinction risk related to climate change. Yet mortality owing to extreme
events, while spectacular and remarkable when noted, may be less important
to the persistence of many species than more subtle interactions and dynamics.
If the emphasis is placed on detecting and documenting direct responses to extreme events, we may find ourselves surprised by the outcomes of less obvious
interactions such as those typified by the examples of trophic mismatch and
altered pathogen life history dynamics described in chapter 3. Nonetheless,
reports of die-offs of entire cohorts of pinniped offspring and of reproductive failure in Darwin's finches during El Niño events (Barber and Chavez
1983; Grant et al. 2000; Grant and Grant 2002), of massive mortality among
migratory birds related to severe weather (Brown and Brown 1998), and of
the relatively sudden death of more than 3,500 individual flying foxes among
nine bat colonies during a heat wave in 2002 in New South Wales, Australia
(Welbergen et al. 2008) serve to communicate the urgency of climate change
consequences for wildlife populations with undisputed effectiveness (Brodie,
Post, and Doak 2012). Designing studies to explicitly account for and detect
the indirect effects of climate change, such as those driven by biotic interactions, will better prepare us for keeping the discipline apace of the changes
we aim to understand.

References

Achard, F., R. DeFries, H. Eva, M. Hansen, P. Mayaux, and H. J. Stibig. 2007. Pan-tropical monitoring of deforestation. *Environmental Research Letters* 2:1–11.

Achard, F., H. D. Eva, P. Mayaux, H. J. Stibig, and A. Belward. 2004. Improved estimates of net carbon emissions from land cover change in the tropics for the 1990s. *Global Biogeochemical Cycles* 18:1–11.

Achard, F., H. D. Eva, H. J. Stibig, P. Mayaux, J. Gallego, T. Richards, and J. P. Malingreau. 2002. Determination of deforestation rates of the world's humid tropical forests. *Science* 297:999–1002.

Adler, P. B., J. Hille Ris Lambers, P. C. Kyriakidis, Q. F. Guan, and J. M. Levine. 2006. Climate variability has a stabilizing effect on the coexistence of prairie grasses. *Proceedings of the National Academy of Sciences of the United States of America* 103:12793–12798.

Adler, P. B., J. Leiker, and J. M. Levine. 2009. Direct and indirect effects of climate change on a prairie plant community. *PLoS One* 4.

Agrawal, A. A., D. D. Ackerly, F. Adler, A. E. Arnold, C. Caceres, D. F. Doak, E. Post, P. J. Hudson, J. Maron, K. A. Mooney, M. Power, D. Schemske, J. Stachowicz, S. Strauss, M. G. Turner, E. Werner, and Ni. 2007a. Filling key gaps in population and community ecology. *Frontiers in Ecology and the Environment* 5:145–152.

Agrawal, A. A., D. D. Ackerly, F. Adler, A. E. Arnold, C. Caceres, D. F. Doak, E. Post, P. J. Hudson, J. Maron, K. A. Mooney, M. Power, J. Stachowicz, S. Strauss, M. G. Turner, and E. Werner. 2007b. In support of observational studies: reply. *Frontiers in Ecology and the Environment* 5:294–295.

Ainsworth, E. A., and S. P. Long. 2005. What have we learned from 15 years of free-air CO_2 enrichment (FACE)? A meta-analytic review of the responses of photosynthesis, canopy. *New Phytologist* 165:351–371.

Akcakaya, H. R. 1992. Population cycles of mammals: evidence for a ratio-dependent predation hypothesis. *Ecological Monographs* 62:119–142.

Albertson, F. W., and J. E. Weaver. 1944a. Effects of drought, dust, and intensity of grazing on cover and yield of short -grass pastures. *Ecological Monographs* 14:1–29.

————. 1944b. Nature and degree of recovery of grassland from the great drought of 1933 to 1940. *Ecological Monographs* 14:393–479.

Albon, S. D., and R. Langvatn. 1992. Plant phenology and the benefits of migration in a temperate ungulate. *Oikos* 65:502–513.

Alley, R. B., J. Marotzke, W. D. Nordhaus, J. T. Overpeck, D. M. Peteet, R. A. Pielke, R. T. Pierrehumbert, P. B. Rhines, T. F. Stocker, L. D. Talley, and J. M. Wallace. 2003. Abrupt climate change. *Science* 299:2005–2010.

Alley, W. M. 1984. The Palmer Drought Severity Index: limitations and assumptions. *Journal of Climate and Applied Meteorology* 23:1100–1109.

Altermatt, F. 2010. Climatic warming increases voltinism in European butterflies and moths. *Proceedings of the Royal Society B: Biological Sciences* 277:1281–1287.

Anders, A. D., and E. Post. 2006. Distribution-wide effects of climate on population densities of a declining migratory landbird. *Journal of Animal Ecology* 75:221–227.

Andrews, R. M. 1991. Population stability of a tropical lizard. *Ecology* 72:1204–1217.

Araújo, M. B., and M. Luoto. 2007. The importance of biotic interactions for modelling species distributions under climate change. *Global Ecology and Biogeography* 16:743–753.

Arens, S. J. T., P. F. Sullivan, and J. M. Welker. 2008. Nonlinear responses to nitrogen and strong interactions with nitrogen and phosphorus additions drastically alter the structure and function of a high arctic ecosystem. *Journal of Geophysical Research-Biogeosciences* 113:10.

Arft, A. M., M. D. Walker, J. Gurevitch, J. M. Alatalo, M. S. Bret-Harte, M. Dale, M. Diemer, F. Gugerli, G. H. R. Henry, M. H. Jones, R. D. Hollister, I. S. Jonsdottir, K. Laine, E. Levesque, G. M. Marion, U. Molau, P. Molgaard, U. Nordenhall, V. Raszhivin, C. H. Robinson, G. Starr, A. Stenstrom, M. Stenstrom, O. Totland, P. L. Turner, L. J. Walker, P. J. Webber, J. M. Welker, and P. A. Wookey. 1999. Responses of tundra plants to experimental warming: meta-analysis of the international tundra experiment. *Ecological Monographs* 69:491–511.

Augustine, D. J., and S. J. McNaughton. 1998. Ungulate effects on the functional species composition of plant communities: herbivore selectivity and plant tolerance. *Journal of Wildlife Management* 62:1165–1183.

Barber, D. G., and J. M. Hanesiak. 2004. Meteorological forcing of sea ice concentrations in the southern Beaufort Sea over the period 1979 to 2000. *Journal of Geophysical Research—Oceans* 109.

Barber, R. T., and F. P. Chavez. 1983. Biological consequences of El Niño. *Science* 222:1203–1210.

Barker, S., G. Knorr, R. L. Edwards, F. Parrenin, A. E. Putnam, L. C. Skinner, E. Wolff, and M. Ziegler. 2012. 800,000 years of abrupt climatic variability. *Science* 334:347–351.

Barnosky, A. D. 2008. Megafauna biomass tradeoff as a driver of Quaternary and future extinctions. *Proceedings of the National Academy of Sciences of the United States of America* 105:11543–11548.

Barnosky, A. D., M. A. Carrasco, and R. W. Graham. 2011. Collateral mammal diversity loss associated with late Quaternary megafaunal extinctions and implications for the future. Geological Society, London, Special Publications 358:179–189.

Barnosky, A. D., E. A. Hadly, J. Bascompte, E. L. Berlow, J. H. Brown, M. Fortelius, W. M. Getz, J. Harte, A. Hastings, P. A. Marquet, N. D. Martinez, A. Mooers, P. Roopnarine, G. Vermeij, J. W. Williams, R. Gillespie, J. Kitzes, C. Marshall, N. Matzke, D. P. Mindell, E. Revilla, and A. B. Smith. 2012. Approaching a state shift in Earth's biosphere. *Nature* 486:52–58.

Barnosky, A. D., E. A. Hadly, and C. J. Bell. 2003. Mammalian response to global warming on varied temporal scales. *Journal of Mammalogy* 84: 354–368.

Barnosky, A. D., P. L. Koch, R. S. Feranec, S. L. Wing, and A. B. Shabel. 2004. Assessing the causes of Late Pleistocene extinctions on the continents. *Science* 306:70–75.

Barry, J. P., C. H. Baxter, R. D. Sagarin, and S. E. Gilman. 1995. Climate-related, long-term faunal changes in a California rocky intertidal community. *Science* 267:672–675.

Barton, B. T. 2011. Local adaptation to temperature conserves top-down control in a grassland food web. *Proceedings of the Royal Society B: Biological Sciences* 278:3102–3107.

Barton, B. T., A. P. Beckerman, and O. J. Schmitz. 2009. Climate warming strengthens indirect interactions in an old-field food web. *Ecology* 90: 2346–2351.

Barton, B. T., and O. J. Schmitz. 2009. Experimental warming transforms multiple predator effects in a grassland food web. *Ecology Letters* 12: 1317–1325.

Beebee, T. J. C. 1995. Amphibian breeding and climate. *Nature* 374:219–220.

Berger, L., R. Speare, P. Daszak, D. E. Green, A. A. Cunningham, C. L. Goggin, R. Slocombe, M. A. Ragan, A. D. Hyatt, K. R. McDonald, H. B. Hines, K. R. Lips, G. Marantelli, and H. Parkes. 1998. Chytridiomycosis causes amphibian mortality associated with population declines in the rain forests of Australia and Central America. *Proceedings of the National Academy of Sciences of the United States of America* 95:9031–9036.

Berteaux, D., M. M. Humphries, C. J. Krebs, M. Lima, A. G. McAdam, N. Pettorelli, D. Reale, T. Saitoh, E. Tkadlec, R. B. Weladji, and N. C. Stenseth. 2006. Constraints to projecting the effects of climate change on mammals. *Climate Research* 32:151–158.

Bertness, M. D., and G. H. Leonard. 1997. The role of positive interactions in communities: Lessons from intertidal habitats. *Ecology* 78:1976–1989.

Betancourt, J. L., T. R. Van Devender, and P. S. Martin, eds. 1990. *Packrat middens: the last 40,000 years of biotic change*. Tucson: University of Arizona Press.

Bhatt, U. S., D. A. Walker, M. K. Raynolds, J. C. Comiso, H. E. Epstein, G. S. Jia, R. Gens, J. E. Pinzon, C. J. Tucker, C. E. Tweedie, and P. J. Webber. 2010. Circumpolar arctic tundra vegetation change is linked to sea ice decline. *Earth Interactions* 14:20.

Billings, W. D. 1987. Carbon balance of Alaskan tundra and taiga ecosystems: past, present, and future. *Quarterly Science Reviews* 6:165–177.

Billings, W. D., K. M. Peterson, J. O. Luken, and D. A. Mortensen. 1984. Interaction of increasing atmospheric carbon-dioxide and soil nitrogen on the carbon balance of tundra microcosms. *Oecologia* 65:26–29.

Billings, W. D., K. M. Peterson, and G. R. Shaver. 1978. Growth, turnover, and respiration rates of roots and tillers in tundra graminoids. *In:* L. Tiezen, ed., *Vegetation and production ecology of an Alaskan arctic tundra*. New York: Springer-Verlag.

Billings, W. D., K. M. Peterson, G. R. Shaver, and A. W. Trent. 1977. Root growth, respiration, and carbon-dioxide evolution in an arctic tundra soil. *Arctic and Alpine Research* 9:129–137.

Bjørnstad, O. N., W. Falck, and N. C. Stenseth. 1995. Geographic gradient in small rodent density fluctuations: a statistical modeling approach. *Proceedings of the Royal Society of London, Series B: Biological Sciences* 262: 127–33.

Bjørnstad, O. N., and B. T. Grenfell. 2001. Noisy clockwork: time series analysis of population fluctuations in animals. *Science* 293:638–643.

Bjørnstad, O. N., R. A. Ims, and X. Lambin. 1999. Spatial population dynamics: analyzing patterns and processes of population synchrony. *Trends in Ecology and Evolution* 14:427–432.

Bjørnstad, O. N., S. M. Sait, N. C. Stenseth, D. J. Thompson, and M. Begon. 2001. The impact of specialized enemies on the dimensionality of host dynamics. *Nature* 409:1001–1006.

Bjørnstad, O. N., N. C. Stenseth, and T. Saitoh. 1999. Synchrony and scaling in dynamics of voles and mice in northern Japan. *Ecology* 80:622–637.

Blinnikov, M. S., B. V. Gaglioti, D. A. Walker, M. J. Wooller, and G. D. Zazula. 2011. Pleistocene graminoid-dominated ecosystems in the Arctic. *Quaternary Science Reviews* 30:2906–2929.

Blois, J. L., J. L. McGuire, and E. A. Hadly. 2010. Small mammal diversity loss in response to late-Pleistocene climatic change. *Nature* 465:771–U775.

Boelman, N. T., M. Stieglitz, K. L. Griffin, and G. R. Shaver. 2005. Interannual variability of NDVI in response to long-term warming and fertilization in wet sedge and tussock tundra. *Oecologia* 143:588–597.

Boelman, N. T., M. Stieglitz, H. M. Rueth, M. Sommerkorn, K. L. Griffin, G. R. Shaver, and J. A. Gamon. 2003. Response of NDVI, biomass, and ecosystem gas exchange to long-term warming and fertilization in wet sedge tundra. *Oecologia* 135:414–421.

Bokhorst, S., J. W. Bjerke, F. W. Bowles, J. Melillo, T. V. Callaghan, and G. K. Phoenix. 2008. Impacts of extreme winter warming in the sub-Arctic: growing season responses of dwarf shrub heathland. *Global Change Biology* 14:2603–2612.

Bonan, G. B. 2008. Forests and climate change: forcings, feedbacks, and the climate benefits of forests. *Science* 320:1444–1449.

Both, C., S. Bouwhuis, C. M. Lessells, and M. E. Visser. 2006. Climate change and population declines in a long-distance migratory bird. *Nature* 441:81–83.

Both, C., J. J. Sanz, A. V. Artemyev, B. Blaauw, R. J. Cowie, A. J. Dekhuizen, A. Enemar, A. Javinen, N. E. I. Nyholm, J. Potti, P. A. Ravussin, B. Silverin, F. M. Slater, L. V. Sokolov, M. E. Visser, W. Winkel, J. Wright, H. Zang, and Mf. 2006. Pied flycatchers Ficedula hypoleuca travelling from Africa to breed in Europe: differential effects of winter and migration conditions on breeding date. *Ardea* 94:511–525.

Both, C., M. van Asch, R. G. Bijlsma, A. B. van den Burg, and M. E. Visser. 2009. Climate change and unequal phenological changes across four trophic levels: constraints or adaptations? *Journal of Animal Ecology* 78:73–83.

Boutin, S., C. J. Krebs, R. Boonstra, M. R. T. Dale, S. J. Hannon, K. Martin, and A. R. E. Sinclair. 1995. Population changes of the vertebrate community during a snowshoe hare cycle in Canada boreal forest. *Oikos* 74:69–80.

Boyce, M. S., C. V. Haridas, and C. T. Lee. 2006. Demography in an increasingly variable world. *Trends in Ecology & Evolution* 21:141–148.

Bradley, N. L., A. C. Leopold, J. Ross, and W. Huffaker. 1999. Phenological changes reflect climate change in Wisconsin. *Proceedings of the National Academy of Sciences of the United States of America* 96:9701–9704.

Brando, P. M., S. J. Goetz, A. Baccini, D. C. Nepstad, P. S. A. Beck, and M. C. Christman. 2010. Seasonal and interannual variability of climate and vegetation indices across the Amazon. *Proceedings of the National Academy of Sciences of the United States of America* 107:14685–14690.

Bråthen, K. A., R. A. Ims, N. G. Yoccoz, P. Fauchald, T. Tveraa, and V. H. Hausner. 2007. Induced shift in ecosystem productivity ? Extensive scale effects of abundant large herbivores. *Ecosystems* 10:773–789.

Bråthen, K. A., and J. Oksanen. 2001. Reindeer reduce biomass of preferred plant species. *Journal of Vegetation Science* 12:473–80.

Brearley, F. Q., J. Proctor, Suriantata, L. Nagy, G. Dalrymple, and B. C. Voysey. 2007. Reproductive phenology over a 10-year period in a lowland evergreen rain forest of central Borneo. *Journal of Ecology* 95:828–839.

Brodie, J., E. Post, and D. Doak, eds. 2012. *Wildlife conservation in a changing climate*. Chicago: University of Chicago Press.

Brodie, J., E. Post, and W. Laurance. 2010. How to conserve the tropics as they warm. *Nature* 468:634.

———. 2012. Climate change and tropical biodiversity: a new focus. *Trends in Ecology & Evolution* 27 (3): 145–150.

Brodie, J. F., and E. Post. 2010. Nonlinear responses of wolverine populations to declining winter snowpack. *Population Ecology* 52:279–287.

Brohan, P., J. J. Kennedy, I. Harris, S. F. B. Tett, and P. D. Jones. 2006. Uncertainty estimates in regional and global observed temperature changes: a new data set from 1850. *Journal of Geophysical Research—Atmospheres* 111.

Brook, B. W., N. S. Sodhi, and P. K. L. Ng. 2003. Catastrophic extinctions follow deforestation in Singapore. *Nature* 424:420–423.

Brooker, R. W., R. M. Callaway, L. A. Cavieres, Z. Kikvidze, C. J. Lortie, R. Michalet, F. I. Pugnaire, A. Valiente-Banuet, and T. G. Whitham. 2009. Don't diss integration: a comment on Ricklefs's Disintegrating Communities. *American Naturalist* 174:919–927.

Brooks, T., and A. Balmford. 1996. Atlantic forest extinctions. *Nature* 380:115. doi:10.1038/380115a0.

Brooks, T. M., R. A. Mittermeier, G. A. B. da Fonseca, J. Gerlach, M. Hoffmann, J. F. Lamoreux, C. G. Mittermeier, J. D. Pilgrim, and A. S. L. Rodrigues. 2006. Global biodiversity conservation priorities. *Science* 313:58–61.

Brooks, T. M., R. A. Mittermeier, C. G. Mittermeier, G. A. B. da Fonseca, A. B. Rylands, W. R. Konstant, P. Flick, J. Pilgrim, S. Oldfield, G. Magin, and C. Hilton-Taylor. 2002. Habitat loss and extinction in the hotspots of biodiversity. *Conservation Biology* 16:909–923.

Brooks, T. M., S. L. Pimm, and J. O. Oyugi. 1999. Time lag between deforestation and bird extinction in tropical forest fragments. *Conservation Biology* 13:1140–1150.

Brown, C. R., and M. B. Brown. 1998. Intense natural selection on body size and wing and tail asymmetry in cliff swallows during severe weather. *Evolution* 52:1461–1475.

Brown, J. H., and E. J. Heske. 1990. Temporal changes in a Chihuahuan desert rodent community. *Oikos* 59:290–302.

Brown, J. L., S. H. Li, and N. Bhagabati. 1999. Long-term trend toward earlier breeding in an American bird: a response to global warming? *Proceedings of the National Academy of Sciences of the United States of America* 96:5565–5569.

Brown, R., C. Derksen, and L. B. Wang. 2010. A multi-data set analysis of variability and change in Arctic spring snow cover extent, 1967–2008. *Journal of Geophysical Research—Atmospheres* 115.

Burgin, A. J., P. M. Groffman, and D. N. Lewis. 2010. Factors regulating denitrification in a riparian wetland. *Soil Science Society of America Journal* 74:1826–1833.

Burrows, M. T., D. S. Schoeman, L. B. Buckley, P. Moore, E. S. Poloczanska, K. M. Brander, C. Brown, J. F. Bruno, C. M. Duarte, B. S. Halpern, J. Holding, C. V. Kappel, W. Kiessling, M. I. O'Connor, J. M. Pandolfi, C. Parmesan, F. B. Schwing, W. J. Sydeman, and A. J. Richardson. 2011. The pace of shifting climate in marine and terrestrial ecosystems. *Science* 334: 652–655.

Cahoon, S. M. P., P. F. Sullivan, E. Post, and J. W. Welker. 2012. Herbivores limit CO_2 uptake and suppress carbon cycle responses to warming in the Arctic. *Global Change Biology* 18:469–479.

Cain, M. L., B. G. Milligan, and A. E. Strand. 2000. Long-distance seed dispersal in plant populations. *American Journal of Botany* 87:1217–1227.

Callaghan, T. V., L. O. Bjorn, F. S. I. Chapin, Y. Chernov, T. R. Christensen, B. Huntley, R. A. Ims, M. Johansson, D. J. Riedlinger, S. Jonasson, N. Matveyeva, W. C. Oechel, N. Panikov, and G. R. Shaver. 2005. Arctic tundra and polar desert ecosystems. *In:* ACIA, ed., *Arctic climate impact assessment*, 243–352. Cambridge: Cambridge University Press.

Canadell, J. G., C. Le Quere, M. R. Raupach, C. B. Field, E. T. Buitenhuis, P. Ciais, T. J. Conway, N. P. Gillett, R. A. Houghton, and G. Marland. 2007. Contributions to accelerating atmospheric $CO(2)$ growth from economic activity, carbon intensity, and efficiency of natural sinks. *Proceedings of the National Academy of Sciences of the United States of America* 104:18866–18870.

Canadell, J. G., and M. R. Raupach. 2008. Managing forests for climate change mitigation. *Science* 320:1456–1457.

Case, T. J., R. D. Holt, M. A. McPeek, and T. H. Keitt. 2005. The community context of species' borders: ecological and evolutionary perspectives. *Oikos* 108:28–46.

Caswell, H. 2000. *Matrix population models: construction, analysis, and interpretation.* New York: Sinauer.

Cattadori, I. M., D. T. Haydon, and P. J. Hudson. 2005. Parasites and climate synchronize red grouse populations. *Nature* 433:737–741.

Chapin, F. S. 1978. Phosphate uptake and nutrient utilization by Barrow tundra vegetation. *In:* L. L. Tieszen, ed., *Vegetation and production ecology of an Alaskan arctic tundra,* 483–508. New York: Springer-Verlag.

Chapin, F. S., S. E. Hobbie, and G. R. Shaver. 1997. Impacts of global change on composition of arctic communities: implications for ecosystem functioning. *In:* W. C. Oechel, T. V. Callaghan, T. Gilmanov, J. I. Holten, B. Maxwell, U. Molau, and B. Sveinbjornsson, eds., *Global change and arctic terrestrial ecosystems,* 221–228. New York: Springer-Verlag.

Chapin, F. S., R. L. Jefferies, J. F. Reynolds, G. Shaver, and J. Svoboda. 1992a. Arctic plant physiological ecology: a challenge for the future. *In:* F. S. Chapin, R. L. Jefferies, J. F. Reynolds, G. Shaver, and J. Svoboda, eds., *Arctic ecosystems in a changing climate: an ecophysiological perspective,* 3–8. New York: Academic Press.

———, eds. 1992b. *Arctic ecosystems in a changing climate: an ecophysiological perspective.* New York: Academic Press.

Chapin, F. S., J. McFarland, A. D. McGuire, E. S. Euskirchen, R. W. Ruess, and K. Kielland. 2009. The changing global carbon cycle: linking plant-soil carbon dynamics to global consequences. *Journal of Ecology* 97: 840–850.

Chapin, F. S., J. T. Randerson, A. D. McGuire, J. A. Foley, and C. B. Field. 2008. Changing feedbacks in the climate-biosphere system. *Frontiers in Ecology and the Environment* 6:313–320.

Chapin, F. S., and G. R. Shaver. 1985. Individualistic growth-response of tundra plant species to environmental manipulations in the field. *Ecology* 66:564–576.

———. 1996. Physiological and growth responses of arctic plants to a field experiment simulating climatic change. *Ecology* 77:822–840.

Chapin, F. S., G. R. Shaver, A. E. Giblin, K. J. Nadelhoffer, and J. A. Laundre. 1995. Responses of arctic tundra to experimental and observed changes in climate. *Ecology* 76:694–711.

Chapin, F. S., B. H. Walker, R. J. Hobbs, D. U. Hooper, J. H. Lawton, O. E. Sala, and D. Tilman. 1997. Biotic control over the functioning of ecosystems. *Science* 277:500–504.

Chapin, F. S., E. S. Zavaleta, V. T. Eviner, R. L. Naylor, P. M. Vitousek, H. L. Reynolds, D. U. Hooper, S. Lavorel, O. E. Sala, S. E. Hobbie, M. C. Mack, and S. Diaz. 2000. Consequences of changing biodiversity. *Nature* 405:234–242.

Chapin, F. S., S. A. Zimov, G. R. Shaver, and S. E. Hobbie. 1996. CO_2 fluctuation at high latitudes. *Nature* 383:585–586.

Chavez, F. P., P. G. Strutton, C. E. Friederich, R. A. Feely, G. C. Feldman, D. C. Foley, and M. J. McPhaden. 1999. Biological and chemical response of the equatorial Pacific Ocean to the 1997–98 El Niño. *Science* 286:2126–2131.

Cheng, T. L., S. M. Rovito, D. B. Wake, and V. T. Vredenburg. 2011. Coincident mass extirpation of neotropical amphibians with the emergence of the infectious fungal pathogen *Batrachochytrium dendrobatidis*. *Proceedings of the National Academy of Sciences of the United States of America* 108:9502–9507.

Chesson, P. 2000. Mechanisms of maintenance of species diversity. *Annual Review of Ecology and Systematics* 31:343–366.

Chesson, P., and N. Huntly. 1989. Short-term instabilities and long-term community dynamics. *Trends in Ecology & Evolution* 4:293–298.

Chitty, D. 1952. Mortality among voles (Microtus agrestis) at Lake Vyrnwy, Montgomeryshire in 1936–9. *Philosophical Transactions of the Royal Society of London, Series B: Biological Sciences* 236:505–552.

Chitty, D., and E. Phipps. 1966. Seasonal changes in survival in mixed populations of 2 species of vole. *Journal of Animal Ecology* 35:313–331.

Clark, J. S. 1998. Why trees migrate so fast: confronting theory with dispersal biology and the paleorecord. *American Naturalist* 152:204–224.

———S. 2007. *Models for ecological data: an introduction*. Princeton, NJ: Princeton University Press.

Clark, J. S., and O. N. Bjørnstad. 2004. Population time series: process variability, observation errors, missing values, lags, and hidden states. *Ecology* 85:3140–3150.

Clements, F. E. 1916. *Plant succession: an analysis of the development of vegetation*. Publication 242. Washington, DC: Carnegie Institution of Washington.

———. 1936. Nature and structure of the climax. *Journal of Ecology* 24: 252–284.

Cochrane, M. A., A. Alencar, M. D. Schulze, C. M. Souza, D. C. Nepstad, P. Lefebvre, and E. A. Davidson. 1999. Positive feedbacks in the fire dynamic of closed canopy tropical forests. *Science* 284:1832–1835.

Condit, R., P. S. Ashton, P. Baker, S. Bunyavejchewin, S. Gunatilleke, N. Gunatilleke, S. P. Hubbell, R. B. Foster, A. Itoh, J. V. LaFrankie, H. S.

Lee, E. Losos, N. Manokaran, R. Sukumar, and T. Yamakura. 2000. Spatial patterns in the distribution of tropical tree species. *Science* 288:1414–1418.

Condit, R., S. P. Hubbell, and R. B. Foster. 1996. Changes in tree species abundance in a Neotropical forest: impact of climate change. *Journal of Tropical Ecology* 12:231–256.

Connell, J. H. 1961. The influence of interspecific competition and other factors on the distribution of the barnacle *Chthamalus stellatus*. Ecology 42:710–723.

Connor, E. F., and E. D. McCoy. 1979. Statistics and biology of the species-area relationship. *American Naturalist* 113:791–833.

Cook, E. R., D. M. Meko, D. W. Stahle, and M. K. Cleaveland. 1999. Drought reconstructions for the continental United States. *Journal of Climate* 12: 1145–1162.

Cramer, W., A. Bondeau, F. I. Woodward, I. C. Prentice, R. A. Betts, V. Brovkin, P. M. Cox, V. Fisher, J. A. Foley, A. D. Friend, C. Kucharik, M. R. Lomas, N. Ramankutty, S. Sitch, B. Smith, A. White, and C. Young-Molling. 2001. Global response of terrestrial ecosystem structure and function to CO2 and climate change: results from six dynamic global vegetation models. *Global Change Biology* 7:357–773.

Crick, H. Q. P., C. Dudley, D. E. Glue, and D. L. Thomson. 1997. UK birds are laying eggs earlier. *Nature* 388:526–526.

Crick, H. Q. P., and T. H. Sparks. 1999. Climate change related to egg-laying trends. *Nature* 399:423–424.

Cross, M. S., and J. Harte. 2007. Compensatory responses to loss of warming-sensitive plant species. *Ecology* 88:740–48.

Crous, K. Y., P. B. Reich, M. D. Hunter, and D. S. Ellsworth. 2010. Maintenance of leaf N controls the photosynthetic CO_2 response of grassland species exposed to 9 years of free-air CO_2 enrichment. *Global Change Biology* 16:2076–2088.

Crozier, L. G., M. D. Scheuerell, and R. W. Zabel. 2011. Using time series analysis to characterize evolutionary and plastic responses to environmental change: a case study of a shift toward earlier migration date in sockeye salmon. *American Naturalist* 178:755–773.

Cushing, D. H. 1983. *Climate and fisheries.* New York: Academic Press.

Dai, A., K. E. Trenberth, and T. T. Qian. 2004. A global dataset of Palmer Drought Severity Index for 1870–2002: relationship with soil moisture and effects of surface warming. *Journal of Hydrometeorology* 5:1117–1130.

Darwin, C. 1859. *On the origin of species by means of natural selection, or, The preservation of favoured races in the struggle for life.* London: John Murray.

Davidson, E. A., and I. A. Janssens. 2006. Temperature sensitivity of soil carbon decomposition and feedbacks to climate change. *Nature* 440:165–173.

Davis, A. J., L. S. Jenkinson, J. H. Lawton, B. Shorrocks, and S. Wood. 1998. Making mistakes when predicting shifts in species range in response to global warming. *Nature* 391:783–786.

Dawson, T. P., S. T. Jackson, J. I. House, I. C. Prentice, and G. M. Mace. 2011. Beyond predictions: biodiversity conservation in a changing climate. *Science* 332:53–58.

DeFries, R. S., T. Rudel, M. Uriarte, and M. Hansen. 2010. Deforestation driven by urban population growth and agricultural trade in the twenty-first century. *Nature Geoscience* 3:178–181.

DeLucia, E. H., J. G. Hamilton, S. L. Naidu, R. B. Thomas, J. A. Andrews, A. C. Finzi, M. Lavine, R. Matamala, J. E. Mohan, G. R. Hendrey, and W. H. Schlesinger. 1999. Net primary production of a forest ecosystem with experimental CO_2 enrichment. *Science* 284:1177–1179.

Denman, K. L., G. Brasseur, A. Chidthaisong, P. Ciais, P. M. Cox, R. E. Dickinson, D. Hauglustaine, C. Heinze, E. Holland, D. Jacob, U. Lohmann, S. Ramachandran, P. L. da Silva Dias, S. C. Wofsy, and X. Zhang. 2007. Couplings between changes in the climate system and biogeochemistry. *In:* S. Solomon, D. Qin, M. Manning, Z. Chen, M. Marquis, K. B. Averyt, M. Tignor, and H. L. Miller, eds. *Climate Change 2007: The physical science basis. Contribution of Working Group 1 to the Fourth Assessment Report of the Intergovernmental Panel on Climate Change.* Cambridge: Cambridge University Press.

Diamond, J. M., and T. J. Case, eds. 1985. *Community ecology.* New York: Harper and Row.

Dixon, R. K., S. Brown, R. A. Houghton, A. M. Solomon, M. C. Trexler, and J. Wisniewski. 1994. Carbon pools and flux of global forest ecosystems. *Science* 263:185–190.

Doak, D., J. Brodie, and E. Post. 2012. What to expect, and how to plan, for wildlife conservation in the face of climate change. *In:* J. Brodie, E. Post, and D. Doak, eds., *Wildlife conservation in a changing climate*, 387–396. Chicago: University of Chicago Press.

Doak, D. F., D. Bigger, E. K. Harding, M. A. Marvier, R. E. O'Malley, and D. Thomson. 1998. The statistical inevitability of stability-diversity relationships in community ecology. *American Naturalist* 151:264–276.

Dobrovolski, R., A. S. Melo, F. A. S. Cassemiro, and J. A. F. Diniz. 2012. Climatic history and dispersal ability explain the relative importance of turnover and nestedness components of beta diversity. *Global Ecology and Biogeography* 21:191–197.

Duehl, A. J., F. H. Koch, and F. P. Hain. 2011. Southern pine beetle regional outbreaks modeled on landscape, climate and infestation history. *Forest Ecology and Management* 261:473–479.

Dukes, J. S., and H. A. Mooney. 1999. Does global change increase the success of biological invaders? *Trends in Ecology & Evolution* 14:135–139.

Dukes, J. S., J. Pontius, D. Orwig, J. R. Garnas, V. L. Rodgers, N. Brazee, B. Cooke, K. A. Theoharides, E. E. Stange, R. Harrington, J. Ehrenfeld, J. Gurevitch, M. Lerdau, K. Stinson, R. Wick, and M. Ayres. 2009. Responses of insect pests, pathogens, and invasive plant species to climate change in the forests of northeastern North America: what can we predict? *Canadian Journal of Forest Research—Revue Canadienne De Recherche Forestiere* 39:231–248.

Dunn, P. O., and D. W. Winkler. 1999. Climate change has affected the breeding date of tree swallows throughout North America. *Proceedings of the Royal Society of London, Series B: Biological Sciences* 266:2487–2490.

Dwyer, G., J. Dushoff, and S. H. Yee. 2004. The combined effects of pathogens and predators on insect outbreaks. *Nature* 430:341–345.

Dymond, C. C., E. T. Neilson, G. Stinson, K. Porter, D. A. MacLean, D. R. Gray, M. Campagna, and W. A. Kurz. 2010. Future spruce budworm outbreak may create a carbon source in eastern Canadian forests. *Ecosystems* 13:917–931.

Ehrhardt-Martinez, K., E. M. Crenshaw, and J. C. Jenkins. 2002. Deforestation and the environmental Kuznets curve: a cross-national investigation of intervening mechanisms. *Social Science Quarterly* 83:226–243.

Ehrlich, P., and A. Ehrlich. 1981. *Extinction: the causes and consequences of the disappearance of species*. New York: Random House.

Ehrlich, P., and B. Walker. 1998. Rivets and redundancy. *BioScience* 48: 387–387.

Eldredge, N. 1995. *Reinventing Darwin: the great debate at the high table of evolutionary theory*. New York: John Wiley.

Eldredge, N., and S. J. Gould. 1972. Punctuated equilibria: an alternative to phyletic gradualism. *In:* Thomas J. M. Schopf, ed., *Models in paleobiology*, 82–115. San Francisco: Freeman, Cooper, and Co.

———. 1997. On punctuated equilibria. *Science* 276:337–341.

Elliot, M., L. Labeyrie, G. Bond, E. Cortijo, J. L. Turon, N. Tisnerat, and J. C. Duplessy. 1998. Millennial-scale iceberg discharges in the Irminger Basin during the last glacial period: relationship with the Heinrich events and environmental settings. *Paleoceanography* 13:433–446.

Ellis, A. M., and E. Post. 2004. Population response to climate change: linear vs. non-linear modeling approaches. *BMC Ecology* 4:2.

Ellison, A. M., N. J. Gotelli, J. S. Brewer, D. L. Cochran-Stafira, J. M. Kneitel, T. E. Miller, A. C. Worley, and R. Zamora. 2003. The evolutionary ecology of carnivorous plants. *Advances in Ecological Research* 33:1–74.

Elmendorf, S. C., G. H. R. Henry, R. D. Hollister, R. G. Bjork, A. D. Bjorkman, T. V. Callaghan, L. S. Collier, E. J. Cooper, J. H. C. Cornelissen, T. A. Day, A. M. Fosaa, W. A. Gould, J. Gretarsdottir, J. Harte, L. Hermanutz, D. S. Hik, A. Hofgaard, F. Jarrad, I. S. Jonsdottir, F. Keuper, K. Klanderud, J. A. Klein, S. Koh, G. Kudo, S. I. Lang, V. Loewen, J. L. May, J. Mercado, A. Michelsen, U. Molau, I. H. Myers-Smith, S. F. Oberbauer, S. Pieper, E. Post, C. Rixen, C. H. Robinson, N. M. Schmidt, G. R. Shaver, A. Stenstrom, A. Tolvanen, O. Totland, T. Troxler, C. H. Wahren, P. J. Webber, J. M. Welker, and P. A. Wookey. 2012. Global assessment of experimental climate warming on tundra vegetation: heterogeneity over space and time. *Ecology Letters* 15:164–175.

Elton, C. 1924. Periodic fluctuations in the numbers of animals: their causes and effects. *British Journal of Experimental Biology* 2:119–163.

Elton, C. S. 1927. *Animal ecology*. London: Sidgwick & Jackson.

———. 1958. *The ecology of invasions by animals and plants*. London: Methuen and Co.

Elton, C. S., and R. S. Miller. 1954. The ecological survey of animal communities: with a practical system of classifying habitats by structural characters. *Journal of Ecology* 42:460–496.

Epstein, P. R. 2001. Climate change and emerging infectious diseases. *Microbes and Infection* 3:747–754.

Faith, J. T., and T. A. Surovell. 2009. Synchronous extinction of North America's Pleistocene mammals. *Proceedings of the National Academy of Sciences of the United States of America* 106:20641–20645.

Fearnside, P. M. 2000. Global warming and tropical land-use change: geenhouse gas emissions from biomass burning, decomposition and soils in forest conversion, shifting cultivation and secondary vegetation. *Climatic Change* 46:115–158.

Ferraz, G., J. D. Nichols, J. E. Hines, P. C. Stouffer, R. O. Bierregaard, and T. E. Lovejoy. 2007. A large-scale deforestation experiment: effects of patch area and isolation on Amazon birds. *Science* 315:238–241.

Fillios, M., J. Field, and B. Charles. 2010. Investigating human and megafauna co-occurrence in Australian prehistory: mode and causality in fossil accumulations at Cuddie Springs. *Quaternary International* 211:123–143.

Finstad, A. G., T. Forseth, B. Jonsson, E. Bellier, T. Hesthagen, A. J. Jensen, D. O. Hessen, and A. Foldvik. 2011. Competitive exclusion along climate gradients: energy efficiency influences the distribution of two salmonid fishes. *Global Change Biology* 17:1703–1711.

Fonseca, C. R., and G. Ganade. 2001. Species functional redundancy, random extinctions and the stability of ecosystems. *Journal of Ecology* 89:118–25.

Forchhammer, M., and D. Boertmann. 1993. The muskoxen, Ovibos moschatus, in north and northeast Greenland: population trends and the influence of abiotic parameters on population dynamics. *Ecography* 16:299–308.

Forchhammer, M. C., and T. Asferg. 2000. Invading parasites cause a structural shift in red fox dynamics. *Proceedings of the Royal Society of London, Series B* 267:779–786.

Forchhammer, M. C., and J. J. Boomsma. 1995. Foraging strategies and seasonal diet optimization of muskoxen in West Greenland. *Oecologia* 104:169–180.

Forchhammer, M. C., T. V. Callaghan, J. Nabe-Nielsen, N. M. Schmidt, M. P. Tamstorf, P. Aastrup, and E. Post. 2012. The biological currency of climate change? Manuscript.

Forchhammer, M. C., T. R. Christensen, B. U. Hansen, M. T. Tamstorf, J. Hinkler, N. M. Schmidt, T. T. Høye, M. Rasch, H. Meltofte, B. Elberling, and E. Post. 2008. Zackenberg in a circumpolar context. *Advances in Ecological Research* 40:499–544.

Forchhammer, M. C., E. Post, T. B. G. Berg, T. T. Hoye, and N. M. Schmidt. 2005. Local-scale and short-term herbivore-plant spatial dynamics reflect influences of large-scale climate. *Ecology* 86:2644–2651.

Forchhammer, M. C., E. Post, and N. C. Stenseth. 1998. Breeding phenology and climate. *Nature* 391:29–30.

———. 2002. North Atlantic Oscillation timing of long- and short-distance migration. *Journal of Animal Ecology* 71:1002–1014.

Forchhammer, M. C., E. Post, N. C. Stenseth, and D. Boertmann. 2002. Long-term responses in arctic ungulate dynamics to variation in climate and trophic processes. *Population Ecology* 44:113–120.

Forchhammer, M. C., N. C. Stenseth, E. Post, and R. Langvatn. 1998. Population dynamics of Norwegian red deer: density-dependence and climatic variation. *Proceedings of the Royal Society of London, Series B* 265:341–350.

Forrest, J., D. W. Inouye, and J. D. Thomson. 2010. Flowering phenology in subalpine meadows: Does climate variation influence community co-flowering patterns? *Ecology* 91:431–40.

Framstad, E., N. C. Stenseth, O. N. Bjørnstad, and W. Falck. 1997. Limit cycles in Norwegian lemmings: tensions between phase-dependence and density-dependence. *Proceedings of the Royal Society of London, Series B* 264:31–38.

Frank, D. A., and P. F. Groffman. 1998. Ungulate vs. landscape control of soil C and N processes in grasslands of Yellowstone National Park. *Ecology* 79:2229–2241.

Fredriksson, G. M., L. S. Danielsen, and J. E. Swenson. 2007. Impacts of El Nino related drought and forest fires on sun bear fruit resources in lowland diptero-carp forest of East Borneo. *Biodiversity and Conservation* 16:1823–1838.

Freed, L. A., and R. L. Cann. 2010. Misleading trend analysis and decline of Hawaiian forest birds. *Condor* 112:213–221.

Freidenburg, L. K., and D. K. Skelly. 2004. Microgeographical variation in thermal preference by an amphibian. *Ecology Letters* 7:369–373.

Friedlingstein, P., P. Cox, R. Betts, L. Bopp, W. Von Bloh, V. Brovkin, P. Cad-ule, S. Doney, M. Eby, I. Fung, G. Bala, J. John, C. Jones, F. Joos, T. Kato, M. Kawamiya, W. Knorr, K. Lindsay, H. D. Matthews, T. Raddatz, P. Rayner, C. Reick, E. Roeckner, K. G. Schnitzler, R. Schnur, K. Strassmann, A. J. Weaver, C. Yoshikawa, and N. Zeng. 2006. Climate-carbon cycle feedback analysis: results from the (CMIP)-M-4 model intercomparison. *Journal of Climate* 19:3337–3353.

Friend, A. D., A. K. Stevens, R. G. Knox, and M. G. R. Cannell. 1997. A process-based, terrestrial biosphere model of ecosystem dynamics (Hybrid v3.0). *Ecological Modelling* 95:249–287.

Fung, I. Y., C. J. Tucker, and K. C. Prentice. 1987. Application of advanced very high resolution radiometer vegetation index to study atmosphere-biosphere ex-change of CO_2. *Journal of Geophysical Research-Atmospheres* 92:2999–3015.

Gaston, K. J. 2000. Global patterns in biodiversity. *Nature* 405:220–227.

Gibbs, H. K., A. S. Ruesch, F. Achard, M. K. Clayton, P. Holmgren, N. Ra-mankutty, and J. A. Foley. 2010. Tropical forests were the primary sources of new agricultural land in the 1980s and 1990s. *Proceedings of the National Academy of Sciences of the United States of America* 107:16732–16737.

Gibbs, J. P., and A. R. Breisch. 2001. Climate warming and calling phenol-ogy of frogs near Ithaca, New York, 1900–1999. *Conservation Biology* 15:1175–1178.

Gienapp, P., C. Teplitsky, J. S. Alho, J. A. Mills, and J. Merila. 2008. Climate change and evolution: disentangling environmental and genetic responses. *Molecular Ecology* 17:167–178.

Gilman, S. E., M. C. Urban, J. Tewksbury, G. W. Gilchrist, and R. D. Holt. 2010. A framework for community interactions under climate change. *Trends in Ecology & Evolution* 25:325–331.

Givnish, T. J. 1994. Does diversity beget stability? *Nature* 371:113–114.

Gleason, H. A. 1926. The individualistic concept of plant association. *Bulletin of the Torrey Botanical Club* 53:1–20.

Goldman, B. D. 2001. Mammalian photoperiodic system: formal properties and neuroendocrine mechanisms of photoperiodic time measurement. *Jour-nal of Biological Rhythms* 16:283–301.

Gompertz, B. 1825. On the nature and function expressive of the law of human mortality, and on a new mode of determining the value of life contingencies. *Philosophical Transactions of the Royal Society of London, Series B: Biological Sciences* 115:513–585.

Goodale, C. L., M. J. Apps, R. A. Birdsey, C. B. Field, L. S. Heath, R. A. Houghton, J. C. Jenkins, G. H. Kohlmaier, W. Kurz, S. R. Liu, G. J. Nabuurs, S. Nilsson, and A. Z. Shvidenko. 2002. Forest carbon sinks in the Northern Hemisphere. *Ecological Applications* 12:891–899.

Goodale, C. L., and E. A. Davidson. 2002. Carbon cycle: uncertain sinks in the shrubs. *Nature* 418:593–594.

Gordon, C., C. Cooper, C. A. Senior, H. Banks, J. M. Gregory, T. C. Johns, J. F. B. Mitchell, and R. A. Wood. 2000. The simulation of SST, sea ice extents and ocean heat transports in a version of the Hadley Centre coupled model without flux adjustments. *Climate Dynamics* 16:147–168.

Gorham, E. 1991. Northern peatlands: role in the carbon cycle and probable responses to climatic warming. *Ecological Applications* 1:182–195.

Grace, J., J. Lloyd, J. McIntyre, A. C. Miranda, P. Meir, H. S. Miranda, C. Nobre, J. Moncrieff, J. Massheder, Y. Malhi, I. Wright, and J. Gash. 1995. Carbon-dioxide uptake by an undisturbed tropical rainforest in southwest Amazonia, 1992–1993. *Science* 270:778–780.

Graham, R. 2007. The role of non-analog biotas in the terminal Pleistocene extinction. *Journal of Vertebrate Paleontology* 27:84A.

Graham, R. W. 2005. Quaternary mammal communities: relevance of the individualistic response and non-analogue faunas. *Paleontological Society Papers* 11:141–158.

Graham, R. W., and E. C. Grimm. 1990. Effects of global climate change on the patterns of terrestrial biological communities. *Trends in Ecology & Evolution* 5:289–292.

Graham, R. W., E. L. Lundelius, M. A. Graham, E. K. Schroeder, R. S. Toomey, E. Anderson, A. D. Barnosky, J. A. Burns, C. S. Churcher, D. K. Grayson, R. D. Guthrie, C. R. Harington, G. T. Jefferson, L. D. Martin, H. G. McDonald, R. E. Morlan, H. A. Semken, S. D. Webb, L. Werdelin, and M. C. Wilson. 1996. Spatial response of mammals to late quaternary environmental fluctuations. *Science* 272:1601–1606.

Graham, R. W., and J. I. Mead. 1987. Environmental fluctuations and evolution of mammalian faunas during the last deglaciation in North America. *In:* W. F. Ruddiman and H. E. Wright, eds., *North America and adjacent oceans during the last deglaciation: the geology of North America.* Denver: Geological Society of America.

Grant, P. R., and B. R. Grant. 2002. Unpredictable evolution in a 30-year study of Darwin's finches. *Science* 296:707–711.

Grant, P. R., B. R. Grant, L. F. Keller, and K. Petren. 2000. Effects of El Nino events on Darwin's finch productivity. *Ecology* 81:2442–57.

Grenfell, B. T., O. F. Price, S. D. Albon, and T. H. Clutton-Brock. 1992. Overcompensation and population cycles in an ungulate. *Nature* 355:823–26.

Grenfell, B. T., K. Wilson, B. Finkenstädt, T. N. Coulson, S. Murray, S. D. Albon, J. M. Pemberton, T. H. Clutton-Brock, and M. J. Crawley. 1998. Noise and determinism in synchronized sheep dynamics. *Nature* 394:674–677.

Grinnell, J. 1917. The niche-relationships of the California thrasher. *Auk* 34:427–433.

Grulke, N. E., G. H. Riechers, W. C. Oechel, U. Hjelm, and C. Jaeger. 1990. Carbon balance in tussock tundra under ambient and atmospheric CO_2. *Oecologia* 83:485–494.

Guisan, A., and W. Thuiller. 2005. Predicting species distribution: offering more than simple habitat models. *Ecology Letters* 8:993–1009.

Guisan, A., and N. E. Zimmermann. 2000. Predictive habitat distribution models in ecology. *Ecological Modelling* 135:147–186.

Guthrie, R. D. 1968. Paleoecology of the large-mammal community in interior Alaska during the Late Pleistocene. *American Midland Naturalist* 79: 346–363.

———. 1982. Mammals of the mammoth steppe as paleoenvironmental indicators. *In:* D. M. Hopkins, J. V. Matthews, Jr., C. E. Schweger, and S. B. Young, eds., *Paleoecology of Beringia,* 307–326. New York: Academic Press.

———. 1989. Mosaics, allelochemicals and nutrients: an ecological theory of late Pleistocene megafaunal extinctions. *In:* P. S. Martin, ed., *Quaternary extinctions: a prehistoric revolution,* 259–298. Tucson: University of Arizona Press.

———. 1992. New paleoecological and paleoethological information on the extinct helmeted muskoxen from Alaska. *Annales Zoologici Fennici* 28: 175–186.

———. 2001. Origin and causes of the mammoth steppe: a story of cloud cover, woolly mammoth tooth pits, buckles, and inside-out Beringia. *Quaternary Science Reviews* 20:549–574.

———. 2003. Rapid body size decline in Alaskan Pleistocene horses before extinction. *Nature* 426:169–171.

———. 2006. New carbon dates link climatic change with human colonization and Pleistocene extinctions. *Nature* 441:207–209.

Hagedorn, F., M. Martin, C. Rixen, S. Rusch, P. Bebi, A. Zurcher, R. T. W. Siegwolf, S. Wipf, C. Escape, J. Roy, and S. Hattenschwiler. 2010. Short-term responses of ecosystem carbon fluxes to experimental soil warming at the Swiss alpine treeline. *Biogeochemistry* 97:7–19.

Haile, J., D. G. Froese, R. D. E. MacPhee, R. G. Roberts, L. J. Arnold, A. V. Reyes, M. Rasmussen, R. Nielsen, B. W. Brook, S. Robinson, M. Demuro, M. T. P. Gilbert, K. Munch, J. J. Austin, A. Cooper, I. Barnes, P. Moller, and E. Willerslev. 2009. Ancient DNA reveals late survival of mammoth and horse in interior Alaska. *Proceedings of the National Academy of Sciences of the United States of America* 106:22352–22357.

Hairston, N. G., F. E. Smith, and L. B. Slobodkin. 1960. Community structure, population control, and competition. *American Naturalist* 94:421–425.

Hall, W. E., T. R. Van Devender, and C. A. Olson. 1988. Late Quaternary arthropod remains from Sonoran Desert packrat middens, southwestern Arizona and northwestern Sonora. *Quaternary Research* 29:277–293.

Hannah, L. 2011. Climate change, connectivity, and conservation success. *Conservation Biology* 25:1139–1142.

Hansen, J., M. Sato, K. P., D. Beerling, R. Berner, V. Masson-Delmotte, M. Pagani, M. Raymo, D. L. Royer, and J. C. Zachos. 2008. Target atmospheric CO2: Where should humanity aim? *Open Atmospheric Science Journal* 2:217–231.

Hansen, J., M. Sato, R. Ruedy, K. Lo, D. W. Lea, and M. Medina-Elizade. 2006. Global temperature change. *Proceedings of the National Academy of Sciences of the United States of America* 103:14288–14293.

Hansen, M. C., S. V. Stehman, P. V. Potapov, T. R. Loveland, J. R. G. Townshend, R. S. DeFries, K. W. Pittman, B. Arunarwati, F. Stolle, M. K. Steininger, M. Carroll, and C. DiMiceli. 2008. Humid tropical forest clearing from 2000 to 2005 quantified by using multitemporal and multiresolution remotely sensed data. *Proceedings of the National Academy of Sciences of the United States of America* 105:9439–944.

Hansen, S. R., and S. P. Hubbell. 1980. Single nutrient microbial competition: qualitative agreement between experimental and theoretically forecast outcomes. *Science* 207:1491–1493.

Hanski, I., L. Hansson, and H. Henttonen. 1991. Specialist predators, generalist predators, and the microtine rodent cycle. *Journal of Animal Ecology* 60:353–367.

Hanski, I., and O. Ovaskainen. 2002. Extinction debt at extinction threshold. *Conservation Biology* 16:666–673.

Hanski, L., and E. Korpimäki. 1995. Microtine rodent dynamics in northern Europe: parameterized models for the predator-prey interacton. *Ecology* 76:840–850.

Hanski, L., P. Turchin, E. Korpimäki, and H. Henttonen. 1993. Population oscillations of boreal rodents: regulation by mustelid predators leads to chaos. *Nature* 364:232–235.

Hansson, L., and H. Henttonen. 1985. Gradients in density variations of small rodents: the importance of latitude and snow cover. *Oecologia* 67:394–402.

Hardy, J. W. 1967. Evolutionary and ecological relationships between 3 species of blackbirds (Icteridae) in central Mexico. *Evolution* 21:196.

Harington, C. R. 2011. Pleistocene vertebrates of the Yukon Territory. *Quaternary Science Reviews* 30:2341–2354.

Harte, J., S. Saleska, and T. Shih. 2006. Shifts in plant dominance control carbon-cycle responses to experimental warming and widespread drought. *Environmental Research Letters* 1.

Harte, J., and R. Shaw. 1995. Shifting dominance within a montane vegetation community: results of a climate-warming experiment. *Science* 267: 876–880.

Harte, J., M. S. Torn, F. R. Chang, B. Feifarek, A. P. Kinzig, R. Shaw, and K. Shen. 1995. Global warming and soil microclimate: results from a meadow-warming experiment. *Ecological Applications* 5:132–150.

Hassell, M. P. 1974. Density-dependence in single-species populations. *Journal of Animal Ecology* 44:283–95.

Hassell, M. P., J. H. Lawton, and R. M. May. 1976. Patterns of dynamical behaviour in single-species populations. *Journal of Animal Ecology* 45: 471–486.

Hastie, T. J., and R. J. Tibshirani. 1999. *Generalized additive models*. London: Chapman & Hall/CRC.

Hebblewhite, M., E. Merrill, and G. McDermid. 2008. A multi-scale test of the forage maturation hypothesis in a partially migratory ungulate population. *Ecological Monographs* 78:141–166.

Hedges, S. B., and S. Kumar. 2009. *The timetree of life*. Oxford: Oxford University Press.

Hedges, S. B., P. H. Parker, C. G. Sibley, and S. Kumar. 1996. Continental breakup and the ordinal diversification of birds and mammals. *Nature* 381: 226–229.

Heikkinen, R. K., M. Luoto, R. Virkkala, R. G. Pearson, and J. H. Korber. 2007. Biotic interactions improve prediction of boreal bird distributions at macro-scales. *Global Ecology and Biogeography* 16:754–763.

Heino, M., V. Kaitala, E. Ranta, and J. Lindström. 1997. Synchronous dynamics and rates of extinction in spatially structured populations. *Proceedings of the Royal Society of London, Series B* 264:481–486.

Heliasz, M., T. Johansson, A. Lindroth, M. Molder, M. Mastepanov, T. Friborg, T. V. Callaghan, and T. R. Christensen. 2011. Quantification of C uptake in subarctic birch forest after setback by an extreme insect outbreak. *Geophysical Research Letters* 38:5.

Helland, I. P., A. G. Finstad, T. Forseth, T. Hesthagen, and O. Ugedal. 2011. Ice-cover effects on competitive interactions between two fish species. *Journal of Animal Ecology* 80:539–547.

Henden, J. A., R. A. Ims, and N. G. Yoccoz. 2009. Nonstationary spatio-temporal small rodent dynamics: evidence from long-term Norwegian fox bounty data. *Journal of Animal Ecology* 78:636–645.

Hendrey, G. R. 1992. Global greenhouse studies: need for a new approach to ecosystem manipulation. *Critical Reviews in Plant Sciences* 11:61–74.

Hendrey, G. R., D. S. Ellsworth, K. F. Lewin, and J. Nagy. 1999. A free-air enrichment system for exposing tall forest vegetation to elevated atmospheric CO2. *Global Change Biology* 5:293–309.

Henry, G. H. R., and U. Molau. 1997. Tundra plants and climate change: the International Tundra Experiment (ITEX). *Global Change Biology* 3:1–9.

Higgins, S. I., and D. M. Richardson. 1999. Predicting plant migration rates in a changing world: the role of long-distance dispersal. *American Naturalist* 153:464–475.

Hik, D. S., and R. L. Jefferies. 1990. Increases in the net aboveground primary production of a salt-marsh forage grass: a test of the predictions of the herbivore-optimization model. *Journal of Ecology* 78:180–195.

Hill, M. P., A. A. Hoffmann, S. Macfadyen, P. A. Umina, and J. Elith. 2012. Understanding niche shifts: using current and historical data to model the invasive redlegged earth mite, Halotydeus destructor. *Diversity and Distributions* 18:191–203.

Hobbie, S. E. 1992. Effects of plant species on nutrient cycling. *Trends in Ecology & Evolution* 7:336–339.

Hobbie, S. E., and F. S. Chapin. 1998. Response of tundra plant biomass, aboveground production, nitrogen, and CO2 flux to experimental warming. *Ecology* 79:1526–1544.

Hobbie, S. E., L. Gough, and G. R. Shaver. 2005. Species compositional differences on different-aged glacial landscapes drive contrasting responses of tundra to nutrient addition. *Journal of Ecology* 93:770–782.

Hobbie, S. E., J. P. Schimel, S. E. Trumbore, and J. R. Randerson. 2000. Controls over carbon storage and turnover in high-latitude soils. *Global Change Biology* 6:196–210.

Holdo, R. M., A. R. E. Sinclair, A. P. Dobson, K. L. Metzger, B. M. Bolker, M. E. Ritchie, and R. D. Holt. 2009. A disease-mediated trophic cascade in the Serengeti and its implications for ecosystem C. *PLoS Biology* 7.

Holt, R. D. 1977. Predation, apparent competition, and structure of prey communities. *Theoretical Population Biology* 12:197–229.

———. 1990. The microevolutionary consequences of climate change. *Trends in Ecology & Evolution* 5:311–315.

————. 2009. Bringing the Hutchinsonian niche into the 21st century: ecological and evolutionary perspectives. *Proceedings of the National Academy of Sciences of the United States of America* 106:19659–19665.

Hooper, D. U., F. S. Chapin, J. J. Ewel, A. Hector, P. Inchausti, S. Lavorel, J. H. Lawton, D. M. Lodge, M. Loreau, S. Naeem, B. Schmid, H. Setala, A. J. Symstad, J. Vandermeer, and D. A. Wardle. 2005. Effects of biodiversity on ecosystem functioning: a consensus of current knowledge. *Ecological Monographs* 75:3–35.

Houghton, J. T., Y. Ding, D. J. Griggs, M. Noguer, P. J. van der Linden, X. Dai, K. Maskell, and C. A. Johnson, editors. 2001. *Climate Change 2001: The scientific basis. Contribution of Working Group I to the Third Assessment Report of the IPCC.* Cambridge: Cambridge University Press.

Houghton, R. A. 1999. The annual net flux of carbon to the atmosphere from changes in land use 1850–1990. *Tellus Series B: Chemical and Physical Meteorology* 51:298–313.

————. 2003. Why are estimates of the terrestrial carbon balance so different? *Global Change Biology* 9:500–509.

Høye, T. T. 2007. Ecological effects of climate change in high-arctic Greenland. PhD diss., University of Copenhagen.

Høye, T. T., E. Post, H. Meltofte, N. M. Schmidt, and M. C. Forchhammer. 2007. Rapid advancement of spring in the High Arctic. *Current Biology* 17:R449–R451.

Hubbell, S. P. 1979. Tree dispersion, abundance, and diversity in a tropical dry forest. *Science* 203:1299–1309.

————. 2001. *The unified neutral theory of biodiversity and biogeography.* Princeton, NJ: Princeton University Press

Huemmrich, K. F., G. Kinoshita, J. A. Gamon, S. Houston, H. Kwon, and W. C. Oechel. 2010. Tundra carbon balance under varying temperature and moisture regimes. *Journal of Geophysical Research—Biogeosciences* 115.

Hulme, P. E. 2012. Invasive species unchecked by climate. *Science* 335: 537–538.

Hutchinson, G. E. 1957. Concluding remarks: population studies, animal ecology, and demography. *Cold Spring Harbor Symposia on Quantitative Biology* 22:415–427.

————. 1959. Homage to Santa Rosalia: or why are there so many kinds of animals? *American Naturalist* 93:145–159.

————. 1961. The paradox of the plankton. *American Naturalist* 95:137–145.

Ims, R. A., and H. P. Andreassen. 2000. Spatial synchronization of vole population dynamics by predatory birds. *Nature* 408:194–196.

Ims, R. A., and E. Fuglei. 2005. Trophic interaction cycles in tundra ecosystems and the impact of climate change. *BioScience* 55:311–322.

Ims, R. A., J. A. Henden, and S. T. Killengreen. 2008. Collapsing population cycles. *Trends in Ecology & Evolution* 23:79–86.

Inouye, D. W. 2008. Effects of climate change on phenology, frost damage, and floral abundance of montane wildflowers. *Ecology* 89:353–362.

Inouye, D. W., B. Barr, K. B. Armitage, and B. D. Inouye. 2000. Climate change is affecting altitudinal migrants and hibernating species. *Proceedings of the National Academy of Sciences of the United States of America* 97:1630–1633.

Inouye, D. W., F. Saavedra, and W. Lee-Yang. 2003. Environmental influences on the phenology and abundance of flowering by Androsace septentrionalis (Primulaceae). *American Journal of Botany* 90:905–910.

Iverson, L. R., and A. M. Prasad. 1998. Predicting abundance of 80 tree species following climate change in the eastern United States. *Ecological Monographs* 68:465–485.

Iwasa, Y., and S. A. Levin. 1995. The timing of life history events. *Journal of Theoretical Biology* 172:33–42.

Jackson, J. B. C., M. X. Kirby, W. H. Berger, K. A. Bjorndal, L. W. Botsford, B. J. Bourque, R. H. Bradbury, R. Cooke, J. Erlandson, J. A. Estes, T. P. Hughes, S. Kidwell, C. B. Lange, H. S. Lenihan, J. M. Pandolfi, C. H. Peterson, R. S. Steneck, M. J. Tegner, and R. R. Warner. 2001. Historical overfishing and the recent collapse of coastal ecosystems. *Science* 293:629–638.

Jackson, S. T., and J. T. Overpeck. 2000. Responses of plant populations and communities to environmental changes of the late Quaternary. *Paleobiology* 26:194–220.

Jackson, S. T., R. S. Webb, K. H. Anderson, J. T. Overpeck, T. Webb, J. W. Williams, and B. C. S. Hansen. 2000. Vegetation and environment in Eastern North America during the Last Glacial Maximum. *Quaternary Science Reviews* 19:489–508.

Jankowski, J. E., S. K. Robinson, and D. J. Levey. 2010. Squeezed at the top: interspecific aggression may constrain elevational ranges in tropical birds. *Ecology* 91:1877–1884.

Janzen, D. H. 1967. Why mountain passes are higher in the tropics. *American Naturalist* 101:233–249.

Jepsen, J. U., S. B. Hagen, K. A. Hogda, R. A. Ims, S. R. Karlsen, H. Tommervik, and N. G. Yoccoz. 2009. Monitoring the spatio-temporal dynamics of geometrid moth outbreaks in birch forest using MODIS-NDVI data. *Remote Sensing of Environment* 113:1939–1947.

Jepsen, J. U., S. B. Hagen, R. A. Ims, and N. G. Yoccoz. 2008. Climate change and outbreaks of the geometrids Operophtera brumata and Epirrita autumnata in subarctic birch forest: evidence of a recent outbreak range expansion. *Journal of Animal Ecology* 77:257–264.

Jeschke, J. M., and D. L. Strayer. 2008. Usefulness of bioclimatic models for studying climate change and invasive species. *Year in Ecology and Conservation Biology 2008* 1134:1–24.

Joel, G., F. S. Chapin, N. R. Chiariello, S. S. Thayer, and C. B. Field. 2001. Species-specific responses of plant communities to altered carbon and nutrient availability. *Global Change Biology* 7:435–450.

Johnson, C. J., M. S. Boyce, R. L. Case, H. D. Cluff, R. J. Gau, A. Gunn, and R. Mulders. 2005. Cumulative effects of human development on arctic wildlife. *Wildlife Monographs* 160:1–36.

Johnson, K. H., K. A. Vogt, H. J. Clark, O. J. Schmitz, and D. J. Vogt. 1996. Biodiversity and the productivity and stability of ecosystems. *Trends in Ecology & Evolution* 11:372–377.

Jorgenson, A. K., and T. J. Burns. 2007. Effects of rural and urban population dynamics and national development on deforestation in less-developed countries, 1990–2000. *Sociological Inquiry* 77:460–482.

Karlsen, S. R., K. A. Hogda, F. E. Wielgolaski, A. Tolvanen, H. Tommervik, J. Poikolainen, and E. Kubin. 2009. Growing-season trends in Fennoscandia 1982–2006, determined from satellite and phenology data. *Climate Research* 39:275–286.

Keeling, C. D., J. F. S. Chin, and T. P. Whorf. 1996. Increased activity of northern vegetation inferred from atmospheric CO2 measurements. *Nature* 382:146–149.

Keith, L. B. 1963. *Wildlife's ten-year cycle.* Madison: University of Wisconsin Press.

Kerby, J., C. C. Wilmers, and E. Post. 2012. Climate change, phenology, and the nature of consumer-resource interactions: advancing the match/mismatch hypothesis. *In:* T. Ohgushi, O. J. Schmitz, and R. D. Holt, eds., *Trait-mediated indirect interactions: ecological and evolutionary perspectives*, 508–525. Cambridge: Cambridge University Press and the British Ecological Society.

Kielland, K., and J. P. Bryant. 1998. Moose herbivory in taiga: effects on biogeochemistry and vegetation dynamics in primary succession. *Oikos* 82: 377–383.

Kielland, K., J. P. Bryant, and R. W. Ruess. 1997. Moose herbivory and carbon turnover of early successional stands in interior Alaska. *Oikos* 80:25–30.

Kielland, K., and F. S. Chapin. 1992. Nutrient absorption and accumulation in arctic plants. *In:* F. S. Chapin, R. L. Jefferies, J. F. Reynolds, G. R. Shaver, and J. Svoboda, eds., *Arctic ecosystems in a changing climate,* 321–326. New York: Academic Press.

Kier, G., H. Kreft, T. M. Lee, W. Jetz, P. L. Ibisch, C. Nowicki, J. Mutke, and W. Barthlott. 2009. A global assessment of endemism and species richness across island and mainland regions. *Proceedings of the National Academy of Sciences of the United States of America* 106:9322–9327.

Klein, D. R. 1968. Introduction, increase, and crash of reindeer on St. Matthew Island. *Journal of Wildlife Management* 32:350–367.

———. 1990. Variation in quality of caribou and reindeer forage plants associated with season, plant part, and phenology. *Rangifer,* Special Issue 3: 123–130.

Koch, P. L., and A. D. Barnosky. 2006. Late Quaternary extinctions: state of the debate. *Annual Review of Ecology Evolution and Systematics* 37:215–250.

Koenig, W. D. 1999. Spatial autocorrelation of ecological phenomena. *Trends in Ecology and Evolution* 14:22–26.

———. 2002. Global patterns of environmental synchrony and the Moran effect. *Ecography* 25:283–288.

Koenig, W. D., and J. M. H. Knops. 1998. Scale of mast-seeding and tree-ring growth. Nature 396:225–226.

———. 2000. Patterns of annual seed production by Northern Hemisphere trees: A global perspective. *American Naturalist* 155:59–69.

Koenig, W. D., J. M. H. Knops, W. J. Carmen, and M. T. Stanback. 1999. Spatial dynamics in the absence of dispersal: acorn production by oaks in central coastal California. *Ecography* 22:499–506.

Körner, C. 2000. Biosphere responses to CO2 enrichment. *Ecological Applications* 10:1590–1619.

Körner, C., M. Diemer, B. Schappi, P. Niklaus, and J. Arnone. 1997. The responses of alpine grassland to four seasons of CO_2 enrichment: a synthesis. *Acta Oecologica—International Journal of Ecology* 18:165–175.

Korpimaki, E., and C. J. Krebs. 1996. Predation and population cycles of small mammals: a reassessment of the predation hypothesis. *BioScience* 46:754–764.

Kozak, K. H., and J. J. Wiens. 2010. Accelerated rates of climatic-niche evolution underlie rapid species diversification. *Ecology Letters* 13:1378–1389.

Krebs, C. J. 1998. Whither small rodent population studies? *Researches on Population Ecology* 40:123–125.

———. 2003. Nonlinear population dynamics: complex population dynamics: a theoretical/empirical synthesis by Peter Turchin. *Trends in Ecology & Evolution* 18:615.

Krebs, C. J., S. Boutin, R. Boonstra, A. R. E. Sinclair, J. N. M. Smith, M. R. T. Dale, K. Martin, and R. Turkington. 1995. Impact of food and predation on the snowshoe hare cycle. *Science* 269:1112–1115.

Krebs, C. J., M. S. Gaines, B. L. Keller, J. H. Myers, and R. H. Tamarin. 1973. Population cycles in small rodents. *Science* 179:35–41.

Krebs, C. J., B. L. Keller, and R. H. Tamarin. 1969. Microtus population biology: demographic changes in fluctuating populations of M. ochrogaster and M. pennsylvanicus in southern Indiana. *Ecology* 50:587–607.

Kumar, S., and S. B. Hedges. 1998. A molecular timescale for vertebrate evolution. *Nature* 392:917–920.

Kurz, W. A., C. C. Dymond, G. Stinson, G. J. Rampley, E. T. Neilson, A. L. Carroll, T. Ebata, and L. Safranyik. 2008. Mountain pine beetle and forest carbon feedback to climate change. *Nature* 452:987–990.

Kutz, S. J., A. P. Dobson, and E. P. Hoberg. 2009. Where are the parasites? *Science* 326:1187–1188.

Kutz, S. J., E. P. Hoberg, L. Polley, and E. J. Jenkins. 2005. Global warming is changing the dynamics of arctic host-parasite systems. *Proceedings of the Royal Society B: Biological Sciences* 272:2571–2576.

Lange, C. B., S. K. Burke, and W. H. Berger. 1990. Biological production off southern California is linked to climatic change. *Climatic Change* 16:319–329.

Laurance, W. F. 1997. Hyper-disturbed parks: edge effects and the ecology of isolated rainforest reserves in tropical Australia. *In:* W. F. Laurance and R. O. Bierregaard, eds., *Tropical forest remnants: ecology, management, and conservation of fragmented communities,* 71–83. Chicago: University of Chicago Press.

———. 1998. A crisis in the making: responses of Amazonian forests to land use and climate change. *Trends in Ecology & Evolution* 13:411–415.

———. 1999. Reflections on the tropical deforestation crisis. *Biological Conservation* 91:109–117.

———. 2004. Forest-climate interactions in fragmented tropical landscapes. *Philosophical Transactions of the Royal Society of London, Series B: Biological Sciences* 359:345–352.

———. 2008. Global warming and amphibian extinctions in eastern Australia. *Austral Ecology* 33:1–9.

Laurance, W. F., S. G. Laurance, and P. Delamonica. 1998. Tropical forest fragmentation and greenhouse gas emissions. *Forest Ecology and Management* 110:173–180.

Laurance, W. F., T. E. Lovejoy, H. L. Vasconcelos, E. M. Bruna, R. K. Didham, P. C. Stouffer, C. Gascon, R. O. Bierregaard, S. G. Laurance, and E. Sampaio. 2002. Ecosystem decay of Amazonian forest fragments: a 22-year investigation. *Conservation Biology* 16:605–618.

Laurance, W. F., K. R. McDonald, and R. Speare. 1996. Epidemic disease and the catastrophic decline of Australian rain forest frogs. *Conservation Biology* 10:406–413.

Laurance, W. F., A. A. Oliveira, S. G. Laurance, R. Condit, H. E. M. Nascimento, A. C. Sanchez-Thorin, T. E. Lovejoy, A. Andrade, S. D'Angelo, J. E. Ribeiro, and C. W. Dick. 2004. Pervasive alteration of tree communities in undisturbed Amazonian forests. *Nature* 428:171–175.

Laurance, W. F., D. C. Useche, L. P. Shoo, S. K. Herzog, M. Kessler, F. Escobar, G. Brehm, J. C. Axmacher, I. C. Chen, L. A. Gamez, P. Hietz, K. Fiedler, T. Pyrcz, J. Wolf, C. L. Merkord, C. Cardelus, A. R. Marshall, C. Ah-Peng, G. H. Aplet, M. D. Arizmendi, W. J. Baker, J. Barone, C. A. Bruhl, R. W. Bussmann, D. Cicuzza, G. Eilu, M. E. Favila, A. Hemp, C. Hemp, J. Homeier, J. Hurtado, J. Jankowski, G. Kattan, J. Kluge, T. Kromer, D. C. Lees, M. Lehnert, J. T. Longino, J. Lovett, P. H. Martin, B. D. Patterson, R. G. Pearson, K. S. H. Peh, B. Richardson, M. Richardson, M. J. Samways, F. Senbeta, T. B. Smith, T. M. A. Utteridge, J. E. Watkins, R. Wilson, S. E. Williams, and C. D. Thomas. 2011. Global warming, elevational ranges and the vulnerability of tropical biota. *Biological Conservation* 144:548–557.

Lauscher, A., and F. Lauscher. 1990. *Phänologie norwegens,* teil IV. Vienna: Eigenverlag F. Lauscher.

Leirs, H., N. C. Stenseth, J. D. Nichols, J. E. Hines, R. Verhagen, and W. Verheyen. 1997. Stochastic seasonality and nonlinear density-dependent factors regulate population size in an African rodent. *Nature* 389:176–180.

Levin, S. A. 1970. Community equilibria and stability, and an extension of the competitive exclusion principle. *American Naturalist* 104:413–423.

———. 1992. The problem of pattern and scale in ecology. *Ecology* 73: 1943–1967.

Levine, J. M., and C. M. D'Antonio. 1999. Elton revisited: a review of evidence linking diversity and invasibility. *Oikos* 87:15–26.

Lewontin, R. C. 1965. Selection for colonizing ability. *In:* H. G. Baker and G. L. Stebbins, eds., *The genetics of colonizing species*, 79–94. New York: Academic Press.

———. 1969. The meaning of stability. *In: Diversity and stability in ecological systems.* Brookhaven Symposium in Biology. Springfield, VA: Brookhaven National Laboratory.

Lincoln, G. A., H. Andersson, and A. Loudon. 2003. Clock genes in calendar cells as the basis of annual timekeeping in mammals: a unifying hypothesis. *Journal of Endocrinology* 179:1–13.

Linder, H. P. 2003. The radiation of the Cape flora, southern Africa. *Biological Reviews* 78:597–638.

Lindström, J., E. Ranta, and H. Lindén. 1996. Large-scale synchrony in the dynamics of capercaillie, black grouse and hazel grouse populations in Finland. *Oikos* 76:221–227.

Lips, K. R., F. Brem, R. Brenes, J. D. Reeve, R. A. Alford, J. Voyles, C. Carey, L. Livo, A. P. Pessier, and J. P. Collins. 2006. Emerging infectious disease and the loss of biodiversity in a Neotropical amphibian community. *Proceedings of the National Academy of Sciences of the United States of America* 103:3165–3170.

Lister, A. M. 1989. Rapid dwarfing of red deer on Jersey in the Last Interglacial. *Nature* 342:539–542.

Livingstone, D. A. 1975. Late Quaternary climatic change in Africa. *Annual Review of Ecology and Systematics* 6:249–280.

Lominicki, A. 1988. *Population ecology of individuals.* Princeton, NJ: Princeton University Press.

Long, S. P. 1991. Modification of the response of photosynthetic productivity to rising temperature by atmospheric CO2 concentrations: has its importance been underestimated? *Plant Cell and Environment* 14:729–739.

Loreau, M. 2000. Biodiversity and ecosystem functioning: recent theoretical advances. *Oikos* 91:3–17.

———. 2010. *From populations to ecosystems: theoretical foundations for a new ecological synthesis.* Princeton, NJ: Princeton University Press.

Loreau, M., S. Naeem, P. Inchausti, J. Bengtsson, J. P. Grime, A. Hector, D. U. Hooper, M. A. Huston, D. Raffaelli, B. Schmid, D. Tilman, and D. A. Wardle. 2001. Biodiversity and ecosystem functioning: current knowledge and future challenges. *Science* 294:804–808.

Lorenzen, E. D., D. Nogues-Bravo, L. Orlando, J. Weinstock, J. Binladen, K. A. Marske, A. Ugan, M. K. Borregaard, M. T. P. Gilbert, R. Nielsen, S. Y. W. Ho, T. Goebel, K. E. Graf, D. Byers, J. T. Stenderup, M. Rasmussen, P. F. Campos, J. A. Leonard, K. P. Koepfli, D. Froese, G. Zazula, T. W. Stafford, K. Aaris-Sorensen, P. Batra, A. M. Haywood, J. S. Singarayer, P. J. Valdes, G. Boeskorov, J. A. Burns, S. P. Davydov, J. Haile, D. L. Jenkins, P. Kosintsev, T. Kuznetsova, X. L. Lai, L. D. Martin, H. G. McDonald, D. Mol, M. Meldgaard, K. Munch, E. Stephan, M. Sablin, R. S. Sommer, T. Sipko, E. Scott, M. A. Suchard, A. Tikhonov, R. Willerslev, R. K. Wayne, A. Cooper, M. Hofreiter, A. Sher, B. Shapiro, C. Rahbek, and E. Willerslev. 2011. Species-specific responses of Late Quaternary megafauna to climate and humans. *Nature* 479:359–U195.

Lovejoy, T. E., R. O. Bierregaard, J. M. Rankin, and H. O. R. Schubart. 1983. Ecological dynamics of forest fragments. *In:* S. L. Sutton, T. C. Whitmore, and A. C. Chadwick, eds., *Tropical rain forest ecology and management,* 377–384. Oxford: Blackwell Scientific.

Lovejoy, T. E., R. O. Bierregaard, A. B. Rylands, J. R. Malcolm, C. Quintela, L. Harper, K. Brown, A. Powell, G. Powell, H. Schubart, and M. Hays. 1986.

Edge and other effects of isolation on Amazon forest fragments. *In:* M. E. Soule, ed., *Conservation biology: the science of scarcity and diversity*, 257–285. New York: Sinauer.

Lu, W. Q., Q. J. Meng, N. J. C. Tyler, K. A. Stokkan, and A. S. I. Loudon. 2010. A circadian clock is not required in an arctic mammal. *Current Biology* 20:533–537.

Luo, Y. Q., J. Melillo, S. L. Niu, C. Beier, J. S. Clark, A. T. Classen, E. Davidson, J. S. Dukes, R. D. Evans, C. B. Field, C. I. Czimczik, M. Keller, B. A. Kimball, L. M. Kueppers, R. J. Norby, S. L. Pelini, E. Pendall, E. Rastetter, J. Six, M. Smith, M. G. Tjoelker, and M. S. Torn. 2011. Coordinated approaches to quantify long-term ecosystem dynamics in response to global change. *Global Change Biology* 17:843–854.

Lyons, S. K. 2003. A quantitative assessment of the range shifts of Pleistocene mammals. *Journal of Mammalogy* 84:385–402.

MacArthur, R. 1955. Fluctuations of animal populations, and a measure of community stability. *Ecology* 36:533–536.

MacArthur, R. H. 1957. On the relative abundance of bird species. *Proceedings of the National Academy of Sciences of the United States of America* 43:293–295.

———. 1972. *Geographical ecology: patterns in the distribution of species.* New York: Harper & Row.

MacArthur, R. H., and R. Levins. 1964. Competition, habitat selection, and character displacement in a patchy environment. *Proceedings of the National Academy of Sciences of the United States of America* 51:1207–1210.

———. 1967. The limiting similarity, convergence, and divergence of coexisting species. *American Naturalist* 101:377–385.

MacArthur, R. H., and E. O. Wilson. 1967. *The theory of island biogeography.* Princeton, NJ: Princeton University Press.

Mack, M. C., E. A. G. Schuur, M. S. Bret-Harte, G. R. Shaver, and F. S. Chapin. 2004. Ecosystem carbon storage in arctic tundra reduced by long-term nutrient fertilization. *Nature* 431:440–443.

Maclean, D. A. 1980. Vulnerability of fir-spruce stands during uncontrolled spruce budworm outbreaks: a review and discussion. *Forestry Chronicle* 56: 213–221.

Maclean, I. M. D., and R. J. Wilson. 2011. Recent ecological responses to climate change support predictions of high extinction risk. *Proceedings of the National Academy of Sciences of the United States of America* 108: 12337–12342.

MacPhee, R. D. E., A. N. Tikhonov, D. Mol, C. D. Maliave, H. Van der Plicht, A. D. Greenwood, C. Flemming, and L. Agenbroad. 2002. Radiocarbon chronologies and extinction dynamics of the Late Quaternary mammalian megafauna of the Taimyr Peninsula, Russian Federation. *Journal of Archaeological Science* 29:1017–1042.

Mahall, B. E., and F. H. Bormann. 1978. Quantitative description of the vegetative phenology of herbs in a northern hardwood forest. *Botanical Gazette* 139:467–481.

Malhi, Y., D. D. Baldocchi, and P. G. Jarvis. 1999. The carbon balance of tropical, temperate and boreal forests. *Plant Cell and Environment* 22:715–740.

Mann, M. E., R. S. Bradley, and M. K. Hughes. 1998. Global-scale temperature patterns and climate forcing over the past six centuries. *Nature* 392: 779–787.

———. 1999. Northern hemisphere temperatures during the past millennium: inferences, uncertainties, and limitations. *Geophysical Research Letters* 26: 759–762.

Mann, M. E., and P. D. Jones. 2003. Global surface temperatures over the past two millennia. *Geophysical Research Letters* 30.

Manseau, M., J. Huot, and M. Crête. 1996. Effects of summer grazing by caribou on composition and productivity of vegetation: Community and landscape level. *Journal of Ecology* 84:503–513.

Margary, I. D. 1926. The Marsham phenological record in Norfolk, 1736–1925, and some others. *Quarterly Journal of the Royal Meteorological Society* 22:27–54.

Marion, G. M., G. H. R. Henry, D. W. Freckman, J. Johnstone, G. Jones, M. H. Jones, E. Levesque, U. Molau, P. Molgaard, A. N. Parsons, J. Svoboda, and R. A. Virginia. 1997. Open-top designs for manipulating field temperature in high-latitude ecosystems. *Global Change Biology* 3:20–32.

Marsham, R. A. 1789. Indications of spring. *Philosophical Transactions of the Royal Society* 79:154–156.

Martin, P. S. 1967. Prehistoric overkill. *In:* P. S. Martin and H. Wright, eds., *Pleistocene extinctions: the search for a cause,* 75–120. New Haven, CT: Yale University Press.

———. 1984. Prehistoric overkill: the global model. *In:* P. S. Martin and K. R.G., eds., *Quaternary extinctions: a prehistoric revolution,* 354–403. Tucson: University of Arizona Press.

Martin, T. E., and J. L. Maron. 2012. Climate impacts on bird and plant communities from altered animal-plant interactions. *Nature Climate Change* 2:195–200.

Martin-Vertedor, D., J. J. Ferrero-García, and L. M. Torres-Vila. 2010. Global warming affects phenology and voltinism of Lobesia botrana in Spain. *Agricultural and Forest Entomology* 12:169–176.

Mastepanov, M., C. Sigsgaard, E. J. Dlugokencky, S. Houweling, L. Strom, M. P. Tamstorf, and T. R. Christensen. 2008. Large tundra methane burst during onset of freezing. *Nature* 456:628–630.

Matheus, P., J. Beget, O. Mason, and C. Gelvin-Reymiller. 2003. Late Pliocene to late Pleistocene environments preserved at the Palisades Site, central Yukon River, Alaska. *Quaternary Research* 60:33–43.

May, R. M. 1972. Limit cycles in predator-prey communities. *Science* 177:900–902.

———. 1973a. *Stability and complexity in model ecosystems.* Princeton, NJ: Princeton University Press.

——— 1973b. Stability in randomly fluctuating versus deterministic environments. *American Naturalist* 107:621–650.

———. 1974. Biological populations with nonoverlapping generations: stable points, stable cycles, and chaos. *Science* 186:645–647.

———. 1975a. Biological populations obeying difference equations: stable points, stable cycles, and chaos. *Journal of Theoretical Biology* 51:511–524.

———. 1975b. Deterministic models with chaotic dynamics. *Nature* 256:165–166.

———, ed. 1976. *Theoretical ecology.* Philadelphia: W. B. Saunders.

———. 1986. Species interactions in ecology. *Science* 231:1451–1452.

May, R. M., G. R. Conway, M. P. Hassell, and T. R. E. Southwood. 1974. Time delays, density-dependence and single-species oscillations. *Journal of Animal Ecology* 43:747–770.

May, R. M., and R. H. MacArthur. 1972. Niche overlap as a function of environmental variability. *Proceedings of the National Academy of Sciences of the United States of America* 69:1109–1113.

Maynard Smith, J. 1974. *Models in ecology.* Cambridge: Cambridge University Press.

McAuliffe, J. R., and T. R. Van Devender. 1998. A 22,000-year record of vegetation change in the north-central Sonoran Desert. *Palaeogeography Palaeoclimatology Palaeoecology* 141:253–275.

McCleery, R. H., and C. M. Perrins. 1998. . . . temperature and egg-laying trends. *Nature* 391:30–31.

McCown, B. H. 1978. The interactions of organic nutrients, soil nitrogen, and soil temperature and plant growth and survival in the arctic environment. *In:* L. L. Tieszen, ed., *Vegetation and production ecology of an Alaska arctic tundra*, 435–456. New York: Springer-Verlag.

McGuire, A. D., L. G. Anderson, T. R. Christensen, S. Dallimore, L. D. Guo, D. J. Hayes, M. Heimann, T. D. Lorenson, R. W. Macdonald, and N. Roulet. 2009. Sensitivity of the carbon cycle in the Arctic to climate change. *Ecological Monographs* 79:523–555.

McInnes, P. F., R. J. Naiman, J. Pastor, and Y. Cohen. 1992. Effects of moose browsing on vegetation and litter of the boreal forest, Isle Royale, Michigan, USA. *Ecology* 73:2059–2075.

McKane, R. B., L. C. Johnson, G. R. Shaver, K. J. Nadelhoffer, E. B. Rastetter, B. Fry, A. E. Giblin, K. Kielland, B. L. Kwiatkowski, J. A. Laundre, and G. Murray. 2002. Resource-based niches provide a basis for plant species diversity and dominance in arctic tundra. *Nature* 415:68–71.

McKendrick, J. D., V. J. Ott, and G. A. Mitchell. 1978. Effects of nitrogen and phosphorous fertilization on carbohydrate and nutrient levels in *Dupontia fisheri* and *Arctagrostis latifolia*. *In:* L. L. Tieszen, ed., *Vegetation and production ecology of an Alaskan arctic tundra*, 509–538. New York: Springer-Verlag.

McLaren, B. E., and R. O. Peterson. 1994. Wolves, moose, and tree rings on Isle Royale. *Science* 266:1555–1558.

McNaughton, S. J. 1976. Serengeti migratory wildebeest: facilitation of energy flow by grazing. *Science* 191:92–94.

———. 1977. Diversity and stability of ecological communities: comment on the role of empiricism in ecology. *American Naturalist* 111:515–525.

———. 1983. Serengeti grassland ecology: the role of composite environmental factors and contingency in community organization. *Ecological Monographs* 53:291–320.

———. 1985. Ecology of a grazing ecosystem: the Serengeti. *Ecological Monographs* 55:259–294.

McNaughton, S. J., M. Oesterheld, D. A. Frank, and K. J. Williams. 1989. Ecosystem-level patterns of primary productivity and herbivory in terrestrial habitats. *Nature* 341:142–144.

McNaughton, S. J., R. W. Ruess, and S. W. Seagle. 1988. Large mammals and process dynamics in African ecosystems. *BioScience* 38:794–800.

McPeek, M. A. 1996. Linking local species interactions to rates of speciation in communities. *Ecology* 77:1355–1366.

———. 1998. The consequences of changing the top predator in a food web: a comparative experimental approach. *Ecological Monographs* 68:1–23.

Mech, L. D. 1966. *The wolves of Isle Royale*. Washington, DC: U.S. Government Printing Office.

Meldgaard, M. 1986. The Greenland caribou: zoogeography, taxonomy, and population dynamics. *Meddelelser om Grønland* 20:1–88.

Melillo, J. M., S. Butler, J. Johnson, J. Mohan, P. Steudler, H. Lux, E. Burrows, F. Bowles, R. Smith, L. Scott, C. Vario, T. Hill, A. Burton, Y. M. Zhou, and J. Tang. 2011. Soil warming, carbon-nitrogen interactions, and forest carbon budgets. *Proceedings of the National Academy of Sciences of the United States of America* 108:9508–9512.

Melillo, J. M., A. D. McGuire, D. W. Kicklighter, B. Moore, C. J. Vorosmarty, and A. L. Schloss. 1993. Global climate change and terrestrial net primary production. *Nature* 363:234–240.

Meltofte, H., T. R. Christensen, B. Elberling, M. C. Forchhammer, and M. Rasch, eds. 2008. *High-Arctic ecosystem dynamics in a changing climate.* New York: Academic Press.

Meltofte, H., and M. Rasch. 2008. The study area at Zackenberg. *Advances in Ecological Research* 40:101–110.

Menzel, A., T. H. Sparks, N. Estrella, and D. B. Roy. 2006. Altered geographic and temporal variability in phenology in response to climate change. *Global Ecology and Biogeography* 15:498–504.

Miller-Rushing, A. J., T. T. Hoye, D. W. Inouye, and E. Post. 2010. The effects of phenological mismatches on demography. *Philosophical Transactions of the Royal Society B: Biological Sciences* 365:3177–3186.

Miller-Rushing, A. J., and D. W. Inouye. 2009. Variation in the impact of climate change on flowering phenology and abundance: an examination of two pairs of closely related wildflower species. *American Journal of Botany* 96:1821–1829.

Miller-Rushing, A. J., and R. B. Primack. 2008. Global warming and flowering times in Thoreau's Concord: a community perspective. *Ecology* 89:332–341.

Milly, P. C. D., K. A. Dunne, and A. V. Vecchia. 2005. Global pattern of trends in streamflow and water availability in a changing climate. *Nature* 438:347–350.

Mitchell, T. D. and P. D. Jones. 2005. An improved method of constructing a database of monthly climate observations and associated high-resolution grids. *International Journal of Climatology* 25:693–712.

Molnár, P. K., A. E. Derocher, T. Klanjscek, and M. A. Lewis. 2010. Predicting climate change impacts on polar bear litter size. *Nature Communications* 2:8.

Molnár, P. K., A. E. Derocher, M. A. Lewis, and M. K. Taylor. 2008. Modelling the mating system of polar bears: a mechanistic approach to the Allee effect. *Proceedings of the Royal Society B: Biological Sciences* 275:217–226.

Molvar, E. M., T. R. Bowyer, and V. Van Ballenberghe. 1993. Moose herbivory, browse quality, and nutrient cycling in an Alaskan treeline community. *Oecologia* 94:472–479.

Mooney, H. A., E. R. Fuentes, and B. I. Kronberg, eds. 1993. *Earth system responses to global change: contrasts between North and South America.* New York: Academic Press.

Moorcroft, P. R., G. C. Hurtt, and S. W. Pacala. 2001. A method for scaling vegetation dynamics: the ecosystem demography model (ED). *Ecological Monographs* 71:557–585.

Moore, J. W., M. McClure, L. A. Rogers, and D. E. Schindler. 2010. Synchronization and portfolio performance of threatened salmon. *Conservation Letters* 3:340–348.

Moran, P. A. P. 1953a. The statistical analysis of the Canadian lynx cycle. I. Structure and prediction. *Australian Journal of Zoology* 1:163–173.

———. 1953b. The statistical analysis of the Canadian lynx cycle. II. Synchronization and meteorology. *Australian Journal of Zoology* 1:291–298.

Moritz, C., J. L. Patton, C. J. Conroy, J. L. Parra, G. C. White, and S. R. Beissinger. 2008. Impact of a century of climate change on small-mammal communities in Yosemite National Park, USA. *Science* 322:261–264.

Moyes, K., D. H. Nussey, M. N. Clements, F. E. Guinness, A. Morris, S. Morris, J. M. Pemberton, L. E. B. Kruuk, and T. H. Clutton-Brock. 2011. Advancing breeding phenology in response to environmental change in a wild red deer population. *Global Change Biology* 17:2455–2469.

Myneni, R. B., C. D. Keeling, C. J. Tucker, G. Asrar, and R. R. Nemani. 1997. Increased plant growth in the northern high latitudes from 1981 to 1991. *Nature* 386:698–702.

Mysterud, A., R. Langvatn, N. G. Yoccoz, and N. C. Stenseth. 2001. Plant phenology, migration and geographical variation in body weight of a large herbivore: the effect of a variable topography. *Journal of Animal Ecology* 70:915–923.

Nadelhoffer, K. J., A. E. Giblin, G. R. Shaver, and A. E. Linkins. 1992. Microbial processes and plant nutrient availability in arctic soils. *In:* F. S. Chapin, R. L. Jefferies, J. F. Reynolds, G. R. Shaver, and J. Svoboda, eds., *Arctic ecosystems in a changing climate,* 281–300. New York: Academic Press.

Nadelhoffer, K. J., G. R. Shaver, A. Giblin, and E. Rastetter. 1997. Potential impacts of climate change on nutrient cycling, decomposition, and productivity in arctic ecosystems. *In:* W. C. Oechel, T. V. Callaghan, T. Gilmanov, J. I. Holten, B. Maxwell, U. Molau, and B. Sveinbjornsson, eds., *Global change and arctic terrestrial ecosystems,* 349–364. New York: Springer-Verlag.

Naeem, S. 1998. Species redundancy and ecosystem reliability. *Conservation Biology* 12:39–45.

———. 2002. Ecosystem consequences of biodiversity loss: the evolution of a paradigm. *Ecology* 83:1537–1552.

Naeem, S., K. Hakansson, J. H. Lawton, M. J. Crawley, and L. J. Thompson. 1996. Biodiversity and plant productivity in a model assemblage of plant species. *Oikos* 76:259–264.

Naeem, S., and S. B. Li. 1997. Biodiversity enhances ecosystem reliability. Nature 390:507–509.

Naeem, S., L. J. Thompson, S. P. Lawler, J. H. Lawton, and R. M. Woodfin. 1994. Declining biodiversity can alter performance of ecosystems. *Nature* 368:734–737.

———. 1995. Empirical evidence that declining diversity may alter the performance of terrestrial ecosystems. *Philosophical Transactions of the Royal Society of London, Series B: Biological Sciences* 347:249–262.

Nager, R. G., C. Ruegger, and A. J. VanNoordwijk. 1997. Nutrient or energy limitation on egg formation: a feeding experiment in great tits. *Journal of Animal Ecology* 66:495–507.

Naiman, R. J., C. A. Johnston, and J. C. Kelley. 1988. Alteration of North American streams by beaver. *BioScience* 38:753–762.

Naiman, R. J., J. M. Melillo, and J. E. Hobbie. 1986. Ecosystem alteration of boreal forest streams by beaver (*Castor canadensis*). *Ecology* 67:1254–1269.

Naiman, R. J., G. Pinay, C. A. Johnston, and J. Pastor. 1994. Beaver influences on the long-term biogeochemical characteristics of boreal forest drainage networks. *Ecology* 75:905–921.

Nepstad, D. C., P. Moutinho, M. B. Dias, E. Davidson, G. Cardinot, D. Markewitz, R. Figueiredo, N. Vianna, J. Chambers, D. Ray, J. B. Guerreiros, P. Lefebvre, L. Sternberg, M. Moreira, L. Barros, F. Y. Ishida, I. Tohlver, E. Belk, K. Kalif, and K. Schwalbe. 2002. The effects of partial throughfall exclusion on canopy processes, aboveground production, and biogeochemistry of an Amazon forest. *Journal of Geophysical Research-Atmospheres* 107.

Nepstad, D. C., A. Verissimo, A. Alencar, C. Nobre, E. Lima, P. Lefebvre, P. Schlesinger, C. Potter, P. Moutinho, E. Mendoza, M. Cochrane, and V. Brooks. 1999. Large-scale impoverishment of Amazonian forests by logging and fire. *Nature* 398:505–508.

Nielsen, M. R. 2009. Is climate change causing the increasing narwhal (Monodon monoceros) catches in Smith Sound, Greenland? *Polar Research* 28:238–245.

Nijs, I. and J. Roy. 2000. How important are species richness, species evenness and interspecific differences to productivity? A mathematical model. *Oikos* 88:57–66.

Niklaus, P. A., P. W. Leadley, B. Schmid, and C. Körner. 2001. A long-term field study on biodiversity × elevated CO_2 interactions in grassland. *Ecological Monographs* 71:341–356.

Nikolskiy, P. A., L. D. Sulerzhitsky, and V. V. Pitulko. 2011. Last straw versus Blitzkrieg overkill: climate-driven changes in the Arctic Siberian mammoth population and the Late Pleistocene extinction problem. *Quaternary Science Reviews* 30:2309–2328.

Nogués-Bravo, D., R. Ohlemuller, P. Batra, and M. B. Araujo. 2010. Climate predictors of Late Quaternary extinctions. *Evolution* 64:2442–2449.

Nogués-Bravo, D., J. Rodriguez, J. Hortal, P. Batra, and M. B. Araujo. 2008. Climate change, humans, and the extinction of the woolly mammoth. *PLoS Biology* 6:685–692.

Norby, R. J., E. H. DeLucia, B. Gielen, C. Calfapietra, C. P. Giardina, J. S. King, J. Ledford, H. R. McCarthy, D. J. P. Moore, R. Ceulemans, P. De Angelis, A. C. Finzi, D. F. Karnosky, M. E. Kubiske, M. Lukac, K. S. Pregitzer, G. E. Scarascia-Mugnozza, W. H. Schlesinger, and R. Oren. 2005. Forest response to elevated CO2 is conserved across a broad range of productivity. *Proceedings of the National Academy of Sciences of the United States of America* 102:18052–18056.

Norby, R. J., P. J. Hanson, E. G. O'Neill, T. J. Tschaplinski, J. F. Weltzin, R. A. Hansen, W. X. Cheng, S. D. Wullschleger, C. A. Gunderson, N. T. Edwards, and D. W. Johnson. 2002. Net primary productivity of a CO2-enriched deciduous forest and the implications for carbon storage. *Ecological Applications* 12:1261–1266.

Norby, R. J., J. Ledford, C. D. Reilly, N. E. Miller, and E. G. O'Neill. 2004. Fine-root production dominates response of a deciduous forest to atmospheric CO2 enrichment. *Proceedings of the National Academy of Sciences of the United States of America* 101:9689–9693.

Norby, R. J., J. M. Warren, C. M. Iversen, B. E. Medlyn, and R. E. McMurtrie. 2010. CO2 enhancement of forest productivity constrained by limited nitrogen availability. *Proceedings of the National Academy of Sciences of the United States of America* 107:19368–19373.

Nordli, O., F. E. Wielgolaski, A. K. Bakken, S. H. Hjeltnes, F. Mage, A. Sivle, and O. Skre. 2008. Regional trends for bud burst and flowering of woody plants in Norway as related to climate change. *International Journal of Biometeorology* 52:625–639.

Odum, E. P. 1953. *Fundamentals of ecology.* Philadelphia: W. B. Saunders.

Oechel, W. C., T. Callaghan, T. Gilmanov, J. I. Holten, B. Maxwell, U. Molau, and B. Sveinbjörnsson, eds. 1997. *Global change and arctic terrestrial ecosystems.* New York: Springer-Verlag.

Oechel, W. C., S. J. Hastings, G. Vourlitis, M. Jenkins, G. Riechers, and N. Grulke. 1993. Recent change of arctic tundra ecosystems from a net carbon dioxide sink to a source. *Nature* 361:520–523.

Oechel, W. C., G. L. Vourlitis, S. J. Hastings, and S. A. Bochkarev. 1995. Change in arctic CO2 flux over 2 decades: effects of climate change at Barrow, Alaska. *Ecological Applications* 5:846–855.

Oechel, W. C., G. L. Vourlitis, S. J. Hastings, R. C. Zulueta, L. Hinzman, and D. Kane. 2000. Acclimation of ecosystem CO2 exchange in the Alaskan Arctic in response to decadal climate warming. *Nature* 406:978–981.

Oliver, J. S., R. G. Kvitek, and P. N. Slattery. 1985. Walrus feeding disturbance: scavenging habits and recolonization of the Berin Sea benthos. *Journal of Experimental Marine Biology and Ecology* 91:233–246.

Olofsson, J., L. Ericson, M. Torp, S. Stark, and R. Baxter. 2011. Carbon balance of Arctic tundra under increased snow cover mediated by a plant pathogen. *Nature Climate Change* 1:220–223.

Olofsson, J., H. Kitti, P. Rautiainen, S. Stark, and L. Oksanen. 2001. Effects of summer grazing by reindeer on composition of vegetation, productivity and nitrogen cycling. *Ecography* 24:13–24.

Olson, J. S., J. A. Watts, and L. J. Allison. 1983. *Carbon in live vegetation of major world ecosystems.* Washington, DC: U.S. Department of Energy.

Oren, R., D. S. Ellsworth, K. H. Johnsen, N. Phillips, B. E. Ewers, C. Maier, K. V. R. Schafer, H. McCarthy, G. Hendrey, S. G. McNulty, and G. G. Katul. 2001. Soil fertility limits carbon sequestration by forest ecosystems in a CO_2-enriched atmosphere. *Nature* 411:469–472.

Ouellet, J. P., S. Boutin, and D. C. Heard. 1994. Responses to simulated grazing and browsing of vegetation available to caribou in the Arctic. *Canadian Journal of Zoology—Revue Canadienne de Zoologie* 72:1426–1435.

Ozgul, A., S. Tuljapurkar, T. G. Benton, J. M. Pemberton, T. H. Clutton-Brock, and T. Coulson. 2009. The dynamics of phenotypic change and the shrinking sheep of St. Kilda. *Science* 325:464–467.

Paine, R. T. 1966. Food web complexity and species diversity. *American Naturalist* 100:65–75.

———. 1969. A note on trophic complexity and community stability. *American Naturalist* 103:91.

———. 1988. Food webs: road maps of interactions or grist for theoretical development? *Ecology* 69:1648–1654.

———. 2010. Macroecology: does it ignore or can it encourage further ecological syntheses based on spatially local experimental manipulations? *American Naturalist* 176:385–393.

Paine, T. D., K. F. Raffa, and T. C. Harrington. 1997. Interactions among scolytid bark beetles, their associated fungi, and live host conifers. *Annual Review of Entomology* 42:179–206.

Pan, Y. D., R. A. Birdsey, J. Y. Fang, R. Houghton, P. E. Kauppi, W. A. Kurz, O. L. Phillips, A. Shvidenko, S. L. Lewis, J. G. Canadell, P. Ciais, R. B. Jackson, S. W. Pacala, A. D. McGuire, S. L. Piao, A. Rautiainen, S. Sitch, and D. Hayes. 2011. A large and persistent carbon sink in the world's forests. *Science* 333:988–993.

Parmesan, C. 1996. Climate and species' range. *Nature* 382:765–766.

———. 2006. Ecological and evolutionary responses to recent climate change. *Annual Review of Ecology Evolution and Systematics* 37:637–669.

———. 2007. Influences of species, latitudes and methodologies on estimates of phenological response to global warming. *Global Change Biology* 13:1860–1872.

Parmesan, C., N. Ryrholm, C. Stefanescu, J. K. Hill, C. D. Thomas, H. Descimon, B. Huntley, L. Kaila, J. Kullberg, T. Tammaru, W. J. Tennent, J. A. Thomas, and M. Warren. 1999. Poleward shifts in geographical ranges of butterfly species associated with regional warming. *Nature* 399:579–583.

Parmesan, C., and G. Yohe. 2003. A globally coherent fingerprint of climate change impacts across natural systems. *Nature* 421:37–42.

Pascual, M., X. Rodo, S. P. Ellner, R. Colwell, and M. J. Bouma. 2000. Cholera dynamics and El Niño–Southern Oscillation. *Science* 289:1766–1769.

Pearson, P. N., and M. R. Palmer. 2000. Atmospheric carbon dioxide concentrations over the past 60 million years. *Nature* 406:695–699.

Pearson, R. G., and T. P. Dawson. 2003. Predicting the impacts of climate change on the distribution of species: are bioclimate envelope models useful? *Global Ecology and Biogeography* 12:361–371.

Pedersen, C., and E. Post. 2008. Interactions between herbivory and warming in aboveground biomass production of arctic vegetation. *BMC Ecology* 8:17.

Pellissier, L., K. A. Brathen, J. Pottier, C. F. Randin, P. Vittoz, A. Dubuis, N. G. Yoccoz, T. Alm, N. E. Zimmermann, and A. Guisan. 2010. Species distribution models reveal apparent competitive and facilitative effects of a dominant species on the distribution of tundra plants. *Ecography* 33:1004–1014.

Petchey, O. L., P. T. McPhearson, T. M. Casey, and P. J. Morin. 1999. Environmental warming alters food-web structure and ecosystem function. *Nature* 402:69–72.

Peterjohn, W. T., J. M. Melillo, P. A. Steudler, K. M. Newkirk, F. P. Bowles, and J. D. Aber. 1994. Responses of trace gas fluxes and N availability to experimentally elevated soil temperatures. *Ecological Applications* 4:617–625.

Peterson, A. T., M. A. Ortega-Huerta, J. Bartley, V. Sanchez-Cordero, J. Soberon, R. H. Buddemeier, and D. R. B. Stockwell. 2002. Future projections for Mexican faunas under global climate change scenarios. *Nature* 416:626–629.

Peterson, A. T., J. Soberon, R. G. Pearson, R. P. Anderson, E. Martinez-Meyer, M. Nakamura, and M. B. Araujo. 2011. *Ecological niches and geographic distributions: a modeling perspective*. Princeton, NJ: Princeton University Press.

Peterson, R. O., N. J. Thomas, J. M. Thurber, J. A. Vucetich, and T. A. Waite. 1998. Population limitation and the wolves of Isle Royale. *Journal of Mammalogy* 79:828–841.

Pettorelli, N., J. M. Gaillard, A. Mysterud, P. Duncan, N. C. Stenseth, D. Delorme, G. Van Laere, C. Toigo, and F. Klein. 2006. Using a proxy of plant productivity (NDVI) to find key periods for animal performance: the case of roe deer. *Oikos* 112:565–572.

Pettorelli, N., J. O. Vik, A. Mysterud, J. M. Gaillard, C. J. Tucker, and N. C. Stenseth. 2005. Using the satellite-derived NDVI to assess ecological responses to environmental change. *Trends in Ecology & Evolution* 20:503–510.

Phillips, O. L., L. Aragao, S. L. Lewis, J. B. Fisher, J. Lloyd, G. Lopez-Gonzalez, Y. Malhi, A. Monteagudo, J. Peacock, C. A. Quesada, G. van der Heijden, S. Almeida, I. Amaral, L. Arroyo, G. Aymard, T. R. Baker, O. Banki, L. Blanc, D. Bonal, P. Brando, J. Chave, A. C. A. de Oliveira, N. D. Cardozo, C. I. Czimczik, T. R. Feldpausch, M. A. Freitas, E. Gloor, N. Higuchi, E. Jimenez, G. Lloyd, P. Meir, C. Mendoza, A. Morel, D. A. Neill, D. Nepstad, S. Patino, M. C. Penuela, A. Prieto, F. Ramirez, M. Schwarz, J. Silva, M. Silveira, A. S. Thomas, H. ter Steege, J. Stropp, R. Vasquez, P. Zelazowski, E. A. Davila, S. Andelman, A. Andrade, K. J. Chao, T. Erwin, A. Di Fiore, E. Honorio, H. Keeling, T. J. Killeen, W. F. Laurance, A. P. Cruz, N. C. A. Pitman, P. N. Vargas, H. Ramirez-Angulo, A. Rudas, R. Salamao, N. Silva, J. Terborgh, and A. Torres-Lezama. 2009. Drought sensitivity of the Amazon rainforest. *Science* 323:1344–1347.

Phillips, O. L., Y. Malhi, N. Higuchi, W. F. Laurance, P. V. Nunez, R. M. Vasquez, S. G. Laurance, L. V. Ferreira, M. Stern, S. Brown, and J. Grace. 1998. Changes in the carbon balance of tropical forests: evidence from long-term plots. *Science* 282:439–442.

Pimm, S. L., and R. A. Askins. 1995. Forest losses predict bird extinctions in eastern North America. *Proceedings of the National Academy of Sciences of the United States of America* 92:9343–9347.

Pimm, S. L., G. J. Russell, J. L. Gittleman, and T. M. Brooks. 1995. The future of biodiversity. Science 269:347–350.

Post, D. D., and D. Pettus. 1966. Variation in *Rana pipiens* (Anura: Ranidae) of eastern Colorado. *Southwestern Naturalist* 11:476–482.

Post, D. D., and D. Pettus. 1967. Sympatry of two members of the *Rana pipiens* complex in Colorado. *Herpetologica* 23:323.

Post, E. 2003a. Climate-vegetation dynamics in the fast lane. *Trends in Ecology & Evolution* 18:551–553.

———. 2003b. Large-scale climate synchronizes the timing of flowering by multiple species. *Ecology* 84:277–281.

———. 2003c. Timing of reproduction in large mammals: climatic and density-dependent influences. *In:* M. D. Schwartz, ed., *Phenology: an integrative environmental science,* 437–499. New York: Kluwer.

———. 2005. Large-scale spatial gradients in herbivore population dynamics. *Ecology* 86:2320–2328.

———. 2013. Erosion of community diversity and stability by herbivore removal under warming. *Proceedings of the Royal Society of London, Series B* 280:20122722.

Post, E., P. S. Bøving, C. Pedersen, and M. A. MacArthur. 2003. Synchrony between caribou calving and plant phenology in depredated and non-depredated populations. *Canadian Journal of Zoology* 81:1709–1714.

Post, E., and J. Brodie. 2012. Improving protected area status in anticipation of novel conservation risks of climate change. Manuscript.

Post, E., J. Brodie, M. Hebblewhite, A. D. Anders, J. A. K. Maier, and C. C. Wilmers. 2009. Global population dynamics and hotspots of response to climate change. *BioScience* 59:489–497.

Post, E., and M. C. Forchhammer. 2001. Pervasive influence of large-scale climate in the dynamics of a terrestrial vertebrate community *BMC Ecology* 5:1.

———. 2002. Synchronization of animal population dynamics by large-scale climate. *Nature* 420:168–171.

———. 2004. Spatial synchrony of local populations has increased in association with the recent Northern Hemisphere climate trend. *Proceedings of the National Academy of Sciences of the United States of America* 101: 9286–9290.

———. 2006. Spatially synchronous population dynamics: an indicator of Pleistocene faunal response to large-scale environmental change in the Holocene. *Quaternary International* 151:99–105.

———. 2008. Climate change reduces reproductive success of an arctic herbivore through trophic mismatch. *Philosophical Transactions of the Royal Society of London, Series B* 363:2369–2375.

Post, E., M. C. Forchhammer, M. S. Bret-Harte, T. V. Callaghan, T. R. Christensen, B. Elberling, A. D. Fox, O. Gilg, D. S. Hik, T. T. Hoye, R. A. Ims, E. Jeppesen, D. R. Klein, J. Madsen, A. D. McGuire, S. Rysgaard, D. E. Schindler, I. Stirling, M. P. Tamstorf, N. J. C. Tyler, R. van der Wal, J. M. Welker, P. A. Wookey, N. M. Schmidt, and P. Aastrup. 2009. Ecological dynamics across the Arctic associated with recent climate change. *Science* 325:1355.

Post, E., M. C. Forchhammer, and N. C. Stenseth. 1999. Population ecology and the North Atlantic Oscillation (NAO). *Ecological Bulletins* 47: 117–125.

Post, E., M. C. Forchhammer, N. C. Stenseth, and T. V. Callaghan. 2001. The timing of life-history events in a changing climate. *Proceedings of the Royal Society of London, Series B* 268:15–23.

Post, E., and D. R. Klein. 1999. Caribou calf production and seasonal range quality during a population decline. *Journal of Wildlife Management* 63: 335–345.

Post, E., and C. Pedersen. 2008. Opposing plant community responses to warming with and without herbivores. *Proceedings of the National Academy of Sciences* 105:12353–12358.

Post, E., C. Pedersen, C. C. Wilmers, and M. C. Forchhammer. 2008a. Phenological sequences reveal aggregate life history response to climatic warming. *Ecology* 89:363–370.

————. 2008b. Warming, plant phenology, and the spatial dimension of trophic mismatch for large herbivores. *Proceedings of the Royal Society of London, Series B* 275:2005–2013.

Post, E., R. O. Peterson, N. C. Stenseth, and B. E. McLaren. 1999. Ecosystem consequences of wolf behavioural response to climate. *Nature* 401: 905–907.

Post, E., and N. C. Stenseth. 1998. Large-scale climatic fluctuation and population dynamics of moose and white-tailed deer. *Journal of Animal Ecology* 67:537–543.

————. 1999. Climatic variability, plant phenology, and northern ungulates. *Ecology* 80:1322–1339.

Post, E., N. C. Stenseth, R. Langvatn, and J.-M. Fromentin. 1997. Global climate change and phenotypic variation among red deer cohorts. *Proceedings of the Royal Society of London, Series B* 264:1317–1324.

Post, E., N. C. Stenseth, R. O. Peterson, J. A. Vucetich, and A. M. Ellis. 2002. Phase dependence and population cycles in a large-mammal predator-prey system. *Ecology* 83:2997–3002.

Post, E. S. 1995. Comparative foraging ecology and social dynamics of caribou (Rangifer tarandus). PhD diss., University of Alaska.

Post, E. S., and D. W. Inouye. 2008. Phenology: response, driver, and integrator. *Ecology* 89:319–320.

Post, E. S., and D. R. Klein. 1996. Relationships between graminoid growth form and levels of grazing by caribou (Rangifer tarandus) in Alaska. *Oecologia* 107:364–372.

Post, W. M., W. R. Emanuel, P. J. Zinke, and A. G. Stangenberger. 1982. Soil carbon pools and world life zones. *Nature* 298:156–159.

Post, W. M., and C. C. Travis. 1979. Quantitative stability in models of ecological communities. *Journal of Theoretical Biology* 79:547–553.

Pounds, J. A., M. R. Bustamante, L. A. Coloma, J. A. Consuegra, M. P. L. Fogden, P. N. Foster, E. La Marca, K. L. Masters, A. Merino-Viteri, R. Puschendorf, S. R. Ron, G. A. Sanchez–Azofeifa, C. J. Still, and B. E. Young. 2006. Widespread amphibian extinctions from epidemic disease driven by global warming. *Nature* 439:161–167.

Pounds, J. A., M. P. L. Fogden, and J. H. Campbell. 1999. Biological response to climate change on a tropical mountain. *Nature* 398:611–615.

Poyry, J., R. Leinonen, G. Soderman, M. Nieminen, R. K. Heikkinen, and T. R. Carter. 2011. Climate-induced increase of moth multivoltinism in boreal regions. *Global Ecology and Biogeography* 20:289–298.

Prideaux, G. J., L. K. Ayliffe, L. R. G. DeSantis, B. W. Schubert, P. F. Murray, M. K. Gagan, and T. E. Cerling. 2009. Extinction implications of a chenopod browse diet for a giant Pleistocene kangaroo. *Proceedings of the National Academy of Sciences of the United States of America* 106: 11646–11650.

Prideaux, G. J., R. G. Roberts, D. Megirian, K. E. Westaway, J. C. Hellstrom, and J. I. Olley. 2007. Mammalian responses to Pleistocene climate change in southeastern Australia. *Geology* 35:33–36.

Pushkina, D., and P. Raia. 2008. Human influence on distribution and extinctions of the late Pleistocene Eurasian megafauna. *Journal of Human Evolution* 54:769–782.

Raich, J. W., and C. S. Potter. 1995. Global patterns of carbon-dioxide emissions from soils. *Global Biogeochemical Cycles* 9:23–36.

Raich, J. W., and W. H. Schlesinger. 1992. The global carbon-dioxide flux in soil respiration and its relationship to vegetation and climate. *Tellus Series B: Chemical and Physical Meteorology* 44:81–99.

Ramankutty, N., H. K. Gibbs, F. Achard, R. Defriess, J. A. Foley, and R. A. Houghton. 2007. Challenges to estimating carbon emissions from tropical deforestation. *Global Change Biology* 13:51–66.

Ranta, E., V. Kaitala, and J. Lindström. 1997. Dynamics of Canadian lynx populations in space and time. *Ecography* 20:454–460.

Ranta, E., V. Kaitala, and P. Lundberg. 1997. The spatial dimension in population fluctuations. *Science* 278:1621–1623.

Rasmusson, E. M. and T. H. Carpenter. 1982. Variations in tropical sea-surface temperature and surface wind fields associated with the Southern Oscillation El Nino. *Monthly Weather Review* 110:354–384.

Ray, G. C., G. L. Hufford, I. I. Krupnik,and J. E. Overland. 2008. Diminishing sea ice. *Science* 321:1443–1444.

Ray, G. C., J. McCormick-Ray, P. Berg, and H. E. Epstein. 2006. Pacific walrus: benthic bioturbator of Beringia. *Journal of Experimental Marine Biology and Ecology* 330:403–419.

Ray, G. C., J. E. Overland, and G. L. Hufford. 2010. Seascape as an organizing principle for evaluating walrus and seal sea-ice habitat in Beringia. *Geophysical Research Letters* 37.

Reale, D., D. Berteaux, A. G. McAdam, and S. Boutin. 2003. Lifetime selection on heritable life-history traits in a natural population of red squirrels. *Evolution* 57:2416–2423.

Reale, D., A. G. McAdam, S. Boutin, and D. Berteaux. 2003. Genetic and plastic responses of a northern mammal to climate change. *Proceedings of the Royal Society of London, Series B: Biological Sciences* 270:591–596.

Reed, T. E., R. S. Waples, D. E. Schindler, J. J. Hard, and M. T. Kinnison. 2010. Phenotypic plasticity and population viability: the importance of environmental predictability. *Proceedings of the Royal Society B: Biological Sciences* 277:3391–3400.

Regehr, E. V., N. J. Lunn, S. C. Amstrup, and L. Stirling. 2007. Effects of earlier sea ice breakup on survival and population size of polar bears in western Hudson Bay. *Journal of Wildlife Management* 71:2673–2683.

Reich, P. B., J. Knops, D. Tilman, J. Craine, D. Ellsworth, M. Tjoelker, T. Lee, D. Wedin, S. Naeem, D. Bahauddin, G. Hendrey, S. Jose, K. Wrage, J. Goth, and W. Bengston. 2001. Plant diversity enhances ecosystem responses to elevated CO2 and nitrogen deposition. *Nature* 410:809–812.

Reich, P. B., M. B. Walters, and D. S. Ellsworth. 1997. From tropics to tundra: global convergence in plant functioning. *Proceedings of the National Academy of Sciences of the United States of America* 94:13730–13734.

Ricklefs, R. E. 1987. Community diversity: relative roles of local and regional processes. *Science* 235:167–171.

———. 2004. A comprehensive framework for global patterns in biodiversity. *Ecology Letters* 7:1–15.

———. 2008. Disintegration of the ecological community. *American Naturalist* 172:741–750.

———. 2009a. A brief response to Brooker et al.'s comment. *American Naturalist* 174:928–931.

———. 2009b. Speciation, extinction and diversity. *In:* R. K. Butlin, J. R. Bridle, and D. Schluter, eds., *Speciation and patterns of diversity,* 257–277. Cambridge: Cambridge University Press.

———. 2011. Applying a regional community concept to forest birds of eastern North America. *Proceedings of the National Academy of Sciences of the United States of America* 108:2300–2305.

Rind, D. 1998. Latitudinal temperature gradients and climate change. *Journal of Geophysical Research-Atmospheres* 103:5943–5971.

Rodó, X., M. Pascual, G. Fuchs, and A. S. G. Faruque. 2002. ENSO and cholera: a nonstationary link related to climate change? *Proceedings of the National Academy of Sciences of the United States of America* 99:12901–12906.

Roe, F. G. 1970. *The North American buffalo.* Toronto: University of Toronto Press.

Roemmich, D., and J. McGowan. 1995. Climatic warming and the decline of zooplankton in the California Current. *Science* 267:1324–1326.

Root, T. L., D. P. MacMynowski, M. D. Mastrandrea, and S. H. Schneider. 2005. Human-modified temperatures induce species changes: joint attribution. *Proceedings of the National Academy of Sciences of the United States of America* 102:7465–7469.

Root, T. L., J. T. Price, K. R. Hall, S. H. Schneider, C. Rosenzweig, and J. A. Pounds. 2003. Fingerprints of global warming on wild animals and plants. *Nature* 421:57–60.

Ropelewski, C. F., and M. S. Halpert. 1987. Global and regional scale precipitation patterns associated with the El Nino Southern Oscillation. *Monthly Weather Review* 115:1606–1626.

Rosenzweig, C., D. Karoly, M. Vicarelli, P. Neofotis, Q. G. Wu, G. Casassa, A. Menzel, T. L. Root, N. Estrella, B. Seguin, P. Tryjanowski, C. Z. Liu, S. Rawlins, and A. Imeson. 2008. Attributing physical and biological impacts to anthropogenic climate change. *Nature* 453:353–U320.

Royama, T. 1977. Population persistence and density dependence. *Ecological Monographs* 47:1–35. .

———. 1984. Population dynamics of the spruce budworm Choristoneura fumiferana. *Ecological Monographs* 54:429–462.

———. 1992. *Analytical population dynamics.* London: Chapman & Hall.

———. 2005. Moran effect on nonlinear population processes. *Ecological Monographs* 75:277–293.

Ryan, M. G. 1991. Effects of climate change on plant respiration. *Ecological Applications* 1:157–167.

Ryan, M. G., and B. E. Law. 2005. Interpreting, measuring, and modeling soil respiration. *Biogeochemistry* 73:3–27.

Sæther, B.-E., J. Tufto, J. Engen, K. Jerstad, O.W. Røstad, and J. E. Skåtan. 2000. Population dynamical consequences of climate change for a small temperate songbird. *Science* 287:854–856.

Saether, B. E., S. Engen, A. P. Moller, E. Matthysen, F. Adriaensen, W. Fiedler, A. Leivits, M. M. Lambrechts, M. E. Visser, T. Anker-Nilssen, C. Both, A. A. Dhondt, R. H. McCleery, J. McMeeking, J. Potti, O. W. Rostad, and D. Thomson. 2003. Climate variation and regional gradients in population dynamics of two hole-nesting passerines. *Proceedings of the Royal Society of London, Series B: Biological Sciences* 270:2397–2404.

Sætre, G.-P., E. Post, and M. Král. 1999. Can environmental fluctuation prevent competitive exclusion in sympatric flycatchers? *Proceedings of the Royal Society of London, Series B* 266:1247–1251.

Sætre, G. P., M. Král, and V. Bicik. 1993. Experimental evidence for interspecific female mimicry in sympatric ficedula flycatchers. *Evolution* 47: 939–945.

Sætre, G. P., T. Moum, S. Bures, M. Král, M. Adamjan, and J. Moreno. 1997. A sexually selected character displacement in flycatchers reinforces premating isolation. *Nature* 387:589–592.

Sagarin, R. 2007. In support of observational studies. *Frontiers in Ecology and the Environment* 5:294–294.

Saleska, S. R., J. Harte, and M. S. Torn. 1999. The effect of experimental ecosystem warming on CO_2 fluxes in a montane meadow. *Global Change Biology* 5:125–141.

Saunders, J. J. 1988. Fossiliferous spring sites in southwest Missouri. *In:* R. S. Laub, N. G. Miller, and D. W. Steadman, eds., *Late Pleistocene and early Holocene paleoecology and archaeology of the Eastern Great Lakes region*, 127–149. Buffalo, NY: Buffalo Society of Natural Sciences.

Scheffer, M., J. Bascompte, W. A. Brock, V. Brovkin, S. R. Carpenter, V. Dakos, H. Held, E. H. van Nes, M. Rietkerk, and G. Sugihara. 2009. Early-warning signals for critical transitions. *Nature* 461:53–59.

Scheffer, M., S. Carpenter, J. A. Foley, C. Folke, and B. Walker. 2001. Catastrophic shifts in ecosystems. *Nature* 413:591–596.

Scheffer, M., and S. R. Carpenter. 2003. Catastrophic regime shifts in ecosystems: linking theory to observation. *Trends in Ecology & Evolution* 18:648–656.

Schimel, D. S., B. H. Braswell, E. A. Holland, R. McKeown, D. S. Ojima, T. H. Painter, W. J. Parton, and A. R. Townsend. 1994. Climatic, edaphic, and biotic controls over storage and turnover of carbon in soils. *Global Biogeochemical Cycles* 8:279–293.

Schimel, D. S., J. I. House, K. A. Hibbard, P. Bousquet, P. Ciais, P. Peylin, B. H. Braswell, M. J. Apps, D. Baker, A. Bondeau, J. Canadell, G. Churkina, W. Cramer, A. S. Denning, C. B. Field, P. Friedlingstein, C. Goodale, M. Heimann, R. A. Houghton, J. M. Melillo, B. Moore, D. Murdiyarso, I. Noble, S. W. Pacala, I. C. Prentice, M. R. Raupach, P. J. Rayner, R. J. Scholes, W. L. Steffen, and C. Wirth. 2001. Recent patterns and mechanisms of carbon exchange by terrestrial ecosystems. *Nature* 414:169–172.

Schindler, D. E., R. Hilborn, B. Chasco, C. P. Boatright, T. P. Quinn, L. A. Rogers, and M. S. Webster. 2010. Population diversity and the portfolio effect in an exploited species. *Nature* 465:609–U102.

Schindler, D. E., D. E. Rogers, M. D. Scheuerell, and C. A. Abrey. 2005. Effects of changing climate on zooplankton and juvenile sockeye salmon growth in southwestern Alaska. *Ecology* 86:198–209.

Schindler, D. W. 2001. The cumulative effects of climate warming and other human stresses on Canadian freshwaters in the new millennium. *Canadian Journal of Fisheries and Aquatic Sciences* 58:18–29.

Schlesinger, W. H. 2010. Translational ecology. *Science* 329:609–609.

Schlesinger, W. H., J. F. Reynolds, G. L. Cunningham, L. F. Huenneke, W. M. Jarrell, R. A. Virginia, and W. G. Whitford. 1990. Biological feedbacks in global desertification. *Science* 247:1043–1048.

Schmid, B., J. Joshi, and F. Schlapfer. 2001. Empirical evidence for biodiversity-ecosystem functioning relationships. *In:* A. P. Kinzig, S. W. Pacala, and D. Tilman, eds., *The functional consequences of biodiversity*, 120–150. Princeton, NJ: Princeton University Press.

Schmitz, O. J. 1998. Direct and indirect effects of predation and predation risk in old-field interaction webs. *American Naturalist* 151:327–342.

———. 2008. Effects of predator hunting mode on grassland ecosystem function. *Science* 319:952–954.

———. 2010. *Resolving ecosystem complexity.* Princeton, NJ: Princeton University Press.

Schmitz, O. J., P. A. Hamback, and A. P. Beckerman. 2000. Trophic cascades in terrestrial systems: a review of the effects of carnivore removals on plants. *American Naturalist* 155:141–153.

Schmitz, O. J., E. Post, C. E. Burns, and K. M. Johnston. 2003. Ecosystem responses to global climate change: moving beyond color mapping. *BioScience* 53:1199–1205.

Schoener, T. W. 1968. The Anolis lizards of Bimini: resource partitioning in a complex fauna. *Ecology* 49:704–726.

———. 1974. Resource partitioning in ecological communities. *Science* 185: 27–39.

Schwartz, M. D., ed. 2003. *Phenology: an integrative environmental science.* Dordrecht: Kluwer Academic Publishers.

Schwartz, M. K., J. P. Copeland, N. J. Anderson, J. R. Squires, R. M. Inman, K. S. McKelvey, K. L. Pilgrim, L. P. Waits, and S. A. Cushman. 2009. Wolverine gene flow across a narrow climatic niche. *Ecology* 90:3222–3232.

Semken, H. A., R. W. Graham, and T. W. Stafford. 2010. AMS ^{14}C analysis of Late Pleistocene non-analog faunal components from 21 cave deposits in southeastern North America. *Quaternary International* 217:240–255.

Shackleton, N. J., A. Berger, and W. R. Peltier. 1990. An alternative astronomical calibration of the lower Pleistocene time-scale based on ODP site 677a. *Transactions of the Royal Society of Edinburgh-Earth Sciences* 81:251–261.

Shapiro, B., A. J. Drummond, A. Rambaut, M. C. Wilson, P. E. Matheus, A. V. Sher, O. G. Pybus, M. T. P. Gilbert, I. Barnes, J. Binladen, E. Willerslev, A. J. Hansen, G. F. Baryshnikov, J. A. Burns, S. Davydov, J. C. Driver, D. G. Froese, C. R. Harington, G. Keddie, P. Kosintsev, M. L. Kunz, L. D. Martin, R. O. Stephenson, J. Storer, R. Tedford, S. Zimov, and A. Cooper. 2004. Rise and fall of the Beringian steppe bison. *Science* 306:1561–1565.

Shaver, G. R., J. Canadell, F. S. Chapin, J. Gurevitch, J. Harte, G. Henry, P. Ineson, S. Jonasson, J. Melillo, L. Pitelka, and L. Rustad. 2000. Global warming and terrestrial ecosystems: a conceptual framework for analysis. *BioScience* 50:871–882.

Shaver, G. R., and F. S. Chapin. 1991. Production-biomass relationships and element cycling in contrasting arctic vegetation types. *Ecological Monographs* 61:1–31.

Shaver, G. R., L. C. Johnson, D. H. Cades, G. Murray, J. A. Laundre, E. B. Rastetter, K. J. Nadelhoffer, and A. E. Giblin. 1998. Biomass and CO_2 flux in wet sedge tundras: responses to nutrients, temperature, and light. *Ecological Monographs* 68:75–97.

Shaver, G. R., and S. Jonasson. 1999. Response of Arctic ecosystems to climate change: results of long-term field experiments in Sweden and Alaska. *Polar Research* 18:245–252.

Shaver, G. R., and J. Kummerow. 1992. Phenology, resource allocation, and growth of arctic vascular plants. *In:* F. S. Chapin III, R. L. Jefferies, J. F. Reynolds, G. R. Shaver, and J. Svoboda, eds., *Arctic ecosystems in a changing climate,* 193–238. New York: Academic Press.

Shaver, G. R., L. E. Street, E. B. Rastetter, M. T. Van Wijk, and M. Williams. 2007. Functional convergence in regulation of net CO_2 flux in heterogeneous tundra landscapes in Alaska and Sweden. *Journal of Ecology* 95:802–817.

Shaw, M. R., and J. Harte. 2001. Response of nitrogen cycling to simulated climate change: differential responses along a subalpine ecotone. *Global Change Biology* 7:193–210.

Sher, A. V., S. A. Kuzmina, T. V. Kuznetsova, and L. D. Sulerzhitsky. 2005. New insights into the Weichselian environment and climate of the East Siberian Arctic, derived from fossil insects, plants, and mammals. *Quaternary Science Reviews* 24:533–569.

Sherry, R. A., X. H. Zhou, S. L. Gu, J. A. Arnone, D. S. Schimel, P. S. Verburg, L. L. Wallace, and Y. Q. Luo. 2007. Divergence of reproductive phenology under climate warming. *Proceedings of the National Academy of Sciences of the United States of America* 104:198–202.

Simberloff, D. 1972. Properties of the rarefaction diversity measurement. *American Naturalist* 106:414–418.

Simpson, G. G. 1944. *Tempo and mode in evolution.* New York: Columbia University Press.

Sjogersten, S., R. van der Wal, and S. J. Woodin. 2008. Habitat type determines herbivory controls over CO_2 fluxes in a warmer arctic. *Ecology* 89: 2103–2116.

Skelly, D. K. 1997. Tadpole communities. *American Scientist* 85:36–45.

Skelly, D. K., and L. K. Freidenburg. 2000. Effects of beaver on the thermal biology of an amphibian. *Ecology Letters* 3:483–486.

Skelly, D. K., L. N. Joseph, H. P. Possingham, L. K. Freidenburg, T. J. Farrugia, M. T. Kinnison, and A. P. Hendry. 2007. Evolutionary responses to climate change. *Conservation Biology* 21:1353–1355.

Skelly, D. K., E. E. Werner, and S. A. Cortwright. 1999. Long-term distributional dynamics of a Michigan amphibian assemblage. *Ecology* 80:2326–2337.

Skogland, T. 1989. Comparative social organization of wild reindeer in relation to food, mates, and predator avoidance. *Advances in Ethology* 29.

Smith, F. A., J. L. Betancourt, and J. H. Brown. 1995. Evolution of body size in the woodrat over the past 25,000 years of climate change. *Science* 270: 2012–2014.

Smith, T. M., and R. W. Reynolds. 2005. A global merged land-air-sea surface temperature reconstruction based on historical observations (1880–1997). *Journal of Climate* 18:2021–2036.

Sobek-Swant, S., J. C. Crosthwaite, D. B. Lyons, and B. J. Sinclair. 2012. Could phenotypic plasticity limit an invasive species? Incomplete reversibility of mid-winter deacclimation in emerald ash borer. *Biological Invasions* 14:115–125.

Soberon, J. 2007. Grinnellian and Eltonian niches and geographic distributions of species. *Ecology Letters* 10:1115–1123.

Soberon, J., and M. Nakamura. 2009. Niches and distributional areas: concepts, methods, and assumptions. *Proceedings of the National Academy of Sciences of the United States of America* 106:19644–19650.

Solberg, E. J., A. Loison, B. E. Saether, and O. Strand. 2000. Age-specific harvest mortality in a Norwegian moose Alces alces population. *Wildlife Biology* 6:41–52.

Solomon, M. E. 1949. The natural control of animal popuations. *Journal of Animal Ecology* 18:1–35.

Southward, A. J., S. J. Hawkins, and M. T. Burrows. 1995. 70 years observations of changes in distribution and abundance of zooplankton and intertidal organisms in the western English Channel in relation to rising sea temperature. *Journal of Thermal Biology* 20:127–155.

Sparks, T. H., and P. D. Carey. 1995. The responses of species to climate over two centuries: an analysis of the Marsham phonological records, 1736–1947. *Journal of Ecology* 83:321–329.

Sparks, T. H., K. Huber, R. L. Bland, H. Q. P. Crick, P. J. Croxton, J. Flood, R. G. Loxton, C. F. Mason, J. A. Newnham, and P. Tryjanowski. 2007. How consistent are trends in arrival (and departure) dates of migrant birds in the UK? *Journal of Ornithology* 148:503–511.

Spiller, D. A., and T. W. Schoener. 1994. Effects of the top and intermediate predators in a terrestrial food web. *Ecology* 75:182–196.

Steadman, D. W., P. S. Martin, R. D. E. MacPhee, A. J. T. Jull, H. G. Mc-Donald, C. A. Woods, M. Iturralde-Vinent, and G. W. L. Hodgins. 2005. Asynchronous extinction of late Quaternary sloths on continents and islands. *Proceedings of the National Academy of Sciences of the United States of America* 102:11763–11768.

Steen, H. 1995. Untangling the causes of disappearance from a local population of root voles, Microtus oeconomus: a test of the regional synchrony hypothesis. *Oikos* 73:65–72.

Steltzer, H., and E. Post. 2009. Seasons and life cycles. *Science* 324:886–887.

Stenseth, N. C. 1995. Snowshoe hare populations: squeezed from below and above. *Science* 269:1061–1062.

———. 1999. Population cycles in voles and lemmings: density dependence and phase dependence in a stochastic world. *Oikos* 87:427–461.

Stenseth, N. C., O. N. Bjørnstad, and W. Falck. 1996. Is spacing behaviour coupled with predation causing the microtine density cycle? A synthesis of current process-oriented and pattern-oriented studies. *Proceedings of the Royal Society of London, Series B: Biological Sciences* 263:1423–1435.

Stenseth, N. C., O. N. Bjørnstad, and T. Saitoh. 1996. A gradient from stable to cyclic populations of Clethrionomys rufocanus in Hokkaido, Japan. *Proceedings of the Royal Society of London, Series B: Biological Sciences* 263: 1117–1126.

Stenseth, N. C., K. S. Chan, E. Framstad, and H. Tong. 1998. Phase- and density-dependent population dynamics in Norwegian lemmings: interaction between deterministic and stochastic processes. *Proceedings of the Royal Society of London, Series B: Biological Sciences* 265:1957–1968.

Stenseth, N. C., K. S. Chan, H. Tong, R. Boonstra, S. Boutin, C. J. Krebs, E. Post, M. O'Donoghue, N. G. Yoccoz, M. C. Forchhammer, and J. W. Hurrell. 1999. Common dynamic structure of Canada lynx populations within three climatic regions. *Science* 285:1071–1073.

Stenseth, N. C., W. Falck, O. N. Bjørnstad, and C. J. Krebs. 1997. Population regulation in snowshoe hare and Canadian lynx: asymmetric food web configurations between hare and lynx. *Proceedings of the National Academy of Sciences of the United States of America* 94:5147–5152.

Stenseth, N. C., W. Falck, K. S. Chan, O. N. Bjørnstad, M. O'Donoghue, H. Tong, R. Boonstra, S. Boutin, C. J. Krebs, and N. G. Yoccoz. 1998. From patterns to processes: phase and density dependencies in the Canadian lynx cycle. *Proceedings of the National Academy of Sciences of the United States of America* 95:15430–15435.

Stralberg, D., D. Jongsomjit, C. A. Howell, M. A. Snyder, J. D. Alexander, J. A. Wiens, and T. L. Root. 2009. Re-shuffling of species with climate disruption: a no-analog future for california birds? *PLoS One* 4:8.

Street, L. E., G. R. Shaver, M. Williams, and M. T. Van Wijk. 2007. What is the relationship between changes in canopy leaf area and changes in photosynthetic CO_2 flux in arctic ecosystems? *Journal of Ecology* 95:139–150.

Stuart, A. J. 2005. The extinction of woolly mammoth (Mammuthus primigenius) and straight-tusked elephant (Palaeoloxodon antiquus) in Europe. *Quaternary International* 126:171–177.

Stuart, A. J., P. A. Kosintsev, T. F. G. Higham, and A. M. Lister. 2004. Pleistocene to Holocene extinction dynamics in giant deer and woolly mammoth. *Nature* 431:684–689.

Sullivan, P. F., and J. M. Welker. 2005. Warming chambers stimulate early season growth of an arctic sedge: results of a minirhizotron field study. *Oecologia* 142:616–626.

Sullivan, P. F., J. M. Welker, S. J. T. Arens, and B. Sveinbjornsson. 2008. Continuous estimates of CO_2 efflux from arctic and boreal soils during the snow-covered season in Alaska. *Journal of Geophysical Research—Biogeosciences* 113.

Surovell, T., N. Waguespack, and P. J. Brantingham. 2005. Global archaeological evidence for proboscidean overkill. *Proceedings of the National Academy of Sciences of the United States of America* 102:6231–6236.

Suttle, K. B., M. A. Thomsen, and M. E. Power. 2007. Species interactions reverse grassland responses to changing climate. *Science* 315:640–642.

Tarnocai, C., J. G. Canadell, E. A. G. Schuur, P. Kuhry, G. Mazhitova, and S. Zimov. 2009. Soil organic carbon pools in the northern circumpolar permafrost region. *Global Biogeochemical Cycles* 23:1–11.

Taylor, K. C., P. A. Mayewski, R. B. Alley, E. J. Brook, A. J. Gow, P. M. Grootes, D. A. Meese, E. S. Saltzman, J. P. Severinghaus, M. S. Twickler, J. W. C. White, S. Whitlow, and G. A. Zielinski. 1997. The Holocene Younger Dryas transition recorded at Summit, Greenland. *Science* 278:825–827.

Taylor, S. W., A. L. Carroll, R. I. Alfaro, and L. Safranyik. 2006. Forest, climate and mountain pine beetle outbreak dynamics in western Canada. *In:* L. Safranyik and B. Wilson, eds., *The mountain pine beetle: a synthesis of biology, management and impacts in lodgepole pine,* 67–94. Victoria, BC: Natural Resources Canada.

Tebaldi, C., K. Hayhoe, J. M. Arblaster, and G. A. Meehl. 2006. Going to the extremes. *Climatic Change* 79:185–211.

Tenow, O., and H. Bylund. 2000. Recovery of a Betula pubescens forest in northern Sweden after severe defoliation by Epirrita autumnata. *Journal of Vegetation Science* 11:855–862.

Thing, H. 1984. Feeding ecology of the West Greenland caribou (*Rangifer tarandus*) in the Sisimiut-Kangerlussuaq region. *Danish Review of Game Biology* 12:1–53.

Thomas, C. D., A. Cameron, R. E. Green, M. Bakkenes, L. J. Beaumont, Y. C. Collingham, B. F. N. Erasmus, M. F. de Siqueira, A. Grainger, L. Hannah, L. Hughes, B. Huntley, A. S. van Jaarsveld, G. F. Midgley, L. Miles, M. A. Ortega-Huerta, A. T. Peterson, O. L. Phillips, and S. E. Williams. 2004. Extinction risk from climate change. *Nature* 427:145–148.

Tian, H. Q., J. M. Melillo, D. W. Kicklighter, A. D. McGuire, J. V. K. Helfrich, B. Moore, and C. J. Vorosmarty. 1998. Effect of interannual climate variability on carbon storage in Amazonian ecosystems. *Nature* 396:664–667.

Tilman, D. 1982. *Resource competition and community structure.* Princeton, NJ: Princeton University Press.

———. 1988. *Plant strategies and the dynamics and structure of plant communities.* Princeton, NJ: Princeton University Press.

———. 1994. Competition and biodiversity in spatially structured habitats. *Ecology* 75:2–16.

———. 1996. Biodiversity: population versus ecosystem stability. *Ecology* 77:350–363.

———. 1999. The ecological consequences of changes in biodiversity: a search for general principles. *Ecology* 80:1455–1474.

Tilman, D., and J. A. Downing. 1994. Biodiversity and stability in grasslands. *Nature* 367:363–365.

Tilman, D., J. Knops, D. Wedin, P. Reich, M. Ritchie, and E. Siemann. 1997. The influence of functional diversity and composition on ecosystem processes. *Science* 277:1300–1302.

Tilman, D., C. L. Lehman, and C. E. Bristow. 1998. Diversity-stability relationships: statistical inevitability or ecological consequence? *American Naturalist* 151:277–282.

Tilman, D., R. M. May, C. L. Lehman, and M. A. Nowak. 1994. Habitat destruction and the extinction debt. *Nature* 371:65–66.

Tilman, D., P. B. Reich, and J. Knops. 2007. Ecology: diversity and stability in plant communities. Reply. *Nature* 446:E7–E8.

Tilman, D., P. B. Reich, and J. M. H. Knops. 2006. Biodiversity and ecosystem stability in a decade-long grassland experiment. *Nature* 441:629–632.

Tingley, M. W., and S. R. Beissinger. 2009. Detecting range shifts from historical species occurrences: new perspectives on old data. *Trends in Ecology & Evolution* 24:625–633.

Tingley, M. W., W. B. Monahan, S. R. Beissinger, and C. Moritz. 2009. Birds track their Grinnellian niche through a century of climate change. *Proceedings of the National Academy of Sciences of the United States of America* 106:19637–19643.

Tong, H. 1990. *Non-linear time series: a dynamical system approach.* Oxford: Oxford University Press.

Trenberth, K. E., and J. W. Hurrell. 1994. Decadal atmosphere-ocean variations in the Pacific. *Climate Dynamics* 9:303–319.

Trenberth, K. E., P. D. Jones, P. Ambenje, R. Bojariu, D. Easterling, A. Klein Tank, D. Parker, F. Rahimzadeh, J. A. Renwick, M. Rusticucci, B. Soden, and P. Zhai. 2007. Observations: surface and atmospheric climate change. *In:* S. Solomon, D. Qin, M. Manning, Z. Chen, M. Marquis, K. B. Averyt, M. Tignor, and H. L. Miller, eds., *Climate Change 2007: The physical science basis. Contribution of Working Group I to the Fourth Assessment Report of the Intergovernmental Panel on Climate Change.* Cambridge: Cambridge University Press.

Tryjanowski, P., S. Kuzniak, and T. H. Sparks. 2005. What affects the magnitude of change in first arrival dates of migrant birds? *Journal of Ornithology* 146:200–205.

Turchin, P. 2003. *Complex population dynamics: a theoretical/empirical synthesis.* Princeton, NJ: Princeton University Press.

Turchin, P., and G. O. Batzli. 2001. Availability of food and the population dynamics of arvicoline rodents. *Ecology* 82:1521–1534.

Turchin, P., P. L. Lorio, A. D. Taylor, and R. F. Billings. 1991. Why do populations of southern pine beetles (Coleoptera, Scolytidae) fluctuate? *Environmental Entomology* 20:401–409.

Turner, M. G. 2010. Disturbance and landscape dynamics in a changing world. *Ecology* 91:2833–2849.

Turney, C. S. M., T. F. Flannery, R. G. Roberts, C. Reid, L. K. Fifield, T. F. G. Higham, Z. Jacobs, N. Kemp, E. A. Colhoun, R. M. Kalin, and N. Ogle. 2008. Late-surviving megafauna in Tasmania, Australia, implicate human involvement in their extinction. *Proceedings of the National Academy of Sciences of the United States of America* 105:12150–12153.

Tyler, N., and M. C. Forchhammer. 2009. Nonlinear effects of climate and density in the dynamics of a fluctuating population of reindeer (vol 89, 1675, 2008). *Ecology* 90:292–292.

Tyler, N. J. C., M. C. Forchhammer, and N. A. Øritsland. 2008. Nonlinear effects of climate and density in the dynamics of a fluctuating population of reindeer. *Ecology* 89:1675–1686.

Tyler, N. J. C., and N. A. Øritsland. 1989. Why don't Svalbard reindeer migrate? *Holarctic Ecology* 12:369–376.

Ulrich, A., and P. L. Gersper. 1978. Plant nutrient limitations of tundra plant growth. *In:* L. L. Tieszen, ed., *Vegetation and production ecology of an Alaskan arctic tundra,* 457–482. New York: Springer-Verlag.

Ulvan, E. M., A. G. Finstad, O. Ugedal, and O. K. Berg. 2011. Direct and indirect climatic drivers of biotic interactions: ice-cover and carbon runoff shaping Arctic char *Salvelinus alpinus* and brown trout *Salmo trutta* competitive asymmetries. *Oecologia.*

Updegraff, K., S. D. Bridgham, J. Pastor, P. Weishampel, and C. Harth. 2001. Response of CO2 and CH4 emissions from peatlands to warming and water table manipulation. *Ecological Applications* 11:311–326.

Urban, M. C., R. D. Holt, S. E. Gilman, and J. Tewksbury. 2011. Heating up relations between cold fish: competition modifies responses to climate change. *Journal of Animal Ecology* 80:505–507.

Urban, M. C., J. J. Tewksbury, and K. S. Sheldon. 2012. On a collision course: competition and dispersal differences create no-analogue communities and cause extinctions during climate change. *Proceedings of the Royal Society B: Biological Sciences* 279:2072–2080.

Valtonen, A., M. P. Ayres, H. Roininen, J. Poyry, and R. Leinonen. 2011. Environmental controls on the phenology of moths: predicting plasticity and constraint under climate change. *Oecologia* 165:237–248.

Van der Wal, R., S. Sjogersten, S. J. Woodin, E. J. Cooper, I. S. Jonsdottir, D. Kuijper, T. A. D. Fox, and A. D. Huiskes. 2007. Spring feeding by pink-footed geese reduces carbon stocks and sink strength in tundra ecosystems. *Global Change Biology* 13:539–545.

van Oort, B. E. H., N. J. C. Tyler, M. P. Gerkema, L. Folkow, A. S. Blix, and K.-A. Stokkan. 2005. Circadian organisation in reindeer. *Nature* 438:1095–1096.

van Oort, B. E. H., N. J. C. Tyler, M. P. Gerkema, L. Folkow, and K.-A. Stokkan. 2007. Where clocks are redundant: weak circadian mechanisms in reindeer living under polar photic conditions. *Naturwissenschaften* 94:183–194.

Veblen, T. T., K. S. Hadley, M. S. Reid, and A. J. Rebertus. 1991. The response of sub-alpine forests to spruce beetle outbreak in Colorado. *Ecology* 72:213–231.

Visser, M. E., and C. Both. 2005. Shifts in phenology due to global climate change: the need for a yardstick. *Proceedings of the Royal Society B: Biological Sciences* 272:2561–2569.

Visser, M. E., A. J. van Noordwijk, J. M. Tinbergen, and C. M. Lessells. 1998. Warmer springs lead to mistimed reproduction in great tits (Parus major). *Proceedings of the Royal Society of London, Series B: Biological Sciences* 265:1867–1870.

Volkov, I., J. R. Banavar, S. P. Hubbell, and A. Maritan. 2003. Neutral theory and relative species abundance in ecology. *Nature* 424:1035–1037.

Vors, L. S., and M. S. Boyce. 2009. Global declines of caribou and reindeer. *Global Change Biology* 15:2626–2633.

Vourlitis, G. L., and W. C. Oechel. 1997. Landscape-scale CO2, H2O vapour and energy flux of moist-wet coastal tundra ecosystems over two growing seasons. *Journal of Ecology* 85:575–590.

———. 1999. Eddy covariance measurements of CO2 and energy fluxes of an Alaskan tussock tundra ecosystem. *Ecology* 80:686–701.

Vrba, E. S., G. H. Denton, T. C. Partridge, and L. H. Burckle, eds. 1996. *Paleoclimate and evolution, with emphasis on human origins*. New Haven, CT: Yale University Press.

Wahren, C. H. A., M. D. Walker, and M. S. Bret-Harte. 2005. Vegetation responses in Alaskan arctic tundra after 8 years of a summer warming and winter snow manipulation experiment. *Global Change Biology* 11: 537–552.

Wake, D. B., E. A. Hadly, and D. D. Ackerly. 2009. Biogeography, changing climates, and niche evolution. *Proceedings of the National Academy of Sciences of the United States of America* 106:19631–19636.

Walker, B. H. 1992. Biodiversity and ecological redundancy. *Conservation Biology* 6:18–23.

Walker, M. D., C. H. Wahren, R. D. Hollister, G. H. R. Henry, L. E. Ahlquist, J. M. Alatalo, M. S. Bret-Harte, M. P. Calef, T. V. Callaghan, A. B. Carroll, H. E. Epstein, I. S. Jonsdottir, J. A. Klein, B. Magnusson, U. Molau, S. F. Oberbauer, S. P. Rewa, C. H. Robinson, G. R. Shaver, K. N. Suding, C. C. Thompson, A. Tolvanen, O. Totland, P. L. Turner, C. E. Tweedie, P. J. Webber, and P. A. Wookey. 2006. Plant community responses to experimental warming across the tundra biome. *Proceedings of the National Academy of Sciences of the United States of America* 103:1342–1346.

Walker, M. D., D. A. Walker, J. M. Welker, A. M. Arft, T. Bardsley, P. D. Brooks, J. T. Fahnestock, M. H. Jones, M. Losleben, A. N. Parsons, T. R. Seastedt, and P. L. Turner. 1999. Long-term experimental manipulation of winter snow regime and summer temperature in arctic and alpine tundra. *Hydrological Processes* 13:2315–2330.

Wallace, A. R. 1876. *The geographical distribution of animals*. London: Harper.

Walther, G. R., E. Post, P. Convey, A. Menzel, C. Parmesan, T. J. C. Beebee, J. M. Fromentin, O. Hoegh-Guldberg, and F. Bairlein. 2002. Ecological responses to recent climate change. *Nature* 416:389–395.

Wardle, D. A. 2002. *Communities and ecosystems: linking the aboveground and belowground components*. Princeton, NJ: Princeton University Press.

Webb, C. O., D. D. Ackerly, M. A. McPeek, and M. J. Donoghue. 2002. Phylogenies and community ecology. *Annual Review of Ecology and Systematics* 33:475–505.

Welbergen, J. A., S. M. Klose, N. Markus, and P. Eby. 2008. Climate change and the effects of temperature extremes on Australian flying-foxes. *Proceedings of the Royal Society B: Biological Sciences* 275:419–425.

Welker, J. M., J. T. Fahnestock, P. Sullivan, and R. A. Chimner. 2005. Leaf mineral nutrition of arctic plants in response to long-term warming and deeper snow in Northern Alaska. *Oikos* 109:167–177.

Wellborn, G. A., D. K. Skelly, and E. E. Werner. 1996. Mechanisms creating community structure across a freshwater habitat gradient. *Annual Review of Ecology and Systematics* 27:337–363.

Wentz, F. J., L. Ricciardulli, K. Hilburn, and C. Mears. 2007. How much more rain will global warming bring? *Science* 317:233–235.

Werner, E. E. and M. A. McPeek. 1994. Direct and indirect effects of predators on 2 anuran species along an environmental gradient. *Ecology* 75:1368–1382.

Wiederholt, R., E. Fernandez-Duque, D. R. Diefenbach, and R. Rudran. 2010. Modeling the impacts of hunting on the population dynamics of red howler monkeys (Alouatta seniculus). *Ecological Modelling* 221:2482–2490.

Wiederholt, R., and E. Post. 2010. Tropical warming and the dynamics of endangered primates. *Biology Letters* 6:257–260.

Wiens, J. A. 1989. Spatial scaling in ecology. *Functional Ecology* 3:385–397.

Wiens, J. A., D. Stralberg, D. Jongsomjit, C. A. Howell, and M. A. Snyder. 2009. Niches, models, and climate change: assessing the assumptions and uncertainties. *Proceedings of the National Academy of Sciences of the United States of America* 106:19729–19736.

Wiens, J. J., and C. H. Graham. 2005. Niche conservatism: Integrating evolution, ecology, and conservation biology. Pages 519–539 Annual Review of Ecology Evolution and Systematics. Annual Reviews, Palo Alto.

Williams, G. C. 1966. *Adaptation and natural selection: a critique of some current evolutionary thought.* Princeton, NJ: Princeton University Press.

Williams, J. W., and S. T. Jackson. 2007. Novel climates, no-analog communities, and ecological surprises. *Frontiers in Ecology and the Environment* 5:475–482.

Williams, J. W., S. T. Jackson, and J. E. Kutzbacht. 2007. Projected distributions of novel and disappearing climates by 2100 AD. *Proceedings of the National Academy of Sciences of the United States of America* 104:5738–5742.

Williams, J. W., D. M. Post, L. C. Cwynar, A. F. Lotter, and A. J. Levesque. 2002. Rapid and widespread vegetation responses to past climate change in the North Atlantic region. *Geology* 30:971–974.

Williams, J. W., B. N. Shuman, T. Webb, P. J. Bartlein, and P. L. Leduc. 2004. Late-quaternary vegetation dynamics in North America: scaling from taxa to biomes. *Ecological Monographs* 74:309–334.

Wills, C., R. Condit, R. B. Foster, and S. P. Hubbell. 1997. Strong density- and diversity-related effects help to maintain tree species diversity in a neotropical forest. *Proceedings of the National Academy of Sciences of the United States of America* 94:1252–1257.

Wilmers, C. C., E. Post, and A. Hastings. 2007a. The anatomy of predator-prey dynamics in a changing climate. *Journal of Animal Ecology* 76:1037–1044.

————. 2007b. A perfect storm: the combined effects on population fluctuations of autocorrelated environmental noise, age structure, and density dependence. *American Naturalist* 169:673–683.

Wilmers, C. C., E. Post, R. O. Peterson, and J. A. Vucetich. 2006. Predator disease out-break modulates top-down, bottom-up and climatic effects on herbivore population dynamics. *Ecology Letters* 9:383–389.

Wittemyer, G., P. Elsen, W. T. Bean, A. C. O. Burton, and J. S. Brashares. 2008. Accelerated human population growth at protected area edges. *Science* 321:123–126.

Wolkovich, E. M., B. I. Cook, J. M. Allen, T. M. Crimmins, J. L. Betancourt, S. E. Travers, S. Pau, J. Regetz, T. J. Davies, N. J. B. Kraft, T. R. Ault, K. Bolmgren, S. J. Mazer, G. J. McCabe, B. J. McGill, C. Parmesan, N. Salamin, M. D. Schwartz, and E. E. Cleland. 2012. Warming experiments underpredict plant phenological responses to climate change. *Nature* 485:494–497.

Woodwell, G. M., and R. H. Whittaker. 1968. Primary production in terrestrial communities. *American Zoologist* 8:19–30.

Wright, S. J. 2005. Tropical forests in a changing environment. *Trends in Ecology & Evolution* 20:553–560.

Yurtsev, B. 1997. Effect of climate change on biodiversity of arctic plants. *In:* W. C. Oechel, T. V. Callaghan, T. Gilmanov, J. I. Holten, B. Maxwell, U. Molau, and B. Sveinbjornsson, eds., *Global change and arctic terrestrial ecosystems,* 229–244. New York: Springer-Verlag.

Zachos, J., M. Pagani, L. Sloan, E. Thomas, and K. Billups. 2001. Trends, rhythms, and aberrations in global climate 65 Ma to present. *Science* 292:686–693.

Zhang, R., E. Jongejans, and K. Shea. 2011. Warming increases the spread of an invasive thistle. *PLoS One* 6.

Zhao, M. S. and S. W. Running. 2010. Drought-induced reduction in global terrestrial net primary production from 2000 through 2009. *Science* 329:940–943.

Zimov, S. A., V. I. Chuprynin, A. P. Oreshko, F. S. Chapin, J. F. Reynolds, and M. C. Chapin. 1995. Steppe-tundra transition: a herbivore-driven biome shift at the end of the Pleistocene. *American Naturalist* 146:765–794.

Zulueta, R. C., W. C. Oechel, H. W. Loescher, W. T. Lawrence, and K. T. Paw U. 2011. Aircraft-derived regional scale CO(2) fluxes from vegetated drained thaw-lake basins and interstitial tundra on the Arctic Coastal Plain of Alaska. *Global Change Biology* 17:2781–2802.

Index

MONOGRAPHS IN POPULATION BIOLOGY
EDITED BY SIMON A. LEVIN AND HENRY S. HORN